Handbook of
Crystal Structures
and
Magnetic Properties
of
Rare Earth Intermetallics

Andrzej Szytuła
Institute of Physics
Jagellonian University
Cracow, Poland

and

Janusz Leciejewicz
Institute of Nuclear Chemistry and Technology
Warsaw, Poland

CRC Press
Taylor & Francis Group
Boca Raton London New York

CRC Press is an imprint of the
Taylor & Francis Group, an **informa** business

Published 1994 by CRC Press
Taylor & Francis Group
6000 Broken Sound Parkway NW, Suite 300
Boca Raton, FL 33487-2742

First issued in paperback 2019

No claim to original U.S. Government works

ISBN 13: 978-0-367-44955-1 (pbk)
ISBN 13: 978-0-8493-4261-5 (hbk)

Visit the Taylor & Francis Web site at
http://www.taylorandfrancis.com

and the CRC Press Web site at
http://www.crcpress.com

Library of Congress Cataloging-in-Publication Data

Szytuła, Andrzej, 1940–
    Handbook of crystal structures and magnetic properties of rare
earth intermetallics / by Andrzej Szytuła and Janusz Leciejewicz.
        p. cm.
    Includes bibliographical references and index.
    ISBN 0-8493-4261-9
    1. Rare earth metal compounds — Magnetic properties.
    2. Intermetallic compounds — Magnetic properties. 3. Crystals — Structure.
    I. Leciejewicz, Janusz, 1928– . II. Title.
    QD172.R2S99 1993
    546'.41 — dc20   93-1549
                                                                              CIP

Library of Congress Card Number 93-1549

THE AUTHORS

Professor Dr. Andrzej Szytuła took his M.S. degree in physics at the Jagellonian University of Cracow (1965). At the same university he defended his doctoral thesis (1970), habilitation thesis (1972), and was named a professor at the Institute of Physics, Jagellonian University (1985). Since 1991 he has been a director of the Institute of Physics of the Jagellonian University and since 1973 a head of the Solid State Department at this institute.

Professor Szytuła is also a leader of a scientific group working on solid state physics research by means of neutron and X-ray diffraction and magnetometry. His scientific activity concerns investigations of structural and magnetic properties of the following materials oxyhydrates (1964–1970), intermetallic compounds of 3d elements (1970–1978), and lanthanides and actinides (1978–1992). He is the author of over 200 publications.

Professor Dr. Janusz Leciejewicz graduated in 1951 from Warsaw Technical University. During the years 1952–1955 he did postgraduate research at the Chair of Inorganic Chemistry, Wrocław Technical University, specializing in X-ray diffraction and magnetochemistry. In 1955 he started research activity at the Institute of Nuclear Research in Świerk developing neutron diffraction studies, first of crystal structures and then of magnetic structures of uranium intermetallic compounds. He received his Ph.D. degree in physics in 1963, D.Sc. degree in 1975, and the title of professor of physics in 1978. Between 1970 and 1981 he served as the head of the Solid State Physics Department. In the years 1983–1990 he was director of the Institute of Nuclear Chemistry and Technology in Warsaw. Now retired, he continues scientific activity in the field of neutron diffraction studies of magnetic properties of f-electron intermetallic systems and X-ray structural studies of actinide coordination compounds. He is the author of 150 scientific communications and review papers.

CONTENTS

Chapter 1
INTRODUCTION 1

Chapter 2
CRYSTAL STRUCTURES OF TERNARY LANTHANIDE
INTERMETALLICS 9
2.1. RTX Phases 9
    2.1.1. Phases With the $MgCu_2$-type Structure 9
    2.1.2. Phases With the MgAgAs-type Structure 10
    2.1.3. Phases With the LaIrSi-(ZrOS)-type Structure 11
    2.1.4. Phases With Fe_2P-type Structure 12
    2.1.5. Phases With the $MgZn_2$-type Structure 12
    2.1.6. Phases With the AlB_2 and Related Structure Types 14
    2.1.7. Phases With the $CaIn_2$-type Structure 15
    2.1.8. Phases With the LaPtSi-type Structure 16
    2.1.9. Phases With PbFCl-type Structure 16
    2.1.10. Compounds With the TiNiSi-type Structure 17
    2.1.11. Phases With $CeCu_2$-type Structure 21
2.2. RTX_2 Phases 22
2.3. RT_2X Phases 24
2.4. $R_2T_3Si_5$ Phases 27
2.5. RT_2X_2 Phases 28
2.6. $ThMn_{12}$-type Structure 44
2.7. $R_2T_{14}X$-type Structure 47
2.8. RT_6X_6 Compounds With the CoSn-derivate Crystal
    Structure Types 52
    2.8.1. $HfFe_6Ge_6$-type Structure 52
2.9. Phases With $NaZn_{13}$-type Structure 54
2.10. Phases With the $Ce_2Ni_{17}Si_5$-type Structures 54
2.11. Phases With $SrNi_{12}B_6$ ($EuNi_{12}B_6$)-type Structures 55
2.12. Crystal Structures Based on $CaCu_5$-type 55
2.13. The $NdCo_4B_4$ Structure Type and its Polytypes 57

Chapter 3
MAGNETIC PROPERTIES 61
3.1. Magnetic Interactions 61
    3.1.1. Lanthanides Elements 61
    3.1.2. Binary Lanthanide −3d− Transition Metal Compounds ... 65
3.2. The Crystal Electric Field (CEF) Model 67
    3.2.1. Introduction 67
    3.2.2. The Interaction of the Free Ion With CEF 70
    3.2.3. Effect of the CEF on Physical Properties 73

Chapter 4
MAGNETIC PROPERTIES OF THE INTERMETALLIC $R_xT_yX_z$
COMPOUNDS WITH y/x≤z 83
4.1.   RTX Phases 83
        4.1.1.   Compounds With the $MgCu_2$-type Structure 83
        4.1.2.   Compounds With the MgAgAs-type Structure 84
        4.1.3.   Compounds With the LaIrSi (ZrOS)-type Structure 84
        4.1.4.   Compounds With the Fe_2P (ZrNiAl)-type Structure 84
        4.1.5.   Compounds With the $MgZn_2$-type Structure 89
        4.1.6.   Compounds With the AlB_2- and Ni_2In-type Structure 90
        4.1.7. R_2RhSi_3 Compounds 93
        4.1.8.   Compounds With the $CaIn_2$-type Structure 94
        4.1.9.   Compounds With the LaPtSi-type Structure 95
        4.1.10. Compounds With the PbFCl-type Structure 95
        4.1.11. Compounds With the TiNiSi-type Structure 96
        4.1.12. Compounds With the $CeCu_2$-type Structure 98
4.2. RTX_2 Phases 99
4.3. RT_2X Phases 105
4.4. RTX_3 Phases 109
4.5. $R_2T_3Si_5$ Phases 110
        4.5.1. $R_2Fe_3Si_5$ Compounds 110
        4.5.2. $R_2Co_3Si_5$ Compounds 113
4.6. RT_2X_2 Phases 114
        4.6.1. RT_2Si_2 and RT_2Ge_2 Phases With R = Pr and Nd 114
        4.6.2. SmT_2X_2 Phases 122
        4.6.3. GdT_2X_2 Phases 122
        4.6.4. RT_2X_2 Compounds With Heavy Lanthanides
                 (R=Tb–Tm) 125
        4.6.5. CeT_2X_2 Phases 140
        4.6.6. EuT_2X_2 Phases 151
        4.6.7. YbT_2X_2 Phases 155
        4.6.8.   Other RT_2X_2 Compounds 156
4.7. RMn_2X_2 Phases 159
4.8.   Conclusions 171
        4.8.1.   Magnetic Moment 171
        4.8.2.   Exchange Interactions 173
        4.8.3.   Crystalline Electric Field 179

Chapter 5
MAGNETIC MATERIALS BASED ON 3d-RICH TERNARY
COMPOUNDS 193
5.1. $RT_{12-x}M_x$ Phases 193
5.2. $R_2T_{14}X$ Phases 202
        5.2.1. $R_2Fe_{14}B$ 202
        5.2.2. $R_2Fe_{14}C$ 207

    5.2.3. $R_2Co_{14}B$ 207
    5.2.4. Substitution Compounds 209
    5.2.5. Hydrides 212
    5.2.6. Magnetic Moments 214
    5.2.7. Model Description of Exchange Interactions 215
    5.2.8. Magnetocrystalline Anisotropy 218
5.3. RT_6X_6 Phases 223
5.4. $RT_{4+x}Al_{8-x}$ Phases 224
5.5. $La(T_{1-x}X_x)_{13}$ (T=Fe, Co or Ni, X=Al, Si) Phases 231
5.6. RT_9Si_2 234
5.7. $RT_{12}B_6$ Series 236
5.8. RT_4B 239
5.9. $R_{1+\epsilon}T_4B_4$ 241

Chapter 6
CONCLUDING REMARKS 243

References 247

Index 271

Chapter 1

INTRODUCTION

The lanthanide elements, which are characterized by unfilled 4f-electron shell, constitute jointly with scandium and yttrium the group of so-called rare earths. Because their chemical properties and reactivities are very similar, serious difficulties were met in the past in separating and obtaining them in a pure state. The first element belonging to this group was discovered in 1787 and was called, at first, Ytterbium, since it was separated from the mineral ores deposited in the neighborhood of the town Ytterby, Sweden. Now, this element bears the name gadolinium. In the course of time, all elements belonging to this group, including prometheum, which was obtained artificially by nuclear reactions, were discovered and obtained in elemental form. Their crystal structures were determined for the first time in the mid-1930s (Klemm and Bommer, 1937) and later studied in more detail, as single crystal samples of high purity became available. Figure 1.1 shows the crystal data for the lanthanide metals and their stability ranges under normal pressure (Beaudry and Gschneidner, Jr., 1978). At high temperatures the cubic face-centered structure (space group Fm3m) was found to be stable, but, as the temperature is lowered, they transform into hexagonal close-packed (space group $P6_3$/mmc) modification. Also in the mid-1930s the magnetic properties of some lanthanide metals were determined: gadolinium metal was found to be ferromagnetic, with the Curie temperature at 293 K and ordered magnetic moment of 7.12 μ_B (Urbain et al., 1935). When large quantities of high purity lanthanide metals became available in the 1950s and '60s, their magnetic properties were extensively studied. Pioneering work was done by Spedding and his group from the University of Iowa, Ames, and by K.P. Belov from the Moscow University. Physical properties, including magnetic, have been studied using classical as well as nuclear methods, like thermal neutron scattering and Mössbauer spectroscopy. Figure 1.2 displays the magnetic ordering schemes discovered in a number of heavy lanthanide metals. More complex magnetic structures were found in light lanthanide metals. For example, in neodymium an incommensurate antiferromagnetic ordering with two-dimensional character was determined to be below $T_N = 19.9$ K (Lebech and Bak, 1979). The above experimental data led to the development of theoretical models of magnetic interactions in lanthanide metals, which were later extended on their binary and ternary intermetallic compounds. These models were essential for the explanation of the observed properties (Ruderman and Kittel, 1954; Kasuya, 1956; Yoshida, 1957; Wallace, 1973; Fulde, 1979) and stimulated the search for new effects and materials. Fascinating physical properties of lanthanide metals turned attention to their alloys, particularly those with transition metals and with elements of the fourth and fifth group

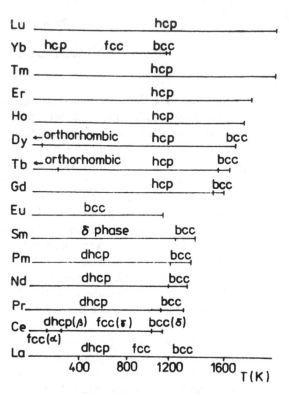

FIGURE 1.1. The crystal structure data for the lathanide metals and their stability ranges under normal pressure (Beaudry and Gschneidner, Jr., 1978).

of the Mendeleyev Table. Large numbers of new phases have been synthetized, exhibiting different stoichiometries and a variety of crystal structure types. In many cases, their physical properties turned out to be of the highest technical importance and these compounds are now manufactured on an industrial scale (e.g., $SmCo_5$).

A good illustration of the wealth of new effects and properties discovered in the binary lanthanide compounds can be exemplified by the magnetic properties of cerium alloys with fifth group elements (cerium pnictides) exhibiting the rock-salt (NaCl) type of crystal structure. Thus, CeN was found to be a mixed valency compound, which remains paramagnetic even at the lowest temperatures (Danan et al., 1969). On the other hand, CeP shows antiferromagnetic long-range ordering of magnetic moments localized on the Ce^{3+} ions. Its magnetic structure is described by the wave vector k = (0, 0, 1) (Hulliger and Ott, 1978). A neutron diffraction study of a single crystal of cerium monoarsenide CeAs revealed a complex antiferromagnetic ordering described by three wave vectors (Rossat-Mignod et al., 1982). Such magnetic structure suggests the presence of competing interactions responsible for the moment alignment along the (001) axis and uniaxial anisotropy which prefers the alignment along the (111) direction. The magnetic structure of CeSb was found to be described by a wave vector (0, 0, k) whose magnitude strongly

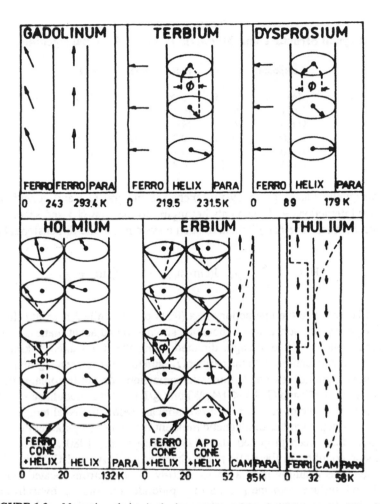

FIGURE 1.2. Magnetic ordering in the heavy rare earths as found by neutron diffraction.

depends on temperature (Rossat-Mignod et al., 1977). Two magnetic phases have been detected in CeBi. At low temperatures the magnetic structure with the wave vector (0, 0, 1/2) has been observed, while above the temperature $T = T_N/2$, a structure with the wave vector (0, 0, 1) has been found to be present. Both structures can be visualized as stacking of ferromagnetic planes along the (0, 0, 1) direction with the sequences $+ + - -$ and $+ - + -$, respectively (Cable and Koechler, 1972).

The magnetic ordering schemes observed in cerium pnictides result from strong magnetic anisotropy as well as from strong f-p hybridization due to the semiconducting character of these phases. The magnetic structures observed in CeAs and CeBi were discussed in terms of the ANNNI model (Bak and von Boehm, 1980) which postulates strong anisotropic interactions in the

TABLE 1.1
Crystal Structure and Magnetic Data for Eu Chalcogenides

Compound	EuO	EuS	EuSe	EuTe
$a(\text{Å})$	5.15	5.96	6.19	6.60
$T_c(K)$	67	16.3	4.3	9.6
$\Theta_p(K)$	76	19	9	-6
$J_1/k_B(K)$	0.58	0.2	$J_1 \sim J_2$	0.03
$J_2/k_B(K)$		-0.1	$J_1 \sim J_2$	-0.15
$\mu(\mu_B)$	6.8	6.87	6.7	

ferromagnetic planes given by the exchange integral $\langle J_0 \rangle$ and considerably weaker interactions with moments belonging to the first $\langle J_1 \rangle$ and second $\langle J_2 \rangle$ adjacent planes. The Hamiltonian of such a system is a one-dimensional Ising

$$H = - \sum_{\gamma=1}^{N} \sum_{r=-\infty}^{+\infty} J(r)m_j m_{j+r} - H \sum_j m_j + \varkappa_0 \qquad (1.1)$$

where H_0 refers to the interactions in a (001) plane, while $J(r)$ is an effective exchange integral between two moments situated on adjacent planes. H is the external magnetic field strength directed along the (001) axis. Taking into account the interactions with the nearest $\langle J_1 \rangle$ and the next-nearest $\langle J_2 \rangle$ neighbors the above hamiltonian yields either ferromagnetic or antiferromagnetic ordering when $J_1 < 0$ and $(J_2/J_1) < 1/4$. When $(J_2/J_1) > 1/4$ a modulated structure with the wave-vector $\cos \pi k = -J_1/4J_2$ appears below the Néel temperature.

Microscopic models of magnetic moment interactions operating in cerium pnictides have been proposed also (Takegahara et al., 1980). Strong ferromagnetic coupling in the (001) plane observed in CeAs and CeSb has been interpreted as resulting from a strong mixing of 4f electrons located close to the Fermi surface with the p band of a pnictide ion, the f-p hybridization. Thus, two types of exchange mechanisms should be considered: long range interactions of RKKY type and short-range superexchange of strong anisotropic character.

It has been observed in experiments that magnetic structures are stable in the presence of strong external magnetic fields or pressures and fully confirm the necessity of considering the above interactions (Rossat-Mignod et al., 1985).

A second illuminating example is provided by the compounds of europium with the elements of the sixth group, i.e., O, S, Se, and Te. The EuX phases exhibit also the rock-salt (NaCl)-type structure. As the lattice parameter arises from EuTe (see Table 1.1), the following transitions to the ordered state of moments localized on Eu ions have been found: EuO and EuS both order ferromagnetically below $T_c = 67$ and 16.3 K, respectively, while EuSe and EuTe show collinear antiferromagnetic ordering with T_N at 4.3 and 9.6 K,

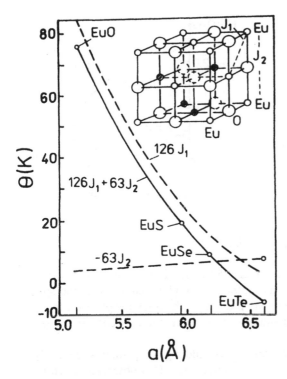

FIGURE 1.3. The changes in the exchange integrals J_1 and J_2 as a function of the lattice constants for the europium chalcogenides (McGuire et al., 1963).

respectively (Methfessel and Mattis, 1968). The magnetism of EuO and EuS can be described as an ideal Heisenberg ferromagnet and EuTe-antiferromagnet. In all cases, the magnetic interactions can be expressed in terms of two exchange integrals (see Table 1.1): $-J_1-$ bound to the interactions with the nearest neighbors and J_2 denoting the interactions with the next-nearest neighbors. In the case of EuTe, the exchange integrals can be evaluated from the equations:

$$k\Theta = 2/3S(S + 1) \sum_i z_i J_i = 126J_1 + 63J_2$$

$$kT_N = 2/3S(S + 1) \; (-z_2 J_2) = -63J_2$$

Θ is the paramagnetic Curie temperature. The J_1 and J_2 exchange integrals derived from the above equations are shown in Figure 1.3 (McGuire et al., 1963).

The molecular field theory shows that in the case of a ferromagnet the Curie temperature T_c and Θ_p are equal, so that only one parameter must be evaluated. The exchange integral J_2 thus describes the superexchange interactions operating at the angle of 180° via the anion. It can be estimated that while passing from EuTe to EuO, the exchange integral is reduced by a factor of 2. Following Goodenough (1963), the exchange integral J_1 describes the

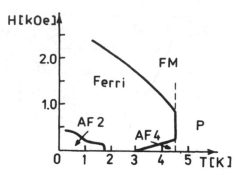

FIGURE 1.4. Magnetic phase diagram for EuSe (Wachter, 1971).

direct interaction between two cations via 5d electrons. The respective exchange integral is given by:

$$J_1 = 2b^2 J_{f-d}/4S^2 U^2$$

b is the overlap integral for the 4f and 5d band electrons of two adjacent cations. J_{f-d} is the 4f-5d interaction integral of a given cation, S is the spin of the 4f shell, and U is the excitation energy. The above expression for J_1 makes it possible to describe quantitatively the interactions in europium pnictides.

EuSe shows anomalous properties. Its ordering temperatures are fairly low, indicating weak Heisenberg-type interactions. This observation can be explained by the magnitude of the lattice parameter a, which is close to a_0. At this value J_1 changes its sign (see Figure 1.3). The magnitude of J_1 thus depends strongly on a; this is shown by the fact that the external pressure of p = 1 kbar raises T_N by 1 K. The small value of J_1 causes an increase in the contributions of the other exchange integrals. The magnetic phase diagram of EuSe is displayed in Figure 1.4 (Wachter, 1971). At the field H = O the Néel point is at 4.6 K. Below it, the antiferromagnetic ordering is given by the wave vector (0, 0, 1/2). At 2.8 K a transition to a ferromagnetic ordering was detected (+ + − sequence). Below T = 1.8 K an antiferrimagnetic structure with the wave vector (0, 0, 1) appears. When an external magnetic field is applied, the ferri- and ferromagnetic structures become stabilized.

In the course of the investigations of the physical properties of binary lanthanide phases, a number of new effects have been either discovered or confirmed. The effects:

- mixed valency
- Kondo lattice
- spin fluctuations
- heavy fermions

were found to be dependent on the electronic structure of the lanthanide ions; in particular, they are strongly related to the position of the 4f electron level

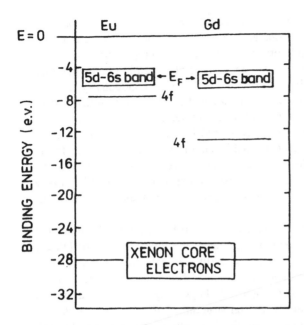

FIGURE 1.5. The energy levels for Eu and Gd (Legvold, 1980).

in respect to the Fermi energy. A typical distance E between the 4f-electron energy and the Fermi level is approximately 20 eV (see Figure 1.5). In the compounds showing either Kondo or mixed valency effects, this gap becomes smaller.

The state of mixed valency results from the action of either external or chemical pressures. The latter can be generated in the crystal lattice by substituting one of the component ions with another of larger (or smaller) ionic radius, leading to positive (or negative) internal pressure. This effect has been observed, for example, in the solid solutions $SmS_{1-x}Se_x$ and $Ce(Rh_{1-x}Pd_x)_3$ (Chazalviel et al., 1984). Owing to this pressure, one of the 4f electrons becomes delocalized. It is transferred from the narrow f-level to the conduction band. Two models have been used to describe this effect:

1. Promotion model — the transfer of a 4f-electron to the conduction band — brings about an unstable state with a limited lifetime.
2. Hybridization model — hybridization leads to widening of the f-level — a band arises.

The Kondo effect is observed in the systems containing diluted magnetic admixtures, which cause anomalies in the temperature dependence of electric resistivity. This effect has been discovered in a number of phases which contain cerium and ytterbium. The mechanism developed for diluted alloys can be adopted for binary intermetallic compounds. In some of them, for example in $CeAl_2$, long-range magnetic ordering appears at low temperatures (3.8 K) (Coqblin, 1982).

The appearance of spin fluctuations is explained as due to the competition between the exchange interactions and the hybridization which leads to the fulfillment of the Stoner condition. At a characteristic temperature, called the spin fluctuation temperature, T_{SF}, an anomaly is observed in the temperature dependence of the specific heat, in thermal and electric conductivity, as well as in magnetic susceptibility. T_{SF} characterizes the degree of hybridization of electron wave functions with the local magnetic moment and band electrons.

The heavy fermion effect was discovered for the first time in $CeCu_2Si_2$ (Steglich et al., 1979). The electron coefficient of the specific heat was found to be 1 J/mol K^2; hence, the presence of an electron with a large effective mass has been deduced. This effect has been subsequently observed in many binary and ternary intermetallic phases containing cerium and uranium (Steglich, 1989).

A wealth of new materials with a large spectrum of physical parameters found among the binary compounds stimulated the investigations of ternary compounds containing, apart from rare earth elements (R), and transition metals (T) and the elements of the fourth and fifth group (X). The $R_nT_mX_p$ phases have been found to exhibit not only large range of compositions (n, m, p), but also a large variety of crystal structure types and properties. For example, in the ternary system Ce–Ni–Si, 21 different phases appear (Bodak et al., 1973). Depending on the composition (n, m, p) they exhibit characteristic crystal structures and different physical properties.

The ternary systems with the crystal structure of $ThCr_2Si_2$ type attracted particular attention during the last few years because all of the most interesting properties which have been observed earlier in binary phases are shown by these phases, including mixed valence, Kondo lattice, and heavy fermion effects. Apart from this, more than 500 compounds with this structure type have been synthesized up to now, so that it became possible to collect a vast amount of data.

The other family of phases, which has been intensively studied lately are $R_2T_{14}B$ and $RT_{12-x}M_x$ compounds with the crystal structures of $Nd_2Fe_{14}B$ and $ThMn_{12}$ types. They have found practical applications as materials for the production of permanent magnets.

The RTX phases, which also show a variety of crystal structure types, have also been studied lately, since the magnetic properties of their uranium analogs turned out to be very intriguing (Tran and Troć, 1991; Sechovsky and Havela, 1992).

The crystal structures of the principal lanthanide (R) ternary systems with transition metals (T) and nonmetals X $R_nT_mX_p$ are described in Chapter 2; Chapter 3 contains the data concerning the exchange interactions and crystal electric field effects operating in these phases. Magnetic properties of $R_nT_mX_p$ compounds are described in detail in Chapter 4. Chapter 5 contains information on the phases which show hard magnetic properties.

Chapter 2

CRYSTAL STRUCTURES OF TERNARY
LANTHANIDE INTERMETALLICS

The basic data concerning the crystal structures of phases with different compositions, relationships among these structures, and the relevant chemical bonding schemes are discussed in this chapter.

2.1. RTX PHASES

A large number of equiatomic ternary rare earth intermetallic compounds with the general formula RTX (R = rare earth, T = transition element, and X = metalloid) have been synthesized and studied (Hovestreydt et al., 1982; Dwight et al., 1968; Bażela, 1987). They crystallize in several different types of structures, such as $MgCu_2$, MgAgAs, ZrOS, Fe_2P, AlB_2, Ni_2In, LaPtSi, LaIrSi, PbFCl, $MgZn_2$, TiNiSi, and $CeCu_2$.

The two crystal lattice parameters a/c and (a+c)/b can be used for the systematics of the various RTX type structures. This is illustrated in Figure 2.1. Each group of RTX metallic phases exhibits different values of these parameters. The exceptions are the hexagonal phases $CaIn_2$, $MgZn_2$, Ni_2In, AlB_2 and R_2RhSi_3-type phases.

It has been found that the a/c ratio contains information on the number of nearest neighbors (Shoemaker and Shoemaker, 1965; Rundqvist and Nawapong, 1966). The length of the "b" lattice parameter and the ratio (a+c)/b have also been found to be factors related to the coordination number. Consequently, the RTX phases can be classified into 11 groups.

2.1.1. PHASES WITH THE $MgCu_2$-TYPE STRUCTURE

The space group is Fd3m. The atomic positions are as follows:

> 8R in 8(a): 0, 0, 0; 1/4, 1/4, 1/4;
> 8T and 8X
> at random in 16(d): 5/8, 5/8, 5/8; 5/8, 7/8, 7/8;
>   7/8, 5/8, 7/8; 7/8, 7/8, 5/8;
>   + face-centering translation

Each R atom is surrounded by 12 (T,X) atoms at a distance of $a\sqrt{11}/8$. This distance amounts to 3.278 Å in NdMnGa. The other compounds with this structure are CeMnGa and NdMnGa (Brabers et al., 1992). The crystal structure of this type is shown in Figure 2.2.

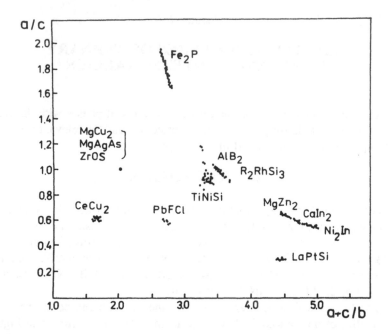

FIGURE 2.1.   Grouping of RTX compounds according to their axial ratios. The lattice constants a, b, and c correspond to the orthorhombic crystal structure. The orthorhombic lattice constants for hexagonal-type phases were taken according to the following way: $a_o = a_h$, $b_o = \sqrt{3}\, a_h$, $c_o = c_h$.

2.1.2. PHASES WITH THE MgAgAs-TYPE STRUCTURE

This structure type has cubic symmetry-space group $F\bar{4}3m$ with atoms

4R in 4(c):	1/4, 1/4, 1/4;
4T in 4(a):	0, 0, 0;
4X in 4(d):	3/4, 3/4, 3/4;
+ face-centering translation		

Each T atom has four closest R neighbors at corners of a tetrahedron $(d_{T-R} = a\sqrt{3}/4)$ and 4X also at corners of a tetrahedron at the same distance

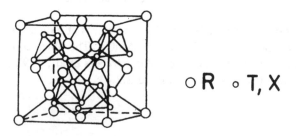

FIGURE 2.2.   The crystal structure of the cubic $MgCu_2$ type.

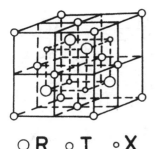

○ R   ○ T   ○ X   **FIGURE 2.3.** The crystal structure of cubic MgAgAs type.

(d = 2.829 Å in DyPdSb). The following phases crystallize in this structure type: RPdSb (Marazza et al., 1980a), RPdBi (Marazza et al., 1980b) RNiSb and RAuSn (Dwight, 1976). The crystal structure of MgAgAs type is shown in Figure 2.3.

2.1.3. PHASES WITH THE LaIrSi- (ZrOS)-TYPE STRUCTURE

The RTSi compounds, in which R is a light, rare earth atom (La–Eu) and T represents Rh or Ir, crystallize in a primitive cubic structure (space group $P2_1 3$) (Chevalier et al., 1982a). In the crystal structure of the LaIrSi (see Figure 2.4), the lanthanum, iridium, and silicon atoms are placed in three fourfold 4(a) sites: x, x, x; 1/2+x, 1/2−x, $\bar{x}$; $\bar{x}$, 1/2+x, 1/2−x; 1/2+x, $\bar{x}$, 1/2+x with the following positional parameters x: x_{La} = 0.635, x_{Ir} = 0.923, and x_{Si} = 0.326.

Each silicon atom is surrounded by four lanthanum and three iridium atoms. The lanthanum atoms are found to be in an elongated silicon tetrahedron with one La–Si distance of 3.392 Å and three of 3.172 Å. The iridium atoms form an equilateral triangle with Ir–Ir distances of 3.975 Å. The silicon atom is just slightly out of the plane of the iridium triangle at a distance of 2.323 Å from each iridium atom.

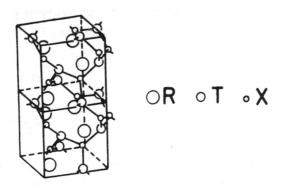

○ R   ○ T   ○ X

FIGURE 2.4. The crystal structure of cubic LaIrSi type.

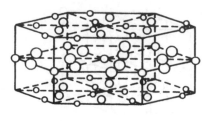

FIGURE 2.5. The crystal structure of the hexagonal
Fe$_2$P type.

2.1.4. PHASES WITH Fe$_2$P-TYPE STRUCTURE

Two series of lanthanide compounds exhibit this structure type: RCuAl
and RNiAl (R = Ce–Lu) (Dwight et al., 1968). The space group is P$\bar{6}$2m.
The atomic sites are:

3Al in 3(f):	x, 0, 0;	0, x, 0;	$\bar{x}$, $\bar{x}$, 0;
3R in 3(g):	x, 0, 1/2;	0, x, 1/2;	$\bar{x}$, $\bar{x}$, 1/2;
2T in 2(c):	1/3, 2/3, 0;	2/3, 1/3, 0;	

The structure has a distinct layer character: hexagonal planes with mixed
atomic composition are stacked along the c axis. The existence of two sets
of interatomic distances between nickel atoms and its neighbors is the dis-
tinctive feature of this structure type: the Ni atom at 0, 0, 1/2 has six Al
neighbors on the adjacent planes with the Ni–Al distance of 2.52 Å (see
Figure 2.5) while each Ni in the 2(c) site coordinates 3 Al in the same plane
with d_{Ni-Al} = 2.81 Å. Cerium atoms form a planar network, each Ce atom
having four Ce neighbors at a distance of 3.62 Å i.e., the sum of Pauling
metallic radii of cerium, which amounts to 3.636 Å. The complicated bonding
scheme operating in this structure type is reflected by a nonlinear variation
of the lattice constants of RCuAl and RNiAl with the R^{3+} ionic radius (Figure
2.6).

2.1.5. PHASES WITH THE MgZn$_2$-TYPE STRUCTURE

This structure type is described by the space group P6$_3$/mmc. Atoms are
located in:

4R in 4(f):	1/3, 2/3, z;		2/3, 1/3, $\bar{z}$;
	2/3, 1/3, 1/2+z;		1/3, 2/3, 1/2−z;
4T and 4X			
at random in 2(a):	0, 0, 0;		0, 0, 1/2;
and in 6(h):	±(x, 2x, 1/4;	2$\bar{x}$, $\bar{x}$, 1/4;	x, $\bar{x}$, 1/4;)

RTAl (T = Fe or Co) phases crystallize in this structure (Oesterreicher,
1977c). The unit cell of this structure type is displayed in Figure 2.7.

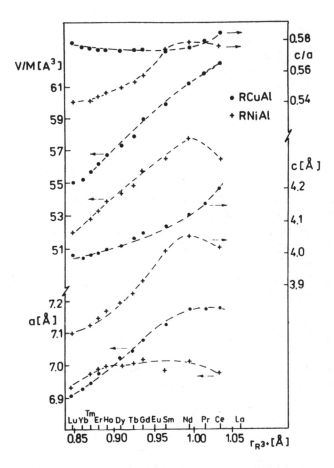

FIGURE 2.6. Variation in the lattice constants a and c, cell volumes per molecule V/M and c/a ratios with the radius $r_{R^{3+}}$ of the rare earth component for RCuAl and RNiAl compounds.

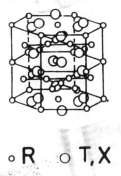

○ R ○ T,X

FIGURE 2.7. The crystal structure of hexagonal MgZn$_2$ type.

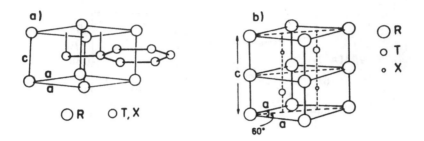

FIGURE 2.8. The crystal structures of hexagonal (a) AlB_2 type and (b) Ni_2In (ZrBeSi) types.

2.1.6. PHASES WITH THE AlB_2 AND RELATED STRUCTURE TYPES

Many ternary equiatomic RTX phases exhibit two closely related structures known as AlB_2 and Ni_2In (ZrBeSi) types. Both structures are displayed schematically in Figure 2.8.

In the AlB_2-type structure, which is described by the space group P6/mmm, the atoms are situated in the following sites:

R in 1(a)   0, 0, 0;

T and X in 2(c)   1/3, 2/3, 1/2;   2/3, 1/3, 1/2;   at random

(Rieger and Parthé, 1969). On the other hand, the Ni_2In (ZrBeSi)-type structure belongs to the space group $P6_3$/mmc with atoms placed at:

2R at 2(a)   0, 0, 0;   0, 0, 1/2;

2T at 2(c)   1/3, 2/3, 1/4;   2/3, 1/3, 3/4;

2X at 2(d)   2/3, 1/3, 1/4;   1/3, 2/3, 3/4;

(Mugnoli et al., 1984; Bażela et al., 1985b). In this case, the doubling of the unit cell along the hexagonal axis is required. RCuSi compounds were found to belong to this structure type. Each R atom has 12 nearest Cu and Si neighbors placed in the two adjacent planes. In addition, it has 6R atoms in the same plane at a distance equal to the lattice parameter a and 2R atoms at a distance c/2. The structure exhibits a layer character: planes composed of Cu and Si atoms and of R atoms are stacked along the hexagonal axis. In HoCuSi, the respective interatomic distances amount to: d_{Ho-Ho} = a = 4.133 Å (six times), d_{Ho-Ho} = c/a = 3.689 Å (twice), d_{Ho-Cu} = d_{Ho-Si} = 3.016 Å. (Bażela et al., 1985b). The Ho–Cu and Ho–Si distances are close to the sum of the Pauling metallic radii of the respective elements for the coordination number 12. The Ho–Ho distances along the hexagonal axis compare well with the sum of the Pauling metallic radius of holmium. Since they are distinctly shorter than Ho–Ho distances in the basal plane, chains of Ho atoms propagating along the hexagonal axis can be distinguished.

The ternary silicides R_2RhSi_3 R = Y,La,Ce,Nd,Sm,Gd–Er crystallize in a hexagonal structure which is derived from the AlB_2 type (Chevalier et al.,

a)   b)

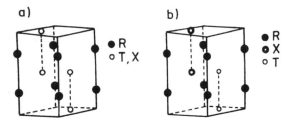

FIGURE 2.9. The crystal structures of hexagonal CaIn₂ (a) and GaGeLi (b) types.

1984). The a and c parameters are approximately doubled values in comparison to disilicides RSi_2. Si and Rh atoms are located inside R_6 distorted trigonal prisms and ordered in planes perpendicular to the c axis. Due to the short Si–Si distances (2.32 Å), the Si atoms form hexagons which are connected via the Rh atoms. The Rh–Si distances which are the same as the Si–Si distances are clearly shorter than the sum of the Si and Rh atomic radii. Since the Si–Si and Rh–Rh distances (3.856 Å) along the c axis are fairly long, Si and Rh atoms form the two-dimensional sublattice. The rare-earth atoms form a three-dimensional framework.

In this structure, which is described by the space group $P\bar{6}2c$, the atoms are situated in the following sites:

R(1) in 2(b)	0, 0, 1/4;	0, 0, 3/4;
R(2) in 6(h)	x, y, 1/4;	etc.
Rh in 4(f)	1/3, 2/3, z_2,	etc.
Si in 12(i)	x_3, y_3, z_3,	etc.

In the case of x = 0.481, y = 0.019, $z_2 \sim 0$, and x_3 = 0.167, y_3 = 0.333, $z_3 \sim 0$ (Chevalier et al., 1984).

2.1.7. PHASES WITH THE CaIn₂-TYPE STRUCTURE

A similar coordination scheme has been found in the structure of a number of RAgSn compounds (Bażela et al., 1992). Their structure can be described as either the CaIn₂ or GaGeLi type. (E. Hovestreydt et al., 1982). The former belongs to the space group $P6_3/mmc$ with:

2R atoms in 2(b):	0, 0, 1/4;	0, 0, 3/4;
2Ag and 2Sn		
atoms at random in 4(f):	1/3, 2/3, z;	2/3, 1/3, $\bar{z}$;
	2/3, 1/3, 1/2+z	1/3, 2/3, 1/2−z;

while the latter to the space group $P6_3mc$ with:

2R atoms in 2(a):	0, 0, z_1;	0, 0, 1/2+z_1;	$z_1 \sim 0.25$
2Ag atoms in 2(b):	1/3, 2/3, z_2;	2/3, 1/3, 1/2+z_2;	$z_2 \sim 0$
2Sn atoms in 2(b):	2/3, 1/3, z_3;	1/3, 2/3, 1/2+z_3;	$z_3 \sim 0$

Figure 2.9 shows both structures.

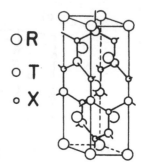

○ R

○ T

○ X

FIGURE 2.10.   The crystal structure of tetragonal LaPtSi type.

In HoAgSn whose structure was described in terms of the GaGeLi-type, the principal interatomic distances are d_{Ho-Ho} = 4.672 Å (six times), d_{Ho-Ho} = 3.659 Å (twice), d_{Ho-Ag} = d_{Ho-Sn} = 3.259 Å. They also amount to the sums of Pauling metallic radii.

2.1.8. PHASES WITH THE LaPtSi-TYPE STRUCTURE

The LaPtSi-type structure is characterized by an ordered distribution of atoms in the sites of the space group I4$_1$md:

4R in 4(a):	0, 0, z_1;	0, 1/2, 1/4 + z_1;
4T in 4(a):	0, 0, z_2;	0, 1/2, 1/4 + z_2;
4X in 4(a):	0, 0, z_3;	0, 1/2, 1/4 + z_3;

+ body-centering translation

where z_1 = 0, z_2 = 0.585, and z_3 = 0.419 (Klepp and Parthé, 1982)

The structure (see Figure 2.10) exhibits a distinct layer character with equiatomic sheets stacked along the tetragonal axis with sequence –R–T–X–R–T–X–R–. In total, there are 12 sheets in a unit cell.

Each R atom has four R-closest neighbors in the same layer with R–R distance of the lattice constant a (d_{R-R} = 4.249 Å in LaPtSi), 4T atoms in the adjacent sheet above (d_{La-Pt} = 3.249 Å), and 4X atoms in the adjacent sheet below (d_{La-Si} = 3.230 Å). These T and X atoms are located on the apexes of a flattened cube. In addition, however, there are more 2T atoms in the next-nearest sheet below (d_{La-Pt} = 3.204 Å) and 2X in the next-nearest sheet above (d_{La-Si} = 3.250 Å). Each T atom is surrounded by 3X atoms at the apexes of a triangle: one in the nearest sheet below (d_{Pt-Si} = 2.410 Å) and two X atoms in the adjacent sheet above (d_{Pt-Si} = 2.450 Å) (Klepp and Parthé, 1982).

2.1.9. PHASES WITH PbFCl-TYPE STRUCTURE

This type of crystal structure has been found in a few lanthanide ternary systems: RCoSi (R = La, Pr, Nd, Sm) and RFeSi (R = La–Ho, Yb) (Bodak et al., 1970). It is described by the space group P4/nmm with atoms distributed among the following positions:

2R in 2(c):	1/4, 1/4, z_1;	3/4, 3/4, $\bar{z}_1$;
2X in 2(c):	1/4, 1/4, z_2;	3/4, 3/4, $\bar{z}_2$;
2T in 2(a):	3/4, 1/4, 0;	1/4, 3/4, 0;

In RTX compounds z_1 and z_2, parameters are usually about 0.7 and 0.2, respectively. The structure is thus built of equiatomic sheets piled up along the tetragonal axis with the following sequence: –T–X–R–R–X–T–.

In the typical compound CeFeSi, the relevant shortest interatomic distances are as follows: each Ce atom is surrounded by 4Si atoms (d_{Ce-Si} = 3.053 Å and two Si atoms at a longer distance of 3.356 Å, forming an elongated octahedron. In addition, there are four iron atoms at a distance of 3.005 Å, so that the coordination polyhedron has a rather complex shape. Each iron atom is surrounded by a flattened tetrahedron of silicon atoms with an Fe-Si distance of 2.349 Å. Apart from this, it has four iron neighbors in the same sheet (z = 0) at a distance amounting to a $\sqrt{2}/2$ = 2.872 Å in CeFeSi. The structure of the PbFCl type is displayed in Figure 2.11.

2.1.10. COMPOUNDS WITH THE TiNiSi-TYPE STRUCTURE

The crystal structure of the TiNiSi type belongs to the space group Pnma. The atoms (four formal molecules) are in the 4(c) sites: x, 1/4, z; $\bar{x}$, 3/4, $\bar{z}$; 1/2 – x, 3/4, 1/2 + z; 1/2 + x, 1/4, 1/2–z. Each kind of atom has its characteristic values of the two free positional parameters. The unit cell of the TiNiSi type is displayed in Figure 2.12. This structure also shows a distinct layer character: multiatomic layers at y = 1/4 and 3/4 are piled up in the direction of the b axis. The R ions form chains running along the a axis, which are interconnected to adjacent chains to and from a three-dimensional network. For example, in HoRhSi (Bażela et al., 1985a) the Ho–Ho distance within the chain is 3.506 Å, while the shortest interchain Ho–Ho distance amounts to 3.513 Å. On the other hand, each Rh atom has two Si neighbors in the same plane with R–Si distances of 2.397 and 2.536 Å. Two other Si atoms belong to the adjacent planes with d_{Rh-Si} of 2.491 Å. The coordination polyhedron around a Rh atom is thus a fairly distorted tetrahedron. The observed mean value of Rh–Si bond length amounts to 2.479 Å, i.e., it is close to the sum of covalent radii of rhodium and silicon and indicates strong chemical

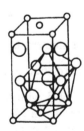

○X   ○R   ○T   FIGURE 2.11.   The crystal structure of tetragonal PbFCl type.

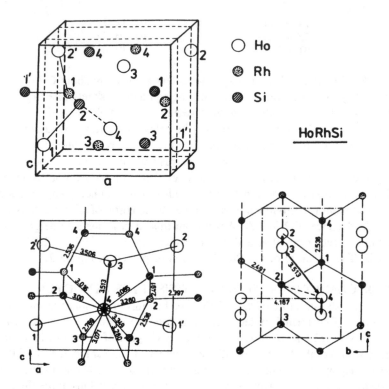

FIGURE 2.12. The crystal structure of TiNiSi type (HoRhSi) and its projections on the a-c and b-c planes.

bonding within the tetrahedron. The coordination tetrahedra share common corners and edges and form a three-dimensional honeycomb framework with Ho atoms inside.

The crystal structure of the TiNiSi type can be derived from the structure of the ZrBeSi type by small displacements of the atoms. The orthorhombic unit cell of TiNiSi is related to the hexagonal cell of the ZrBeSi (Ni_2In) type by the following relation:

$$a_o \sim c_h; \quad c_o \sim \sqrt{3}a_h; \quad b_o \sim a_h; \quad \quad (2.1)$$

To characterize the distortion, an orthorhombicity parameter: $\delta = \dfrac{c_o}{\sqrt{3}b_o} - 1$ has been introduced. It indicates the degree of deformation of the hexagonal basal plane from its perfect shape. The variation of δ with lanthanide atomic radius for a number of RTX compounds is shown in Figure 2.13. The deviation from a perfect hexagonal lattice is thus dependent on the lanthanide atomic radius. It follows from Figure 2.13 that the structures of HoNiSi and GdPtSi can be alternatively considered as being of the ZrBeSi type, since in both cases δ is zero.

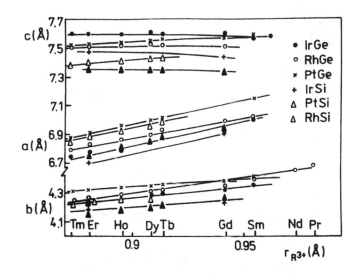

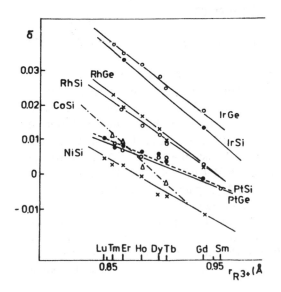

FIGURE 2.13. Variation in the lattice constants a, b, and c with the radius of the rare earth component for RTX phases with the TiNiSi-type crystal structure (Bażela, 1987).

Orthorhombic unit cell parameters of a number of RTX phases plotted against the atomic radii of the lanthanide elements (see Figure 2.14) reveal that the c parameter does not depend on the kind of lanthanide element; however, the a and b parameters' values rise with r_R. This effect reflects the R–R bonding which operates in the chain propagating along the a and b axes. To analyze more in detail the influence of the bonding scheme on the lattice

FIGURE 2.14. Variation of the orthorhombicity parameter δ with the radius r_{R3+} for RTX compounds (Bażela, 1987).

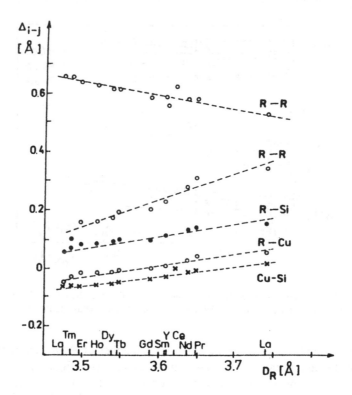

FIGURE 2.15.   Plots of observed Δ values against radius $r_{R^{3+}}$ for RCuSi compounds (Bażela, 1987).

parameters of RRhSi (TiNiSi type structure) and RCuSi (ZrBeSi-type structure), the following parameter has been proposed:

$$\Delta_{i-j} = d_{i-j} - (r_i + r_j) \qquad (2.2)$$

(Pearson and Villers, 1984). Here, d_{i-j} is observed in the experiment distance between the i and j atoms, r_i and r_j being their ionic radii, respectively. In both groups of compounds Δ_{i-j} change linearly with $r_{R^{3+}}$ (see Figure 2.15). Therefore, an additional diagnostic parameter f_{ij} has been introduced, related to Δ_{i-j} in the following way: $\Delta_{i-j} = f_{i-j} R + k$. f_{ij} can be determined from the plot displayed in Figure 2.15. Table 2.1. lists the f_{i-j} for RCuSi and RRhSi phases. In the former, f_{ij} is zero for all R–Cu, R–Si, and Cu–Si bonds, so that the corresponding Δ_{R-Cu}, Δ_{R-Si}, and Δ_{Cu-Si} parameters are also zero. Hence, $d_{R-Cu} = r_R + r_{Cu}$, etc.: the observed interatomic distances are equal to the sums of respective ionic radii. It can thus be concluded that, in the RCuSi phases, the particular i–j contacts control the unit-cell dimensions. In the case of RRhSi phases $f_{Rh-Si} = +0.05$ for Rh and Si atoms in the same plane, but $f_{Rh-Si} = +0.02$ when Si atoms lie in the adjacent planes. For the R–Rh bonds

TABLE 2.1
Values of the f_{i-j} Parameter for RCuSi and RRhSi Compounds[a]

Compound	f_{R-R}	f_{R-T}	f_{R-X}	f_{T-X}	f_{T-T}
RCuSi	−0.492	0.376	0.376	0.293	
	+1.094				
RRhSi					
Atoms in neighboring ac planes	−0.50	−0.04	0.02	−0.18	0.23
		0.29	0.19		
Atoms in the same plane	−0.64	−0.32	−0.32	0.05	
		−0.54		0.55	

[a] From Bażela, 1987.

$f_{R-Rh} = -0.04$. These data suggest that the bonding system operating in the structure of TiNiSi type shows heterodesmic character.

2.1.11. PHASES WITH CeCu$_2$-TYPE STRUCTURE

The structure of this type belongs to the space group Imma. The atomic positions are:

4R in 4(e):	0, 1/4, z_1;	0, 3/4, $\bar{z}_1$;
4T and 4X	1/2, 3/4, 1/2+z_1;	1/2, 1/4, 1/2 − z_1;
at random in 8(h):	0, 1/2 − y, z_2;	0, $\bar{y}$, $\bar{z}_2$;
	1/2, 1/2 + y, 1/2 + z_2;	1/2, y, 1/2 − z_2;
	0, y, z_2;	0, 1/2 + y, z_2;
	1/2, y, 1/2 + z_2;	1/2, 1/2 − y, 1/2 − z_2;

where $z_1 = 0.54$ and $z_2 = 0.167$ (Hovestreydt et al., 1982).

The structure (see Figure 2.16) can be considered as composed of sheets stacked along the b axis, since the experimentally found values of y are close to zero. The sequence of sheets is: −R–T, X–R–T, X−. The R atoms form a system of interpenetrating zig-zag chains propagating in the direction of the

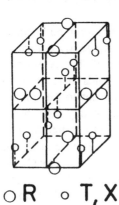

○ R ○ T, X

FIGURE 2.16. The crystal structure of CeCu$_2$ type.

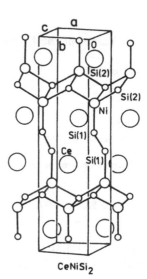

FIGURE 2.17.   The crystal structure of CeNiSi$_2$ type.   CeNiSi$_2$

a axis (the R–R distance is 3.868 Å in PrPtGe) and the b-axis (d_{R-R} = 3.701 Å). Each Ce atom has 6 + 6 (Pt$_{0.5}$Ge$_{0.5}$) atoms located on the adjacent planes with d(Ce − Pt$_{0.5}$Ge$_{0.5}$) distances ranging from 3.092 to 3.233 Å. The interatomic distances within the (Pt, Ge) planes range from 2.557 to 2.991 Å — each statistical (Pt, Ge) atom has four neighbors at corners of a distorted parallelogram. This structure type has been discovered in the R(Pd$_{0.5}$Ge$_{0.5}$)$_2$, (R = La–Tm) and R(Pt$_{0.5}$Ge$_{0.5}$)$_2$ (R = Ce, Pr, Nd) (Hovestreydt et al., 1982).

2.2. RTX$_2$ PHASES

The RTX$_2$ compounds crystallize in different orthorhombic structures. A large number of them crystallize in the orthorhombic CeNiSi$_2$-type structure (see Figure 2.17) (Bodak and Gladyshevskii, 1970). The space group is Cmcm; each atom, i.e., R, T, X$_1$, and X$_2$ occupies the 4(c) site with the following coordinates: 0, y, 1/4; 0, $\bar{y}$, 3/4; 1/2, 1/2+y, 1/4; 1/2, 1/2−y, 3/4.

The characteristic feature of this structure is the coordination polyhedron of the transition metal atoms which are located inside a slightly deformed square pyramid with X$_1$ atoms at its corners. The base of this pyramid is formed by X$_2$ atoms; the top corner is an X$_1$ atom. The pyramids share common corners of X$_2$ and form two-dimensional layers perpendicular to the b-axis. The layers are interconnected by X$_1$–X$_1$ bonds. R atoms occupy large holes in this framework having four X$_1$, four T, and four X$_2$ atoms below. The bond lengths determined for HoCoSi$_2$ (Szytuła et al., 1989) can be considered as characteristic: the five Co–Si distances inside the pyramid amount to: Co–Si 2.291 Å (two bonds); Co–Si$_2$: 2.22 Å (two bonds); Co–Si$_1$: 2.218Å (one bond) (average 2.248 Å). These values compare fairly well with the sum of the Pauling metallic radii for respective elements corrected for the coordination number 5 for Co and 4 for silicon. On the other hand, the Si_2–Si_2 and Si_1–Si_2 distances of 2.769 and 3.819 Å, respectively, are larger than the Si–Si bonds amounting to 2.34 Å in elemental silicon.

The pyramids have common corners of Si_2 which are located close to the planes with y = 1/4 and 3/4, while the top corners of Si_1 point alternately up and down along the y axis resulting in two-dimensional layers normal to the b axis. The Si_1–Si_1 bond are equal 2.366 Å.

Ho^{3+} ions occupy large holes in this framework having four Si_1 atoms above and four Co and four Si_2 atoms below, all at the closest distance of about 3 Å.

The $CeNiSi_2$ structure can be considered as closely related to the $ThCr_2Si_2$-type structure (space group I4/mmm). In both structures, coordination polyhedra with a 3d atom inside form layers perpendicular to the longest axis. The layers are interconnected by Si–Si bonds. R atoms are located in holes of the layers. The sequence of atomic planes perpendicular to the longest axis is the following: R–T–Si_2–T–R–Si_1–Si_1–R–T–Si_2–T–R in the case of $RTSi_2$ compounds and R–Si–T–Si–R–Si–T–Si–R in RT_2Si_2 systems.

Figure 2.18 shows the dependence of the lattice parameters and the unit cell volumes of $RNiSi_2$ and $RNiGe_2$ on the lanthanide ionic radius r_{R3+} (Gil et al., 1993b). A deviation is observed in the case of $CeNiSi_2$, in which a mixed valency state has been detected (Pecharsky et al., 1991).

The following compounds exhibit this structure type:

1. $RCoSi_2$, R = Y, Ce, Nd, Sm, Gd–Tm (Pellizone et al., 1982)
2. $RNiSi_2$, R = La–Lu (Bodak and Gladyshevskii, 1970)
3. $RRhSi_2$, R = La–Sm, Gd (Chevalier et al., 1984b)
4. $RIrSi_2$, R = La–Sm, Gd, Tb (Chevalier et al., 1984b)

A large number of germanides and stannides (T = Mn, Fe, Co, Ni, and Cu) also crystallize in this type of crystal structure. Defected structures have been also observed in these phases (François et al., 1990; Gil et al., 1993b; Weitzer et al., 1992; Venturini et al., 1990).

$RMnSi_2$ (R = La–Sm) and $RFeSi_2$ compounds crystallize in the $TbFeSi_2$-type structure (Yarovetz and Gorelenko, 1981), which is also described by the space group Cmcm with the atoms in 4(c) positions. The position parameters determined for $LaMnSi_2$ are given in Table 2.2 (Malaman et al., 1990a). The $TbFeSi_2$-type structure is very similar to $CeNiSi_2$. In both structure types, all of the R, T, and Si atoms are located in alternating layers stacked in the sequence: R–T–Si–T–R–Si–Si–R–T–Si–T–R for the $CeNiSi_2$ type and R–Si–Si–R–Si–T–T–Si–R for the $TbFeSi_2$ type

$NdRuSi_2$ was found to crystallize in a monoclinic structure, which is described as a distorted variant of the $CeNiSi_2$-type (Cenzual et al., 1992). It belongs to the space group $P2_1m$ with the following atomic positions:

TABLE 2.2
Crystal Structure Data for CeNiSi$_2$, HoCoSi$_2$, TbMn$_x$Ge$_2$, LaMnSi$_2$, and PrFeSi$_2$

Compound	CeNiSi$_2$[a]	HoCoSi$_2$[b]	TbMn$_x$Ge$_2$[c]	LaMnSi$_2$[d]	PrFeSi$_2$[e]
Type of crystal structure	CeNiSi$_2$	CeNiSi$_2$	CeNiSi$_2$	TbFeSi$_2$	TbFeSi$_2$
a(Å)	4.141(2)	3.932(3)	4.112(3)	4.191(3)	4.103(3)
b(Å)	16.418(10)	16.052(24)	15.870(28)	17.88(1)	17.04(3)
c(Å)	4.068(2)	3.894(5)	3.907(9)	4.073(3)	4.016(3)
y_R	0.1070(2)	0.1080(10)	0.094(2)	0.1018(5)	0.1044(6)
y_T	0.3158(8)	0.3199(10)	0.313(7)	0.7510(9)	0.7509(4)
y_{X1}	0.4566(2)	0.4581(18)	0.457(1)	0.4651(12)	0.4646(7)
y_{X2}	0.7492(4)	0.7466(17)	0.752(3)	0.3448(9)	0.3195(8)

[a] Bodak and Gladyshevskii, 1970.
[b] Szytuła et al., 1989.
[c] Gil et al., 1993.
[d] Malaman et al., 1990a.
[e] Malaman et al., 1990b.

2.3. RT$_2$X PHASES

The family of compounds with 1:2:1 stoichiometry crystallizes in two groups of different structure types: in the structure found in the Heusler alloys and in the orthorhombic Fe$_3$C-type structure.

The rare-earth-based Heusler alloys RT$_2$X have cubic symmetry (space group Fm3m) (see Figure 2.20) (Heusler, 1934). This highly ordered structure may be described either as body-centered cubic with a face-centered super-structure or in terms of four interpenetrating f.c.c. sublattices A,B,C,D with the following coordinates.

A	B	C	D
0 0 0	1/4 1/4 1/4	1/2 1/2 1/2	3/4 3/4 3/4
0 1/2 1/2	1/4 3/4 3/4	1/2 0 0	3/4 1/4 1/4
1/2 0 1/2	3/4 1/4 3/4	0 1/2 0	1/4 3/4 1/4
1/2 1/2 0	3/4 3/4 1/4	0 0 1/2	1/4 1/4 3/4

In RT$_2$X compounds the A and C sites are occupied by Cu, the B sites by R, and the D sites by X atoms.

The Heusler phases occur with rare earths in the following systems: RAg$_2$In (Galera et al., 1984), RAu$_2$In (Besnus et al., 1985b) RCu$_2$In (Felner, 1985), as well as in some RPd$_2$Sn (Malik et al., 1985a, 1986) and RNi$_2$Sn (Skolozdra et al., 1982) systems.

In the case of TbPd$_2$Sn compounds, a structural transformation has been observed at low temperatures (Umarji et al., 1985).

TABLE 2.3
Atomic Positions for YIrGe$_2$ (Space Group Immm)[a]

Atoms	Positions	Symmetry	x	y	z
Y$_1$	4(i)	mm	0	0	0.2607(3)
Y$_2$	4(h)	mm	0	0.2057(2)	0.5
Ir	8(1)	m	0.5	0.1459(1)	0.2494(1)
Ge$_1$	4(h)	mm	0	0.0785(3)	0.5
Ge$_2$	4(g)	mm	0.5	0.0770(3)	0.5
Ge$_3$	8(1)	m	0.5	0.2998(2)	0.3474(3)

Note: a = 4.2635(9) Å, b = 15.9786(21) Å, c = 8.8134(16) Å.

[a] François, 1986.

The RPd$_2$Si (R = La–Lu) and RPd$_2$Ge (R = Gd–Er) (Jorda et al., 1983) compounds crystallize in the orthorhombic Fe$_3$C-type structure (Moreau et al., 1982). This structure, which belongs to the Pnma space group, can be described as built from trigonal prisms whose corners are occupied by two rare-earth atoms (4c site) and four Pd atoms (8d site). In its center is a Si atom.

In this structure R and Si atoms are in 4(c) positions x, 1/4, z; $\bar{x}$, 3/4, $\bar{z}$; 1/2 − x, 3/4, 1/2 + z; 1/2 + x, 1/4, 1/2 − z whereas Pd atoms in 8(d) position: x, y, z; 1/2 + x, 1/2 − y, 1/2 − z; $\bar{x}$, 1/2 + y, $\bar{z}$; 1/2 − x, $\bar{y}$, 1/2 + z; $\bar{x}$, $\bar{y}$, $\bar{z}$; 1/2 − x, 1/2 + y, 1/2 + z; x, 1/2 − y, z; 1/2 + x, y, 1/2 − z, where

	x	y	z
Y	0.0303	0.5	0.144
Pd	0.1767	0.0517	0.5928
Si	0.362	0.5	0.853

The segments of these structures are presented in Figure 2.21. The variation of the lattice parameters with the trivalent ionic radii of the rare earth

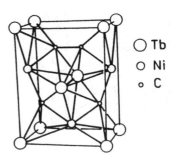

○ Tb
○ Ni
∘ C

FIGURE 2.19. The crystal structure of TbNiC$_2$.

TABLE 2.4
Interatomic Distances in the TbNiC$_2$ (in Å)[a]

C bond lengths			Metal atom distances		
C–C	1.375	1x	Tb–Ni	2.961	6x
C–Ni	1.947	4x	Tb–Ni	2.964	8x
C–Ni	2.000	2x	Tb–Tb	3.594	5x
C–Tb	2.653	8x	Tb–Tb	3.778	8x
C–Tb	2.672	8x	Ni–Ni	3.778	2x

[a]   Yakinthos et al., 1989.

is shown in Figure 2.22. This plot indicates that the rare-earth ions in these compounds are trivalent, except for europium, which appears to be either divalent or in the intermediate valence state. While the b parameter has been found to be almost independent of the ionic radii of the rare-earth element, the a and c parameters show strong dependence due to anisotropy of the bonding schemes.

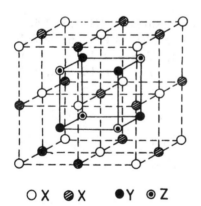

○ X   ◐ X   ●Y   ◉ Z

FIGURE 2.20.   The Heusler alloy type structure.

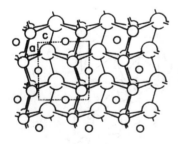

Height R Pd Ge
≈0.05
1/4
≈0.45

FIGURE 2.21.   The crystal structure of the YPd$_2$Si (the Fe$_3$C-type) derivative structure.

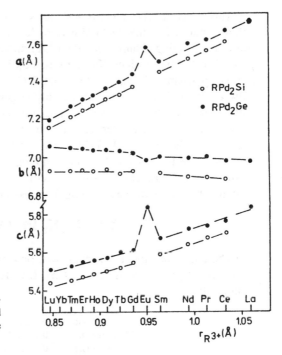

FIGURE 2.22. The variation of the lattice parameters a, b and c of RPd_2Si and RPd_2Ge compounds with the radii of the trivalent rare earth ions.

2.4. $R_2T_3Si_5$ PHASES

A number of rare-earth-ion silicides crystallize in the tetragonal $Sc_2Fe_3Si_5$-type structure space group (P4/mnc) (Bodak et al., 1977a). The unit cell is presented in Figure 2.23. The atoms occupy the following positions:

Sc atoms in	8(h) site	(0.072	0.236	0)
Fe_1	8(h) site	(0.377	0.358	0)
Fe_2	4(d) site	(0	0.5	0.25)
Si_1	8(g) site	(0.175	0.675	0.25)
Si_2	4(e) site	(0	0	0.246)
Si_3	8(h) site	(0.185	0.976	0)

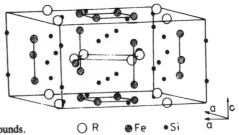

FIGURE 2.23. The unit cell of $R_2T_3Si_5$ compounds.   ○ R   ◉ Fe   ● Si

This structure type can be described as an arrangement of one-dimensional columns parallel to the (001) direction. It gives rise to two different sets of iron sites, forming chains along the (001) direction (Fe_2) and isolated squares parallel to the basal plane (Fe_1). In Sc_2Fe_3Si, S the Fe–Fe distance in both sets is approximately equal (2.64 and 2.67 Å), while the nearest distance between the different sets is considerably larger (4.10 Å). Thus, the transition metals form two types of clusters, although not in the most stringent sense, since the Fe–Fe, intercluster distances are larger than in iron metal (2.48 Å); Fe(1) is octahedrally and Fe(2) tetrahedrally surrounded by silicon atoms (d_{Fe-Si} = 2.28 to 2.57 Å). The Fe(1) squares alternate with scandium squares along the (001) direction forming distorted square antiprisms centered by silicon. The rare-earth nearest-neighbor distance is 3.70 Å.

2.5. RT$_2$X$_2$ PHASES

The tetragonal $BaAl_4$ structure was originally determined more than 50 years ago (Andress and Alberti, 1935). It is described in the space group I4/mmm with

Ba in 2(a) site:	0, 0, 0
Al in 4(d) site:	0 1/2 1/4, 1/2 0 1/4
Al in 4(e) site:	0 0 z, 0 0 $\bar{z}$
	+ body-centering translation

In $BaAl_4$ the free parameter z was determined to be 0.375. From this structure the following three types have been derived: $ThCr_2Si_2$ (Ban and Sikirica, 1965), $CaBe_2Ge_2$ and $BaNiSn_3$ (Dorrscheidt and Schafer, 1978).

The $ThCr_2Si_2$ structure is described by the space group I4/mmm with the following site occupation:

M (Th) at 2(a) site:	0, 0, 0;
T (Cr) at 4(d) site:	0, 1/2, 1/4 1/2, 0, 1/4;
X (Si) at 4(e) site:	0, 0, z_x 0, 0, $\bar{z}_x$;
	+ body-centering translation

On the other hand, the $CaBe_2Ge_2$ structure is described by the space group P4/nmm with site occupations as follows:

M (Ca) at 2(c) site:	1/4, 1/4, z_M 3/4, 3/4, $\bar{z}_M$;
T1(Be1) at 2(a) site:	3/4, 1/4, 0 1/4, 3/4, 0;
T2(Be2) at 2(c) site:	1/4, 1/4, z_T 3/4, 3/4, $\bar{z}_T$;
X1(Ge1) at 2(b) site:	3/4, 1/4, 1/2 1/4, 3/4, 1/2;
X2(Ge2) at 2(c) site:	1/4, 1/4, z_X 3/4, 3/4, $\bar{z}_X$

The $BaNiSn_3$ type structure is described by the space group I4/mm with the following site occupation:

R (Ba) at 2(a) site: 0, 0, z_1; $z_1 = 0.3455$
T (Ni) at 2(a) site: 0, 0, 0;
X (Si) at 2(a) site: 0, 0, z_2; $z_2 = 0.7599$
X (Si) at 4(b) site: 0, 1/2; z_3; $z_3 = 0.1088$

                          + body-centering translation

In $BaAl_4$, aluminium atoms occupy two types of sites, with tetrahedral coordination by four aluminium atoms (and four barium atoms) and with pyramidal coordination by five aluminium atoms (and five barium atoms). The tetrahedral (t) and pyramidal (p) sites are arranged in layers perpendicular to the (001) direction in the sequence ptp ptp. Planes of barium atoms are intercalated between ptp groups of aluminium layers.

In the three ternary derivative structures, the aluminium sites are occupied by transition metal T and X atoms in an ordered fashion, forming three-dimensional T-X networks. The three structures distinguish themselves by the different distribution of T and X atoms at the p and t layers: XTX XTX in $ThCr_2Si_2$, TXT XTX in $CaBe_2Ge_2$ and TXX TXX in $BaNiSn_3$ (Braun et al., 1983; Engel et al., 1983). All three structure types have been found in the La–Ir–Si system (Engel et al., 1983).

The $CaBe_2Ge_2$-type and $ThCr_2Si_2$-type structures are observed as high and low temperature modifications of $LaIr_2Si_2$ (Braun et al., 1983). The transformation temperature is at about 1720° C.

The Ir–Si distances are short 2.40(4) Å and correspond to the sum of the respective covalent radii. All other interatomic distances, in particular the homonuclear Si–Si distance (2.61 Å), are longer than the sums of the respective covalent radii. The intralayer distance between two lanthanum atoms is equal to the crystallographic a axis dimension (4.1 to 4.3 Å); the interlayer La–La distances are larger (5.7 to 5.9 Å).

Band structure calculations performed for both phases are in good agreement with the experimental results and indicate strong Ir–Si hybridization effects in the high temperature phase with mainly d character at E_F (Braun et al., 1985).

The majority of 1:2:2 systems exhibits the structure of the $ThCr_2Si_2$ type. This structure can be visualized as layers of Si tetrahedra with transition metal atoms inside, piled up in the direction of the tetragonal axis. In the empty space between adjacent layers lanthanide atoms are located. This is displayed in Figure 2.24. The X–X distance is usually close to the sum of the covalent radii of X atoms, similar to the T–X contacts. Strong chemical interactions are thus expected within the layers composed of 4X tetrahedra. The above bond lengths are critically dependent on the magnitude of the z parameter and the c/a ratio (a, c — the lattice constants). In the case of the regular tetrahedra

$$z = 1/4 + 1/2\sqrt{2}(a/c) = 3/8 \qquad (2.3)$$

	x	y	z
Si$_1$ in 2(e) site	0.0364(8)	1/4	0.0907(5)
Ru in 2(e) site	0.1179(2)	1/4	0.3869(2)
Nd in 2(e) site	0.4130(2)	1/4	0.7990(9)
Si$_2$ in 2(e) site	0.6657(9)	1/4	0.4913(5)

RIrGe$_2$ (R = Y, Pr, Nd, Sm, Gd–Er) and RPtGe$_2$ (R = Y, Ce, Pr, Nd, Tm) compounds belong to space group Immm. The atomic position parameters determined for YIrGe$_2$ are listed in Table 2.3. Also, this type of crystal structure is related to the CeNiSi$_2$ type (François, 1986).

RTC$_2$ carbides crystallize in the orthorhombic structure described by space group Amm2 (Jeitschko and Gerss, 1986) with the following atomic positions:

R atoms in 2(a) site	0 0 0
T atoms in 2(b) site	1/2 0 z
C atoms in 4(e) site	1/2 y z

For example, the values of the positional parameters in TbNiC$_2$ are z(Ni) = 0.610(2), y(C) = 0.152(2) and z(C) = 0.301(2) (Yakinthos et al., 1989). The crystallographic unit cell is displayed in Figure 2.19.

The interatomic distances in the TbNiC$_2$ unit cell are listed in Table 2.4.

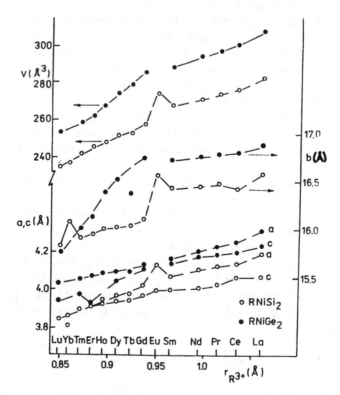

FIGURE 2.18. The lattice parameters and the unit cell volumes in RNiSi$_2$ and RNiGe$_2$ series of compounds vs. the R^{3+} ionic radius.

I-RT₂X₂   P-RT₂X₂   RTX₃

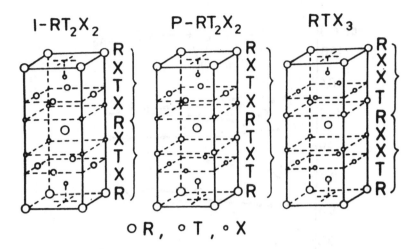

○ R, ○ T, ○ X

FIGURE 2.24. Crystal structure of I–RT₂X₂ (ThCr₂Si₂-type), P–RT₂X₂ (CaBe₂Ge₂-type) and RTX₃ (BaNiSn₃-type) compounds.

Hence, c/a is $2\sqrt{2}$, a much larger value than the usually observed c/a ratio of about 2.5. This means that the tetrahedra deviate from the regular shape, however, with only a small influence on the observed bond lengths within the tetrahedron.

Some z values determined in the course of X-ray and neutron-diffraction studies are displayed in Figure 2.25. They oscillate around the value of 3/8 (0.375), with the exception of Pd- and Cu-containing compounds, for which z > 0.375, in contrast to ruthenium-containing compounds for which z values are smaller than 0.375.

The a and c lattice constants and c/a ratios in many RT₂X₂ phases (see Figures 2.26 and 2.27) exhibit a regular rise with the ionic radius of R^{3+};

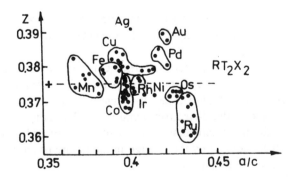

FIGURE 2.25. Stability diagram for ternary RT₂X₂ silicides and germanides. The parameters of a regular tetrahedron are marked +.

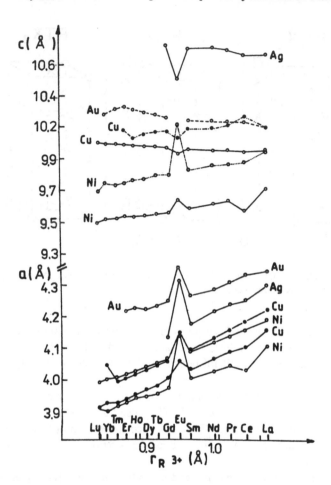

FIGURE 2.26. Lattice parameters of $R(Ni,Cu,Ag,Au)_2Si_2$ silicides (○) and $R(Ni,Cu)_2Ge_2$ germanides (●) vs. the radius of R^{3+} ion.

however, the c/a ratio simultaneously falls. The anomalous behavior observed in the case of Ce, Eu, Yb compounds is connected with the occurrence of divalent or tetravalent states.

Experiments have shown that the lattice constants and the unit cell volumes are sensitive to the kind of transition metal T. The variation of unit cell volume is displayed in Figure 2.28. It suggests that the unit cell volume depends on the size of the d-electron metal atom. The minima observed for Co, Rh, Ir-containing systems may be brought about either by different valency states of the T ions or by the change in the nature of chemical bonding operating in the $T-X$ tetrahedra.

The results of the measurements of a and c lattice performed at different temperatures are available (Meyer and Felner, 1972; Szytuła et al., 1986a).

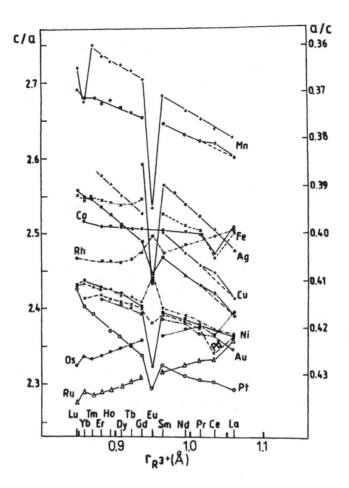

FIGURE 2.27. The c/a ratios of RT_2Si_2 ($\circ,\bullet,x,\triangle$) and RT_2Ge_2 ($+$) vs. the radius of R^{3+} ion.

In the case of $GdRh_2Si_2$ and $TbRh_2Si_2$ (Szytuła et al., 1986a), the variation of the lattice constants in the temperature range 78 to 300 K is described by a third degree polynominal: $a(T) = a_0 + a_1T + a_2T^2 + a_3T^3$ and $c(T) = c_0 + c_1T + c_2T^2 + c_3T^3$.

Interatomic distances, which are directly available from X-ray and neutron diffraction experiments, provide important information about the chemical interactions (chemical bonding) operating in crystalline solids. In $ThCr_2Si_2$-type compounds, the shortest X–T distance, given by

$$d_{x-T} = \sqrt{(a/2)^2 + (z - 1/4)^2c^2} \qquad (2.4)$$

is approximately equal to the sum of the covalent radii of the respective atoms (see Table 2.4). The distance between two X atoms belonging to adjacent X

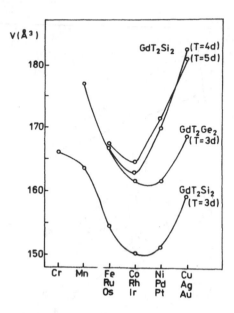

FIGURE 2.28.   Change in the unit cell volume for RT_2Si_2 and RT_2Ge_2 compounds (T = 3d-, 4d-, 5d- electron metals).

layers is $d_{x-x} = (1 - 2z)c$. The value of this distance is mainly responsible for keeping the layers of the tetrahedra together. A lanthanide atom is co-ordinated by 8X and by 8T atoms at distances:

$$d_{R-X} = \sqrt{a^2/2 + (1/2 - z)^2 c^2} \qquad (2.5)$$

and

$$d_{R-T} = \sqrt{(a/2)^2 + (c/4)^2} \qquad (2.6)$$

Interatomic distance calculated from crystallographic data for some RT_2X_2 compounds are listed in Tables 2.5 and 2.6. The analysis of bond lengths, listed in these tables, leads to the following conclusions:

- The Si–Si distances in the Si_4 tetrahedra are close to 2.34 Å, i.e., to the Si–Si distance in silicon (diamond type structure); occurrence of strong covalent bonds can thus be assumed; short T–Si bonds in (TSi_4) tetrahedra are also close to the sums of respective covalent radii, suggesting strong chemical interactions.
- R–Si distances are larger than the sum of covalent, however, smaller than the sum of ionic radii.

- R–T, R–R, T–T distances are large enough to exclude any direct chemical links. The system of chemical bonds operating in the structure of ThCr$_2$Si$_2$-type thus exhibits a heterodesmic character with three kinds of bonds:
- covalent Si–Si and T–Si
- ionic or covalent R–Si
- metallic R–T and T–T

The different sizes of Si and Ge ions reflect themselves in the lattice parameters of RT$_2$Si$_2$ and RT$_2$Ge$_2$, as well as in their unit-cell volumes. The latter are larger in RT$_2$Ge$_2$. The influence of the atomic diameter of the R ion (D$_R$) on the behavior of a and c lattice parameters was described by the following function:

$$\Delta_{i-j} = 1/2(D_i + D_j) - d_{i-j}; \text{ (Pearson and Villars, 1984)} \quad (2.7)$$

TABLE 2.5
Interatomic Distances in RCo$_2$Si$_2$ Compounds (in Å)

R	d$_{R-R}$	d$_{R-T}$	d$_{R-X}$	d$_{T-T}$	d$_{T-X}$	d$_{X-X}$
La	4.011	3.217	3.115	2.836	2.351	2.575
Ce	3.957	3.146	3.066	2.798	2.311	2.505
Pr	3.968	3.175	3.080	2.806	2.324	2.539
Nd	3.960	3.169	3.074	2.800	2.320	2.534
Gd	3.913	3.137	3.039	2.767	2.294	2.512
Tb	3.893	3.124	3.024	2.753	2.283	2.502
Dy	3.883	3.116	3.016	2.746	2.277	2.496
Ho	3.872	3.108	3.008	2.738	2.271	2.490
Er	3.867	3.109	2.998	2.734	2.265	2.488
Tm	3.853	3.096	2.995	2.726	2.262	2.482

TABLE 2.6
Interatomic Distance in EuT$_2$Si$_2$ and DyT$_2$Si$_2$ Compounds (in Å)

Compounds	d$_{R-R}$	d$_{R-T}$	d$_{R-T}$	d$_{T-T}$	d$_{T-x}$	d$_{X-X}$
EuFe$_2$Si$_2$	3.970	3.215	3.079	2.807	2.353	2.530
EuCo$_2$Si$_2$	3.921	3.147	3.033	2.772	2.314	2.462
EuNi$_2$Si$_2$	4.008	3.133	3.079	2.834	2.338	2.409
EuCu$_2$Si$_2$	4.118	3.201	3.124	2.869	2.376	2.477
DyMn$_2$Si$_2$	3.925	3.226	3.078	2.775	2.315	2.687
DyFe$_2$Si$_2$	3.903	3.164	3.049	2.760	2.289	2.587
DyCo$_2$Si$_2$	3.890	3.118	3.027	2.749	2.269	2.511
DyNi$_2$Si$_2$	3.953	3.099	3.058	2.795	2.284	2.438
DyCu$_2$Si$_2$	3.980	3.215	3.084	2.814	2.357	2.525

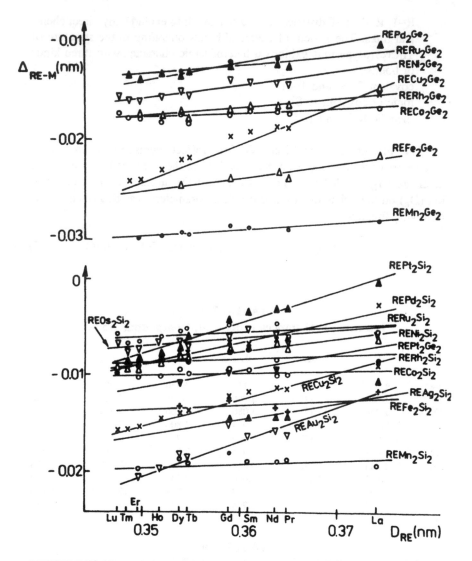

FIGURE 2.29.   Plots of the observed Δ_{i-j} values against the radius of R^{3+} ion for RT_2X_2 phases.

D_i and D_j are the atomic diameters of ion i and j, respectively; d_{1-j} is the calculated interatomic distance between them. The values of Δ_{i-j} vary linearly with D_R for RT_2X_2 compounds (T = 3d transition metal) (see Figure 2.29).

A diagnostic parameter f_{i-j} was introduced by the following relation:

$$\Delta_{i-j} = f_{i-j}D_R + k \tag{2.8}$$

In Table 2.7, the f_{i-j} values are listed for a number of RT_2X_2 systems. The magnitudes of f_{i-j} may be considered as a measure of the stability of the

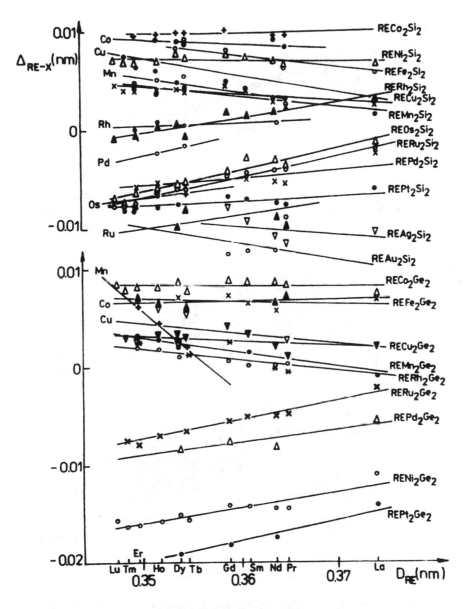

FIGURE 2.29. Continued.

crystal structure framework, indicating which links are responsible for it. In the manganese-containing compounds, the T–T contacts control the magnitudes of a and c lattice parameters, whereas in the cases of iron-, nickel-, and copper-containing phases, the R–T bonds are more important. In RCo_2X_2 compounds, both f_{R-X} and f_{R-T} are approximately zero despite the fact that the Δ_{R-X} values are positive and Δ_{T-X} are negative.

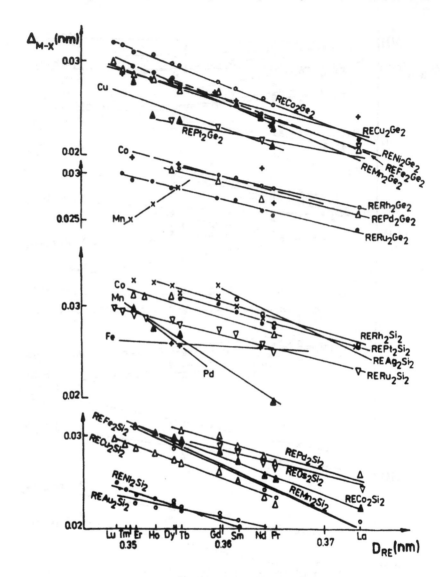

FIGURE 2.29.   Continued.

Figure 2.30 shows a diagram on which the dependence of a on D_R is plotted against the dependence of c on D_R. The following regularities can be noticed:

- The RCo_2Si_2 and RCo_2Ge_2 phases place themselves on the lines $f_{R-T} = 0$ and $f_{R-X} = 0$.
- RMn_2Si_2 and RMn_2Ge_2 compounds behave uniquely, since they fit the $f_{R-T} = 0$ line. It means that their unit-cell dimensions are controlled

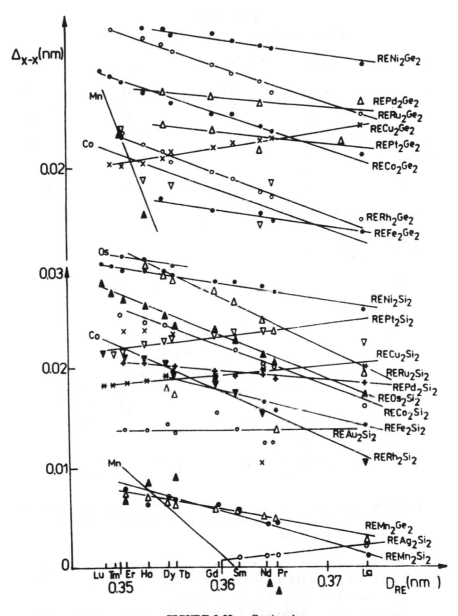

FIGURE 2.29. Continued.

by the R–Mn contacts. An examination of Δ_{R-T} values shows that they are strongly negative, i.e., there is, in fact, no physical contact between the R and Mn ions via their ionic volumes. A change in valency state as compared to the elemental state is possible.

TABLE 2.7
Calculated Values of f_{i-j} for RT_2X_2 Systems[a]

System	R–T	R–X	T–X	T–T
RMn_2Si_2	0	−0.119	−0.444	−0.300
RCo_2Si_2	−0.005	0.005	−0.376	−0.388
RNi_2Si_2	0.150	0	−0.350	−0.156
RCu_2Si_2	0.281	−0.045	−0.349	0.064
RMn_2Ge_2	0	−0.190	−0.483	−0.255
RFe_2Ge_2	0.154	0	−0.347	−0.158
RNi_2Ge_2	0.129	0	−0.354	−0.192
RCu_2Ge_2	0.321	−0.050	−0.344	0.123

[a]   Pearson and Villars, 1984.

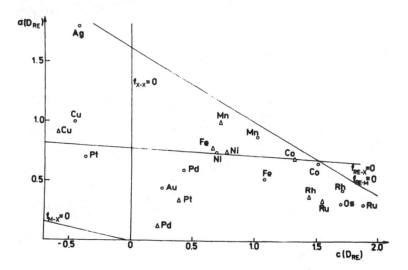

FIGURE 2.30.   Lines calculated using $z = 0.375$ for the conditions $f_{R-T} = 0$ $f_{R-X} = 0$ and $f_{T-X} = 0$ on a plot of the dependence of a on the diameter D_R against the dependence of c on D_R (O = silicides, Δ = germanides).

- The unit-cell dimensions of iron-, nickel-, and copper-containing compounds are controlled to large extent by the R–T contacts, since they place themselves along the $f_{R-X} = 0$ line.
- The unit-cell parameters of RT_2Si_2 (T = Pd, Pt, Au) systems, which lie closest to the line with $f_{R-X} = 0$, are mostly controlled by the R–Si contacts. In the case of RRh_2Si_2 compounds, however, the R–Rh contacts exert stronger control on the dependence of a and c lattice parameters on D_R since they place themselves close to the $f_{R-T} = 0$ line. Moreover, the observed Δ_{R-T} values are only slightly negative, indicating that the R and Rh ions are in contact, and hence, they can control the unit-cell dimensions.

• The position of RCo_2B_2 phases on the diagram shows that the R–B contacts control the dependence of a and c lattice parameters on D_R.

The above features permit the following conclusions to be drawn:

1. The number and disposition of d electrons of the 3d transition metal T appear to be deciding factors for the dependence of a and c lattice parameters on D_R. In this respect, manganese-containing compounds are an exception.

2. The large dependence of a on D_R observed in copper-containing compounds, which requires a negative dependence of c on D_R, is probably due to the filled d shell, in good agreement with electron spectroscopy data (Buschow et al., 1977).

3. The angles determined for the T-4X tetrahedra, are, in general, different from 109°, implying that they are not an important factor of the structure framework stability. (For a perfect tetrahedron, the z value must be 3/8, hence c/a = $2\sqrt{2}$ and c/a = 2.820.) The observed c/a values for the majority of RT_2X_2 compounds are significantly smaller. They are in the range from 2.36 to 2.73 (Figure 2.27).

4. RT_2X_2 compounds containing 4d and 5d transition metals exhibit different behavior from those containing 3d metals.

The anisotropic temperature factors determined for $CeOs_2Si_2$ and $HoRu_2Si_2$ (Horvath and Rogl, 1983, 1985) and for $DyCo_2Si_2$ and $TbRu_2Si_2$ (Ban et al., 1986), using X-ray single crystal data, indicate that the thermal vibration amplitudes of all atoms are small in the basal planes, in comparison to those along the c axis, suggesting, in general, stronger chemical interactions in the planes than in the direction of the c axis.

The values of chemical shifts and the quadrupole splittings determined by Mössbauer spectroscopy contain also information on the electronic structure of the Mössbauer atom and the influence of its environment. Quadrupole interaction is due to the competition between lattice and conduction electron contribution. This effect was studied for RFe_2Si_2 systems (Umarji et al., 1983). Local symmetry of the Fe atoms influences the quadrupole interactions if Si atoms form a regular tetrahedron around an iron ion. They will not produce lattice or conduction-electron contributions to the quadrupole interactions at the spin-paired Fe atom. However, deviations from this regularity are permitted by the space group (z_{Si} free parameter). Consequently, small lattice and conduction electron contributions are possible. The second nearest neighbors are R ions, which form a distorted tetrahedron around Fe ion. They may also contribute to the quadrupole interaction determined by the experiment. Zero quadrupole splitting was found in $GdFe_2Si_2$ and $ErFe_2Si_2$, while it is larger than zero in $LaFe_2Si_2$ and $CeFe_2Si_2$, indicating a probable influence of the tetravalent state of cerium (Umarji et al., 1983; Noakes et al., 1983).

The values of isomer shift 0.36 (1) mms^{-1} relative to ^{57}Co(Cr) and the absence of quadrupole splitting are in favor of the low spin divalent iron with the 3d^6 configuration for Fe atoms in RFe$_2$Si$_2$ (Görlich et al., 1982).

The NMR studies of RCu$_2$Si$_2$ and RCu$_2$Ge$_2$ (Sampathkumaran et al., 1979a) show that the quadrupole coupling constants of ^{63}Cu and the Knight shift of copper and silicon are constant for different rare-earth ions. The anomaly of the value of e^{2}qQ for Yb is probably the result of a mixed valence state.

The Mössbauer effect measurements on EuT$_2$Ge$_2$, GdT$_2$Ge$_2$, and DyT$_2$Si$_2$ have shown that:

- In the case of DyT$_2$Si$_2$ a large and monotonic decrease of the hyperfine as well as of the field gradient at ^{161}Dy nuclei occur, while the values of isomer shift remain constant as the d metal changes. The ^{161}Dy Mössbauer data indicate strong dependence of isomer shift values and the hyperfine fields on the number 3d, 4d, or 5d electrons (Nowik et al., 1983).

- In the case of EuT$_2$Ge$_2$ and GdT$_2$Si$_2$, the isomer shift, which measures the charge density inside the Mössbauer nucleus is proportional to the change of unit cell volumes (Felner and Nowik, 1978). The analysis of the crystal structure and a calculation of the electric field gradient based on the point charge model show that the effective charge on the T ion is close to zero. Thus, the T ion is either neutral or strongly screened.

- The ^{151}Eu Mössbauer data for EuRu$_2$Ge$_2$ indicate strong bonding between Eu and Ge atoms along the a axis and only weak bonds with other rare-earths ions in the basal plane (Nowik and Felner, 1985).

- In the case of GdT$_2$Si$_2$ (T = 3d, 4d, 5d elements) the values of isomer shift and quadrupole interaction show a tendency to rise with atomic number of 3d, 4d, and 5d metals. A conclusion can be drawn that the total s-electron density at the Gd nuclei decreases with the rise of atomic number of the respective d element. The strength of electric quadrupole coupling grows for the lighter elements of 4d and 5d transition groups in T positions and becomes dominant over the magnetic coupling for T = Ru and Os, respectively (Latka, 1989).

X-ray absorption spectroscopy (XAS) studies of charge transfer were also performed for the RT$_2$X$_2$ compounds (R = Nd, Sm, Gd, T = Mn, Fe, Co, Ni, and Cu, X = Si or Ge). The chemical shifts measured for R, T, and X atoms are positive. This result suggests that, unlike in the binary systems, all of the constituents in the ternary intermetallics RT$_2$X$_2$ contribute electrons to the conduction band (Darshan et al., 1984).

XPS studies of the valence band and of the 4f region performed on EuCu$_2$Si$_2$ and YCu$_2$Si$_2$ samples have shown that the copper 3d photoelectrons contribute to the spectrum located about 2 eV below the Fermi level E$_F$, and the Eu ions appear in Eu^{2+} and Eu^{3+} states (Buschow et al., 1977).

Photoemission spectra obtained for CeT_2Si_2 compounds (T = Cu, Ag, Au, Pd) indicate hybridization of 4f electrons of Ce with the d states of T ions (Parks et al., 1983). The band structures of $CeCu_2Si_2$ and $LaCu_2Si_2$ were calculated using the semirelativistic linear orbital method (LMTO) (Jarlborg et al., 1983). The results, analyzed in terms of energy levels, charge distribution, and Fermi surface properties show that the 4f levels of Ce are situated mainly above the E_F. The density of states at E_F is large and concentrated at the 4f band of Ce. Large Fermi surface anisotropies are pointed out in both of the above compounds.

The calculations of the electronic state of the Fe atoms were performed using the approximation of the semiempirical variant of the MALCAO method (Cotton, 1983). They yielded the state of $4s^{0.17}$ $4p^{0.12}$ $3d^{8.5}$ (Koterlin and Luciv, 1978). This confirms the assumption of electronic density transfer from the nearest neighbors of the Fe atoms.

In the $CaBe_2Ge_2$ type of crystal structure in which the high temperature phases of RPt_2Si_2 and RIr_2Si_2 crystallize, the situation is more complicated. Nonregular dependence of the lattice constants as the functions of the R ion radius is observed (see Figure 2.31). The interatomic distances are given by following relations

$$d_{R-X_1} = \sqrt{(a/2)^2 + [(z_R - 1/2)c]^2}$$

$$d_{R-X_2} = \sqrt{a^2/2 + [(1 - z_X - z_R)c]^2}$$

$$d_{R-T_1} = \sqrt{(a/2)^2 + [(1 - z_R)c]^2}$$

$$d_{R-T_2} = \sqrt{a^2/2 + [(z_R + z_T - 1)c]^2} \tag{2.9}$$

These distances indicate four neighbors, but it is not known, *a priori,* which of these distances is the shortest.

The atomic parameters were determined only for two compounds: $LaIr_2Si_2$ (Braun et al., 1983) and $HoPt_2Si_2$ (Leciejewicz et al., 1984a).

RPt_2Ge_2 (R = Ca, Y, La–Dy) compounds crystallize in a monoclinic structure which is a deformation variant of the tetragonal $CaBe_2Ge_2$-type structure. The space group is $P2_1$ (Venturini et al., 1989). The atomic parameters for $LaPt_2Ge_2$ are the following.

	x	y	z
La at 2(a) site	0.2631(39),	1/4,	0.7449(13)
Pt_1 at 2(a) site	0.2559(28),	0.2168(34),	0.3790(13)
Pt_2 at 2(a) site	0.7320(20),	0.3001(28),	0.9984(9)
Ge_1 at 2(a) site	0.2647(79),	0.2651(55),	0.1267(21)
Ge_2 at 2(a) site	0.7273(56),	0.1964(61),	0.4951(22)

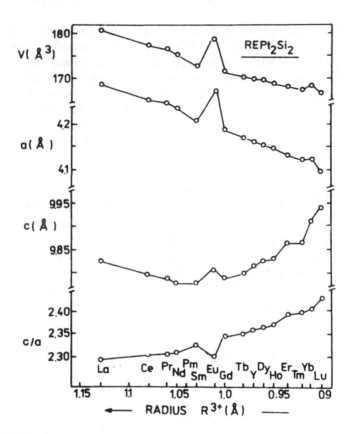

FIGURE 2.31. Lattice parameters, unit cell volumes and c/a ratio of RPt_2Si_2 silicides vs. the ionic radius R^{3+} (Hiebl and Rogl, 1985).

A comparison of the above parameters with those for $LaIr_2Si_2$ shows that y coordinates of platinum and germanium are mainly affected by the monoclinic deformation and the loss of the mirror plane, while x and z coordinates remain close to the corresponding values in the P4/nmm space group of $CaBe_2Ge_2$.

2.6. ThMn₁₂-TYPE STRUCTURE

The structure of the $ThMn_{12}$ type compounds has been known for a long time. However, it has turned out that binary compounds of lanthanides exist only with manganese and zinc, i.e., RMn_{12} and RZn_{12} (Wang and Gilfrich, 1966; Kirchmayr, 1969; Iandelli and Palenzone, 1969; Stewart and Coles, 1974).

The addition of a third element stabilizes the $ThMn_{12}$ phase in many systems containing f-electron metal-3d-transition metal, another transition metal, and Si/Al.

TABLE 2.8
Atomic Position Parameters for NdFe$_{10}$Mo$_2$[a]

Atomic position	x	y	z
Nd in 2(a)	0	0	0
Mo,Fe in 8(i)	0.3568	0	0
Fe in 8(j)	0.2728	0.5	0
Fe in 8(f)	0.25	0.25	0.25

Note: The lattice constants are a = 8.611 Å and c = 4.802 Å, the space group is I4/mmm.

[a] De Mooij and Buschow, 1988.

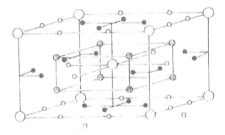

○ 2a ◉ 8 f ○ 8 i ● 8 j   **FIGURE 2.32.** The crystal structure of ThMn$_{12}$ type.

X-ray and neutron diffraction studies of these systems indicate the tetra-gonal ThMn$_{12}$-type crystal structure (space group I4/mmm). The determined values of the atomic position parameters obtained for NdFe$_{10}$Mo$_2$ are listed in Table 2.8 (De Mooij and Buschow, 1988).

The crystal structure of the ThMn$_{12}$ type is shown in Figure 2.32. The actinide or rare earth atoms occupy the corners and the centers of tetragonal prisms. The 3d and other atoms are usually located in one or more of the positions 8i, 8j and 8f.

The numbers of nearest-neighbor sites for 8i are (5, 4, 4, 1); for 8f (4, 2, 4, 2); and for 8j (4, 4, 2, 2), where the numbers in parentheses refer to 8i, 8f, 8j, and 2a neighbors, respectively.

This type of crystal structure is characteristic of two systems: RT$_{12-x}$Al$_x$ (T = Cr, Mn, Fe) and RT$_{12-x}$M$_x$ (T = Fe, Co, Ni, M = Ti, V, Cr, Mo, W, Re, Si).

In the first system YFe$_{12-x}$Al$_x$ the stability of the ThMn$_{12}$-type structure is observed for $4 < x < 6$ (Felner and Nowik, 1986). The occupation of the various sites by Fe and Al atoms was determined by the Mössbauer effect and by neutron diffraction experiments (see Table 2.9).

For x = 4, the site occupancy is the following: R in 2(a): (0, 0, 0); Fe in 8(f): (1/4, 1/4, 1/4), Al in 8(i): (x$_1$, 0, 0); and in 8(j): (x$_2$, 1/2, 0), with x$_1$ = 0.335 and x$_2$ = 0.276. When $x > 4$, the transition metals occupy the

TABLE 2.9
The Occupation of the Various Sites by T and
Al Atoms in $RFe_{12-x}Al_x$ Compounds

Compound	8i	8j	8f	Ref.
RMn_4Al_8	Al	Al	Mn	Moze et al., 1990a
RCr_4Al_8	Al	Al	Cr	Moze et al., 1990a
RFe_4Al_8	Al	Al	Fe	Moze et al., 1990b
RFe_6Al_6	Al	Al,Fe	Fe	Moze et al., 1990b

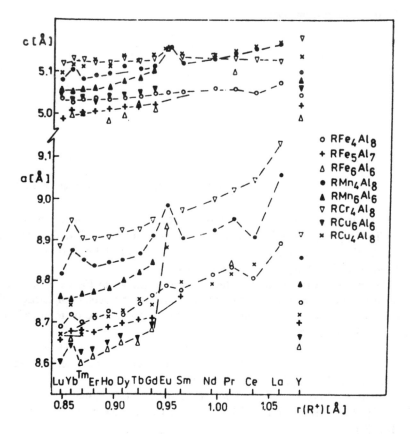

FIGURE 2.33. Dependence of the lattice parameters a and c as a function of ionic radius of the rare earth elements for $RT_{12-x}Al_x$ for x = 6, 7, 8 and T = Cr, Mn, Fe, Cu.

8j positions also. The dependence of the lattice parameters as a function of radius R of rare-earth elements is shown in Figure 2.33. As the radius of R rises, the a axis increases, while the c axis remains constant. For Ce-, Eu-, and Yb-containing compounds, anomalies are observed which suggest the mixed-valence state.

TABLE 2.10

The Occupation of the Various Sites in $RT_{12-x}M_x$ Compounds

Compound	8i	8j	8f	Ref.
$RFe_{11}Ti$	Ti,Fe	Fe	Fe	Ohashi et al., 1987
$RFe_{10}V_2$	V,Fe	Fe	Fe	Helmholdt et al., 1988
$RFe_{10}Cr_2$	Cr,Fe	Cr,Fe	Cr,Fe	Buschow, 1991
$RFe_{10}Mo_2$	Mo,Fe	Fe	Fe	Buschow, 1988
$RFe_{10}Si_2$	Fe	Si,Fe	Si,Fe	Buschow, 1988
$RCo_{11}Ti$	Ti,Fe	Co	Co	Yang et al., 1990
$RCo_{10}Mo_2$	Mo,Co	Co	Co	Lin et al., 1991
$RNi_{10}Si_2$	Ni	Ni	Si,Ni	Moze et al., 1991

$RFe_{12-x}M_x$ compounds form the $ThMn_{12}$-type of structure in limited range of concentrations, e.g., in $RCo_{12-x}V_x$ for x between 1.6 and 3.5. In this concentration region, the lattice constants rise with an increase of x (Jurczyk, 1990).

The neutron diffraction (Helmholdt et al., 1988) and Mössbauer effect data also point to the preferential substitution of atoms. The determined distribution of atoms in some of these compounds is listed in Table 2.10. The transition metal (Ti, V, Mo) prefers the 8i site, while the sp element atoms (Si) occupy preferentially the 8f and 8j sites (Buschow, 1991).

Attempts have been made to explain the variance in site occupations by means of size effects on enthalpy effects (Buschow, 1988; Li and Coey, 1991).

The variation of the lattice constants as a function of the radius of rare-earth atom is presented in Figure 2.34. The a-parameters lattice increases linearly with R, whereas the c-parameter is practically constant.

The $RTiFe_{11-x}Co_x$ systems exist for x = 8 (R = Pr, Nd, Er) and over whole concentration range (x = 11) for R = Tb, Dy, and Ho. With an increase of Co concentration, a fall of the lattice constant values is observed (Cheng et al., 1990). $RFe_{12-x}X_xHy$ hydrides with a $ThMn_{12}$-type structure have also been discovered (Zhang et al., 1990). The hydrogen absorption leads to an increase of the lattice parameters a and c. The a/c ratio rises upon hydrogenation for each compound indicating that hydrogen expands the lattice more along the a axis than along the c direction. The increase of the unit cell volume was found to be from 0.4 to 10%, depending on the nature of the R component (Zhang et al., 1990).

Neutron diffraction study of $YTiFe_{11}N_x$ indicates that this compound also exhibits the $ThMn_{12}$-type crystal structure with the nitrogen atoms in the interstitial 2b sites (Yang et al., 1991).

2.7. $R_2T_{14}X$-TYPE STRUCTURE

$Nd_2Fe_{14}B$ emerged as the representative of $R_2T_{14}X$ compounds. The $R_2Fe_{14}B$ compounds have been reported to exist for R = Y, La, Ce, Pr, Nd, Sm, Gd,

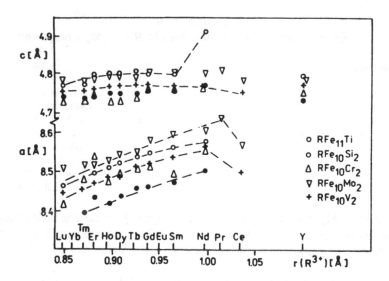

FIGURE 2.34.   Dependence of the lattice parameters a and c as a function of the ionic radius R^{3+} for $RTiFe_{11}$ and $RFe_{10}M_2$, where M = Si, Cr, Mo, V.

Tb, Dy, Ho, Er, Tm, Yb and Lu (Sinnema et al., 1984; Burlet et al., 1986). The $R_2Co_{14}B$ compounds were reported to exist for the "light" rare-earth elements Y, La, Ce, Pr, Nd, Sm, Gd and Tb (Buschow et al., 1985). The $R_2Fe_{14}C$ compounds are formed more easily with the "heavier" rare-earth elements (Gueramian et al., 1987; de Boer et al., 1988). It turned out that for most of the light lanthanides (La, Ce, Pr, and Nd) it is difficult or even impossible to obtain the $R_2Fe_{14}C$ phases, so far. $Nd_2Fe_{14}C$ was eventually obtained in single-phase form (Buschow et al., 1988b), but $Pr_2Fe_{14}C$ could only be stabilized by substituting small amounts of Mn for Fe (Buschow et al., 1988).

All existing $R_2Fe_{14}B$, $R_2Co_{14}B$, and $R_2Fe_{14}C$ compounds crystallize in the same tetragonal structure. It had first been determined by neutron powder diffraction (Herbst et al., 1984) and then confirmed by X-ray single crystal study (Givord et al., 1984; Shoemaker et al., 1984). These experiments were performed on $Nd_2Fe_{14}B$, and its structure is now known as the $Nd_2Fe_{14}B$-type of structure, space group $P4_2/mnm$. It is shown in Figure 2.35. Each unit cell contains four formula units or 68 atoms. Table 2.11 lists the atomic positions for $Nd_2Fe_{14}B$ (Herbst et al., 1985), and Table 2.12 contains interatomic distances computed from them.

All rare-earth and boron atoms, but only 4 of the 56 iron atoms, reside in the z = 0 and z = 1/2 planes. The remaining 52 iron atoms per unit cell form hexagonal nets between these planes. The rare-earth atoms occupy two different positions, the f and g sites, each with occupancy 4. The boron atoms reside on the g site, while the k_1, k_2, j_1, j_2, e, and c sites are occupied by the iron atoms with occupancy factors of 16, 16, 8, 8, 4, and 4, respectively.

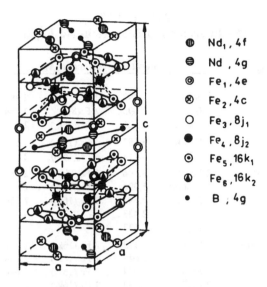

⦿	Nd_1, 4f	
⊖	Nd , 4g	
◎	Fe_1, 4e	
⊗	Fe_2, 4c	
○	Fe_3, $8j_1$	
●	Fe_4, $8j_2$	
⊙	Fe_5, $16k_1$	
◭	Fe_6, $16k_2$	
•	B , 4g	

FIGURE 2.35. Tetragonal unit cell of $Nd_2Fe_{14}B$ the prototype structure for the $R_2T_{14}X$ compounds.

TABLE 2.11
Atomic Sites and Coordinates x,y,z for $Nd_2Fe_{14}B$[a]

Atom	Site	x	y	z
Nd_1	4f	0.2679(5)	0.2679(5)	0
Nd_2	4g	0.1403(4)	−0.1403(4)	0
Fe	4c	0	0.5000	0
Fe	4e	0.5000	0.5000	0.1139(5)
Fe	$8j_1$	0.0979(3)	0.0979(3)	0.2045(2)
Fe	$8j_2$	0.3167(3)	0.3167(3)	0.2464(3)
Fe	$16k_1$	0.2234(3)	0.5673(3)	0.1274(2)
Fe	$16k_2$	0.0375(3)	0.3698(3)	0.1758(2)
B	4g	0.3711(9)	0.3711(9)	0

Note: In units of the lattice constants for $Nd_2Fe_{14}B$ obtained from analysis of room temperature neutron diffraction data

[a] Herbst et al., 1985.

This structure is built up of an alternating sequence of layers, one of Fe atoms and the other containing R and Fe.

Figure 2.36 shows the lattice parameters a and c as a function of the radius of rare earth element. For all systems, the increase of the radius of R element is accompanied by a linear rise of the c-lattice constant. Such correlation in respect to the a lattice parameters is observed for heavy rare earths.

TABLE 2.12
The Interatomic Distances in $Nd_2Fe_{14}B$

Site	Interatomic distances (Å)			
$Fe_1(4e)$	2 $Nd_2(4g)$	3.192	2 $Fe_3(8j_1)$	2.491
	2 B	2.095	2 Fe(8j$_2$)	2.754
	1 $Fe_1(4e)$	2.826	4 $Fe_5(16k_1)$	2.496
$Fe_2(4c)$	2 $Nd_1(4f)$	3.382	4 $Fe_5(16k_1)$	2.573
	2 $Nd_2(4g)$	3.118	4 $Fe_6(16k_2)$	2.492
$Fe_3(8j_1)$	2 $Nd_1(4f)$	3.306	1 $Fe_4(8j_2)$	2.784
	1 $Nd_2(4g)$	3.296	2 $Fe_5(16k_1)$	2.587
	1 $Fe_1(4e)$	2.491	2 $Fe_6(16k_2)$	2.396
	1 $Fe_3(4f_1)$	2.433	—	—
	2 $Fe_4(8j_2)$	2.633	—	—
$Fe_4(8j_2)$	1 $Nd_1(4f)$	3.143	2 $Fe_5(16k_1)$	2.748
	1 $Nd_2(4g)$	3.049	2 $Fe_5(16k_1)$	2.734
	1 $Fe_1(4e)$	2.754	2 $Fe_6(16k_2)$	2.640
	2 $Fe_3(8j_1)$	2.633	2 $Fe_6(16k_2)$	2.662
	1 $Fe_3(8j_1)$	2.784	—	—
$Fe_5(16k_1)$	1 $Nd_1(4f)$	3.066	1 $Fe_4(8j_2)$	2.734
	1 $Nd_2(4g)$	3.060	1 $Fe_4(8j_2)$	2.748
	1 B	2.096	1 $Fe_5(16k_1)$	2.592
	1 $Fe_1(4e)$	2.496	1 $Fe_6(16k_2)$	2.527
	1 $Fe_2(4c)$	2.573	1 $Fe_6(16k_2)$	2.536
$Fe_6(16k_2)$	1 $Nd_1(4f)$	3.279	1 $Fe_5(16k_1)$	2.527
	1 $Nd_2(4g)$	3.069	1 $Fe_5(16k_1)$	2.462
	1 $Fe_2(4c)$	2.492	1 $Fe_5(16k_1)$	2.536
	1 $Fe_3(8j_1)$	2.396	1 $Fe_6(16k_2)$	2.542
	1 $Fe_3(8j_2)$	2.662	2 $Fe_6(16k_2)$	2.549
	1 $Fe_4(8j_2)$	2.640	—	—

The anomalously small lattice parameters of the $Ce_2Fe_{14}B$ compound are associated with the tetravalent state of cerium.

The effect of substitutions by other atoms was also studied. Thus, Fe was substituted by Co, Si, Ru, Ga, Cu, and Al; R elements were substituted and interchanged; and B was replaced by C, Si, Ga, Al, and B (Pędziwiatr et al., 1986, 1987; Sagawa et al., 1984). Substitution of R, Fe, or B does not change the type of crystal structure, but it results in changes of the lattice parameters. Usually only geometric (size) factors produce changes of the lattice parameters. When smaller atoms are introduced into the crystal lattice a decrease in lattice constants is observed (as in the case of Fe replacement by the smaller Co or Si). When atoms with larger ionic radii are incorporated, there is an increase in a and c observed (for example, in the $R_{2-x}Th_xFe_{14}B$ system. Vegard's law is obeyed for most substituted systems, especially for low

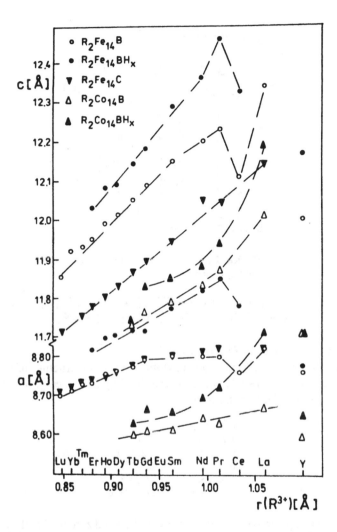

FIGURE 2.36. Dependence of the lattice parameters a and c as a function of ionic radius of R^{3+} for $R_2Fe_{14}B$, $R_2Fe_{14}BH_x$, $R_2Fe_{14}C$, $R_2Co_{14}B$, and $R_2Co_{14}BH_x$.

concentrations of substituted atoms. In the case $R_2Fe_{14-x}Co_xB$, deviation from Vegard's law and observed changes in c/a ratios indicate preference of substitution (Pędziwiatr, 1988).

The introduction of hydrogen to the lattice of $R_2T_{14}B$ compounds leads to an increase in unit-cell volume. The change in unit-cell volume was found to be 3.8 to 4.2% in $R_2Fe_{14}B$ and 1.5 to 3.0% in $R_2Co_{14}B$, depending on the nature of R element (Pourarian, 1990).

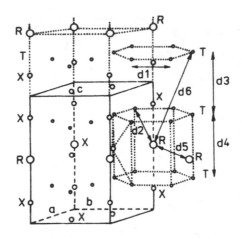

FIGURE 2.37. The crystal structure of RMn_6Sn_6 compounds.

2.8. RT_6X_6 COMPOUNDS WITH THE CoSn-DERIVATE CRYSTAL STRUCTURE TYPES

2.8.1. $HfFe_6Ge_6$-TYPE STRUCTURE

A number of RT_6X_6 compounds exhibit the crystal structure of $HfFe_6Ge_6$ type (Olenitch et al., 1981), which is described by the space group P6/mmm with atoms in the following positions:

R in 1(b):	0, 0, 1/2;		
T in 6(i):	1/2, 0, z_1;	0, 1/2, z_1;	1/2, 1/2, z_1;
	0, 1/2, $\bar{z}_1$;	1/2, 0, $\bar{z}_1$;	1/2, 1/2, $\bar{z}_1$;
X_1 in 2(e):	0, 0, z_2;	0, 0, $\bar{z}_2$;	
X_2 in 2(d):	1/3, 2/3, 1/2;	2/3, 1/3, 1/2;	
X_3 in 2(c):	1/2, 2/3, 0;	2/3, 1/3, 0;	

Since the z_1 parameter is usually close to 0.25 and z_2 to 0.125, the structure can be visualized as composed of sheets stacked along the hexagonal axis. There are four kinds of atomic sheets:

- Hexagonal (H) planes with R and X_2 atoms, all with z = 1/2.
- Planes at z about 1/4 and z about 3/4 with T atoms forming Kagome nets (K) separated by a double layer of X_1 and X_2 on the one side and the hexagonal plane (H) on the other (see Figure 2.37).

The coordination polyhedra with respective interatomic distances have been determined for $TbMn_6Sn_6$ (Chafik El Idrissi et al., 1991a) to be as follows: each Tb atoms has six Sn_2 neighbors in the H plane at the corners of a regular hexagon with $d_{Tb-Sn} = 3.1927(4)$ Å and two Sn_1 atoms below and above it with $d_{Tb-Sn} = 3.046(1)$ Å. On the other hand, each Mn atom

TABLE 2.13
Crystallographic Data for RFe$_6$Sn$_6$[a]

R	q	a	b	c	S.G.
Y	0.375	$2c_h$	$8\sqrt{3}a_h$	a_h	Cmcm
Gd	0.5	$2c_h$	$2\sqrt{3}a_h$	a_h	Cmcm
Tb	0.5	$2c_h$	$2\sqrt{3}a_h$	a_h	Cmcm
Dy	0.4	$2c_h$	$5\sqrt{3}a_h$	a_h	Ammm
Ho	0.333	$2c_h$	$3\sqrt{3}a_h$	a_h	Immm
Er	0.25	$2c_h$	$4\sqrt{3}a_h$	a_h	Cmcm

Note: R = Y, Gd–Er; (a_h and c_h refer to the hexagonal subcell of YCo$_6$Ge$_6$ type structure

[a] Chafik El Idrissi et al., 1991b.

has two nearest Sn$_2$ atoms (d_{Mn-Sn} = 2.745(1) Å), two Sn$_3$ (d_{Mn-Sn3} = 2.782(1) Å) and two Sn$_1$ at 2.8820(7) Å forming a deformed tetrahedron with the Mn atom inside it.

The structure of the HfFe$_6$Ge$_6$-type turned out to be fairly common among the RT$_6$X$_6$ compounds — the following show this structure: (1) RCo$_6$Sn$_6$ (R = Tb, Dy, Ho, Er, Tm, Lu) (Skolozdra and Koretskaya, 1981): RFe$_6$Sn$_6$ (R = Tm, Lu) (Chafik El Idrissi et al., 1991b): RMn$_6$Sn$_6$ (R = Gd, Tb, Dy, Ho, Er, Tm, Lu) (Chafik El Idrissi et al., 1991a); RCo$_6$Ge$_6$ (R = Hf, Gd, Tb, Dy, Ho, Er, Tm, Yb, Lu) (Bucholz and Schuster, 1981), and (2) YCo$_6$Ge$_6$-type structure.

Another derivate of the CoSn-type is the structure of YCo$_6$Ge$_6$ (Bucholz and Schuster, 1982) and RCo$_6$Ge$_6$ phases (Olenitch et al., 1981). R is Gd, Tb, Dy, Ho, Er, Tm, Yb, Lu in the latter case. The space group is P6/mmm. The atoms are in the following positions:

0.5Y at random in 1(a)	0,0,0;		
3Co in 3(g)	1/2, 0, 1/2;	0, 1/2, 1/2;	1/2, 1/2, 1/2;
2Ge in 2(c)	1/3, 2/3, 0;	2/3, 1/3, 0;	
0.5Ge at random in 2(e)	0, 0, z;	0, 0, $\bar{z}$;	z~0.154(2)

The Y atoms are located at random in the centers of hexagonal bipyramids with Ge atoms at their apexes. The cobalt atoms form a fully ordered network on the plane with z = 1/2, with the shortest Co–Co distance of 2.537 Å.

In a number of previously reported YCo$_6$Ge$_6$ type phases, new structures related to this type have been found (Chafik El Idriss et al., 1991b). Table 2.13 lists the superstructures discovered in the RFe$_6$Sn$_6$ compounds and their crystallographic parameters. They are given in terms of orthorhombic unit cells with dimensions a,b,c, which are related to the a_h and c_h hexagonal lattice parameters of a subcell of YCo$_6$Ge$_6$ type in a manner shown in Table 2.13. The corresponding space groups and the appropriate wave vectors/0, q, 0/ are also listed.

2.9. PHASES WITH NaZn$_{13}$-TYPE STRUCTURE

The NaZn$_{13}$-type crystal structure exhibits the space group Fm3c, the atoms being located in the following sites:

8Na in 8(a):	1/4, 1/4, 1/4;	3/4, 3/4, 3/4;
8Zn$_I$ in 8(b):	0, 0, 0;	1/2, 1/2, 1/2;
96Zn$_{II}$ in 96(i):	0, y, z;	etc
	+ face-centered translation	

This type of crystal structure was detected in La(Fe,Al)$_{13}$ and in La(Fe,Si)$_{13}$ phases (Kripyakevich et al., 1968). The occupation of the 8(b) and 96(i) positions by Fe and Al atoms does not proceed in a random way. Neutron diffraction study of La(Fe$_x$Al$_{1-x}$)$_{13}$ has shown that the 8(b) site is fully occupied by irons, whereas the remaining iron atoms Al are distributed randomly in the 96(i) site. The substitution of Fe by Al or Si brings about a linear rise of the lattice parameter a$_1$ (Palstra et al., 1983). The La and Fe$_I$ atoms form a cubic body centered sublattice (CsCl type): each La atom has eight closest neighbors. In addition, there are 24Fe$_{II}$ atoms at the apexes of an icosahedron. Each Fe$_I$ atom is, in turn, surrounded by 12Fe$_{II}$ atoms, while an Fe$_{II}$ atom has 9Fe$_{II}$ and 1Fe$_I$ nearest neighbors. A part of the unit cell of NaZn$_{13}$-type is displayed in Figure 2.38.

2.10. PHASES WITH THE Ce$_2$Ni$_{17}$Si$_5$-TYPE STRUCTURES

A number of phases with the general composition RT$_9$X$_2$ crystallize in the BaCd$_{11}$-type derivate structure (Bodak and Gladyshevskii, 1969) with partial disorder of the Ni and Si atoms on the 4(a) Cd site. The space group is I4$_1$/amd, and the site occupancy is as follows:

32Ni in 32(i):	x, y, z;
8Si in 8(d):	0, 0, 0;
4Ce in 4(b):	0, 1/4, 3/8;
4Z in 4(a):	0, 3/4, 1/8;
Z = 0.6 Ni + 0.4 Si (at random)	
(Origin shifted by 0, 0, 1/2;)	
The composition of the title compound is thus	
Ce$_4$Ni$_{32}$(Ni$_{2.4}$Si$_{1.6}$)Si$_8$ − 4CeNi$_8$(Ni$_{0.6}$Si$_{0.4}$)Si$_2$	

Each Ce atom is surrounded by a complex polyhedron of 22 Ni and Si atoms. This structure type has been found in RCo$_9$Si$_2$ (R = Nd, Sm, Eu, Gd, Tb, Dy, Ho, Y) RNi$_9$Si$_2$ (R = La, Ce, Pr, Nd, Sm, Eu) (Bodak and Gladyshevskii, 1969), RFe$_{10}$SiC$_{0.5}$ (R = Ce, Pr, Nd, Sm), and RCo$_{10}$SiC$_{0.5}$ (R = Nd, Sm) (Le Roy et al., 1987; Allemand et al., 1989).

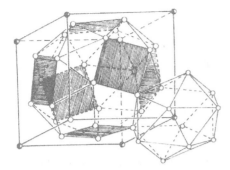

FIGURE 2.38. Part of the LaFe₁₃ unit cell.

2.11. PHASES WITH
SrNi₁₂B₆(EuNi₁₂B₆)-TYPE
STRUCTURES

The space group is $R\bar{3}m$, atoms in the following sites:

18Ni(I) in 18(h):	x_1, y_1, z_1;
18B in 18(h):	x_2, y_2, z_2;
18Ni(II) in 18(g):	$x_3, 0, 1/2$;
3Sr in 3(a):	0, 0, 0;

All boron atoms are located in the centers of deformed trigonal prisms composed of Ni atoms. The prisms share two edges of a rectangular prism with two adjacent prisms and thus form a six-membered ring around every Sr atom. Nickel atoms form prism rings above and below the Sr atom and surround it in the form of an octahedron. Interatomic distances show the absence of direct Sr–B interactions, but only the Ni–B and Ni–Sr bonds.

This structure type has been found in: $RCo_{12}B_6$ (R = La, Ce, Pr, Nd, Sm, Eu, Gd, Tb, Dy, Ho, Er, Y) $RNi_{12}B_6$ (R = La, Eu, Gd, Tb, Dy, Y) (Kuźma et al., 1981), and $NdFe_{12}B_6$ (Niihara and Yajima, 1972).

2.12. CRYSTAL STRUCTURES BASED ON
CaCu₅-TYPE STRUCTURE

Ternary lanthanide borides RFe_4B (R = Er, Tm, Lu) (Spada et al., 1984; van Noort et al., 1985; Hong et al., 1988), RCo_4B [Kuźma et al., 1979], and RNi_4B (Niihara et al., 1973; Chernyak et al., 1982) crystallize in the structure of the $CeCo_4B$ type, which is a derivate of the $CaCu_5$ type. Its space group is P6/mmm; the atoms are in the positions:

1Ce in 1(a):	0, 0, 0		
1Ce in 1(b):	0, 0, 1/2;		
2Co in 2(c):	1/3, 2/3, 0;	2/3, 1/3, 0;	
6Co in 6(i):	1/2, 0, z;	0, 1/2, z;	1/2, 1/2, z;
	1/2, 0, $\bar{z}$;	0, 1/2, $\bar{z}$;	1/2, 1/2, $\bar{z}$;
			z = 0.287
2B in 2(d):	1/3, 2/3, 1/2;	2/3, 1/3, 1/2;	

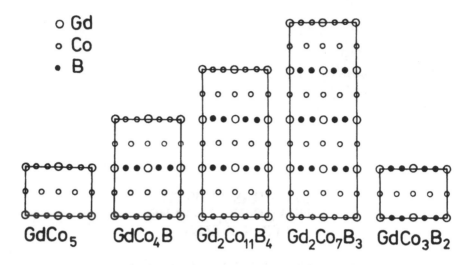

FIGURE 2.39. Schematic representation of the structures derived from the CaCu₅ type for Gd(Co₁₋ₓBₓ)₅ series.

The structure bears a layer character. In the plane with $z = 0$ each Ce atom coordinates six cobalt atoms at a distance of 2.89 Å, while the Ce atom in the plane with $z = 1/2$ has six boron closest neighbors at apexes of a hexagon with $d_{Ce-B} = 2.89$ Å. These sheets are separated by a plane composed of Co atoms.

Each Co atom within this plane has four nearest neighbors at a distance of $a/2 = 2.50$ Å and two other Co neighbors ($d_{Co-Co} = 2.46$ Å) and two boron atoms ($d_{Co-B} = 2.07$ Å), located on the adjacent planes.

The structures of a number of borides can be described as derivatives of the CaCu₅ type, which belongs also to the space group P6/mmm with:

1Ca in 1(a):	0, 0, 0;		
2Cu in 2(c):	1/3, 2/3, 0;	2/3, 1/3, 0;	
3Cu in 3(g):	1/2, 0, 1/2;	0, 1/2, 1/2;	1/2, 1/2, 1/2;

Its projection on the bc plane of an orthohexagonal unit cell (b = a_h $\sqrt{3}$; c = c_h) is presented in Figure 2.39.

The CeCo₄B unit cell is constructed by replacing the Cu atoms on one sheet, which automatically brings about the doubling of the unit cell in the c direction (see Figure 2.39) and a shift of the atoms of the 3(g) sheet towards the boron-containing sheet. This process is accompanied by the change of the z parameter: $z_{3(g)} = 0.25$ in the doubled undistorted CaCu₅ unit cell to $z_{6(i)} = 0.287$ in the CeCo₄B-type unit cell. Further replacement of Co atoms by borons leads to a unit cell tripled in the c direction (Figure 2.39). This is the

TABLE 2.14
Lattice Parameters (in Å) of a Number of CaCu₅-type Derivates

Compound	a	c
$CeCo_5$	4.926	4.020
$CeCo_4B$	5.005	6.932
$Ce_3Co_{11}B_4$	5.045	9.925
$Ce_2Co_7B_3$	5.053	12.97

structure of $Ce_3Co_{11}B_4$-type, space group P6/mmm, atoms in the following sites:

1Ce in 1(a):	0, 0, 0;		
2Ce in 2(e):	0, 0, z_1;	0, 0, $\bar{z}_1$;	$z_1 \sim 0.333$
2Co in 2(c):	1/2, 2/3, 0;	2/3, 1/3, 0;	
3Co in 3(g):	1/2, 0, 1/2;	0, 1/2, 1/2;	1/2, 1/2, 1/2;
6Co in 6(i):	1/2, 0, z_2;	0, 1/2, z_2;	1/2, 1/2, z_2;
	1/2, 0, $\bar{z}_2$;	0, 1/2, $\bar{z}_2$;	1/2, 1/2, $\bar{z}_2$;
			$z_2 = 0.2$
4B in 4(h):	1/3, 2/3, z_3;	2/3, 1/3, z_3;	
	1/3, 2/3, $\bar{z}_3$;	2/3, 1/3, $\bar{z}_3$;	$z_3 = 0.035$

The Ce atom in the 2(e) site coordinates 6Co atoms from the 3(g) positions at a distance of 3.02 Å, six borons at a distance of 2.92 Å and 6Co [6(i) site] with $d_{Ce-Co} = 2.85$ Å. The Co atom [6(i) site] has four boron neighbors at a short distance of 2.08 Å, while the Co [6(i) site] has two borons at the same distance.

A third phase occurring in the ternary Ce–Co–B system has the composition $Ce_2Co_7B_3$. Its structure is schematically displayed in Figure 2.39. The lattice parameters of the compounds belonging to this homolog series are given in Table 2.14.

2.13 THE NdCo₄B₄-STRUCTURE TYPE AND ITS POLYTYPES

The crystal structure of $NdCo_4B_4$ belongs to the space group $P4_2/n$ (Kuźma et al., 1979). The atoms occupy the following sites (origin at $\bar{1}$):

2Nd in 2(a):	±(1/4, 1/4, 1/4)	
8Co in 8(g):	±(x, y, z;	1/2 − x, 1/2 − y, z;
	$\bar{y}$, 1/2 + x, 1/2 + z;	1/2 + y, $\bar{x}$, 1/2 + z)
8B in 2(g):	as above	

The projection of this structure on the (001) plane is shown in Figure 2.40. Cobalt atoms form tetrahedra linked by common edges, giving rise to infinite chains propagating along the tetragonal axis. The Co–Co distance is

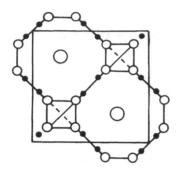

O Nd   o Co   • B

FIGURE 2.40. Projection of the NdCo₄B₄ structure on the (001) plane.

2.89 Å. Since the distance between two cobalt atoms belonging to the adjacent chains are rather short (d_{Co-Co} = 2.46 Å), chemical bonding operating between them can be postulated. The chains are also linked by pairs of boron atoms (d_{B-B} = 1.81 Å), so that an atomic framework arises with octagonal tubes filled with strings of Nd atoms (d_{Nd-Nd} = 3.822 Å).

This type of structure has been found in a number of ternary borides: RCo_4B_4 (R = La, Pr, Nd, Sm) (Kuźma et al., 1979), $LaRu_4B_4$ (Grüttner and Yvon, 1979). ROs_4B_4 (R = La, Pr, Nd, Sm) (Rogl et al., 1982).

The structure of $NdCo_4B_4$ has been found to constitute the basic structure unit in a number of ternary borides with the general formula $R_xT_4B_4$ (1 < x < 1.15) (R = Ce, Pr, Nd, Sm, Gd, Tb).

Although their structures were at first considered in terms of incommensurate polytypes, single crystal X-ray studies made it possible to describe some of them in commensurate unit cells, usually with long periodicity in the direction of the c axis. Thus, the structure of $Nd_{1.11}Fe_4B_4$ has been solved in an orthorhombic unit cell, space group Pccn, with a = b = 7.117 Å, c = 35.07 Å and four formal molecules of $Nd_5Fe_{18}B_{18}$ in the unit cell. The projection onto the (110) plane is displayed in Figure 2.41. In this structure, each Fe atom has five nearest neighbors in the same chain of tetrahedra and one Fe neighbor in the nearest chain, while pairs of boron atoms help to connect the different chains. The Nd–Fe and Nd–B distances amount to 2.834 to 3.125 Å and 2.621 to 3.102 Å, respectively. Two borons forming a pair are 1.744 Å apart (Givord et al., 1985).

On the other hand, single crystal X-ray study of $Sm_{1.13}Fe_4B_4$ permitted to solve its structure in a commensurate tetragonal unit cell (space group P4₂/n), however, with a periodicity of 58.89 Å along the tetragonal axis. The atomic pattern is very similar to that of $Nd_5Fe_{18}B_{18}$, but the chains of iron atom tetrahedra show a periodic twist modulation along the c axis, with boron pairs probably following the same modulation. Thus, the unit cell of $Sm_{1.13}Fe_4B_4$ contains two $Sm_{17}(Fe_4B_4)_{15}$ formula units. The average Fe–Fe bond distances are 2.5 Å (shared tetrahedra edges) and 2.7 Å (between different chains of

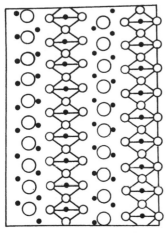

○ Nd   ○ Fe   ● B

FIGURE 2.41. Projection of the structure of $Nd_5Fe_{18}B_{18}$ on the (110) plane (Givord et al., 1985).

TABLE 2.15
$R_{1+\epsilon}$ Fe_4B_4 Phases-Lattice Parameters (in Å)

Compounds	ϵ	a	c
$Ce_{37}(Fe_4B_4)_{33}$	0.1212	7.090	129.04
$Pr_{21}(Fe_4B_4)_{19}$	0.1053	7.158	74.18
$Gd_{33}(Fe_4B_4)_{29}$	0.1379	7.073	113.73
$Tb_{31}(Fe_4B_4)_{27}$	0.1481	7.049	105.81
$NdCo_4B_4$	0	7.07	3.822

tetrahedra). The Sm-Fe distances range from 2.90 to 3.19 Å, while the shortest Sm-Sm contacts operating within the strings along the tetragonal axis amount to 3.46 Å (Bezinge et al., 1985). The other $R_{1-\epsilon}$ Fe_4B_4 compounds with the structures described in terms of commensurate tetragonal unit cells (space group $P4_2/n$) with the corresponding lattice parameters are listed in Table 2.15.

Chapter 3

MAGNETIC PROPERTIES

Intermetallic compounds which are described in this book contain two kinds of transition elements: f-electron (lanthanide) and d-electron transition elements. Each of these groups of elements exhibits different properties due to the localized behavior of "magnetic electrons". While the 3d-electron elements never show local moment behavior, the 4f-electron lanthanide elements almost go through the complete localized behavior.

Figure 3.1 shows the d- and f-electron elements arranged according to the onset of localization (Smith and Kmetko, 1983). The d- and f-electron wave functions shrink across their series and may finally fall within the inert gas core, which allows an atomic-like local moment in the condensed state.

The magnetic properties and magnetic interactions of these two groups elements are discussed later in this chapter.

3.1. MAGNETIC INTERACTIONS

3.1.1. LANTHANIDE ELEMENTS

All rare-earth elements, with the exception of Ce, Eu, and Yb, are in trivalent state. Spin, orbital and total angular momentum of the singlet rare-earth ions are determined by Hund's rule. For the free R^{3+} ions, the magnetic moment is proportional to $g_J \mu_B \sqrt{J(J+1)}$ in the paramagnetic and to $g_J J \mu_B$ in the ordered state. The Hamiltonian which describes the magnetic behavior of lanthanide ions is usually in the form.

$$H_{tot} = H_{Coul} + H_{exch} + H_{cef} + H_{ms} + H_{ext} \qquad (3.1)$$

where the term H_{Coul} represents the Coulomb interactions, H_{exch} describes the exchange interactions, H_{cef} is the crystal electric field term, H_{ms} accounts for magnetostriction effects, and H_{ext} takes account of the interaction in an external magnetic field.

The available experimental data clearly indicate that the terms H_{exch} and H_{ef} play dominant roles in the description of the magnetic properties of the R ions in a crystal. The magnetic coupling energy between two localized moments is usually assumed to be proportional to $\vec{S}_i \cdot \vec{S}_j$.

In the case of rare earths, two mechanisms have been proposed in which the 4f moments can interact in an indirect way. In the first, called the RKKY interaction, the magnetic coupling proceeds by means of spin polarization of the s conduction electrons.

In the second mechanism, the spin polarization of the appreciably less localized rare-earth 5d electrons plays an important role. The essential form

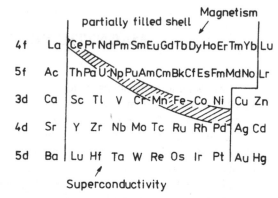

FIGURE 3.1. A nearly periodic table of the d- and f-electron elements arranged to show where the crossover to localized behavior occurs (Smith and Kmetko, 1983).

of the indirect interaction between localized moments was introduced by Ruderman and Kittel (1954) to describe the hyperfine interactions between nuclear moments. Later it was applied to the exchange interaction between localized moments (Kasuya, 1956; Yosida, 1957). The spin polarization of localized spin moment involves exchange interactions,

$$H = -2J_{sf}\vec{s} \cdot \vec{S} \qquad (3.2)$$

where s represents the spin of the Fermi surface conduction electrons, S denotes the localized rare-earth spin, and J_{sf} the exchange integral.

The conduction electron spin polarization interacts with a localized spin moment on a neighbor ion at a distance R_{ij} from the scattering center. The total exchange energy of the indirect interaction between i and j is

$$E = \frac{18\pi n^2}{E_F}J_{sf}^2 \vec{S}_i \cdot \vec{S}_j F(2k_F R_{ij}) \qquad (3.3)$$

Substituting this interaction energy into the molecular field expression (Bleaney and Bleaney, 1965) of T_c or Θ_p in ferromagnets, it can be shown that

$$T_c = \Theta_p = -\frac{3\pi n^2}{k_B E_F}J_{sf}^2(g_J - 1)^2 J(J + 1)\sum_{i \neq o}F(2k_F R_{oi}) \qquad (3.4)$$

where Θ_p is proportional to the molecular field at the central ion O resulting from the interaction with all neighbors i at a distance R_{oi}. Of course, this type of summation is valid only when all magnetic ions are crystallographically identical.

In antiferromagnetic materials, there may exist various types of spin arrangements represented by a propagation vector k = 0. Here, k may represent, e.g., the wave vector of some spin spiral structure whose wave length λ is equal to $2\pi/k$. Conical spin arrangements need not be considered here,

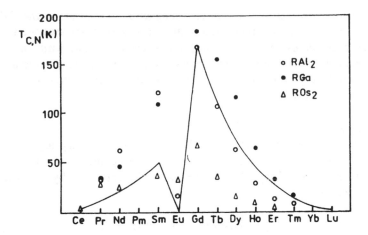

FIGURE 3.2. Magnetic ordering temperatures of some RX_2 compounds as a function of the De Gennes factor.

since they are only stabilized by the presence of anisotropy. The stable structure *(described by k_o)* is that for which the energy of the system is minimum.
Hence (see Mattis, 1965):

$$T_N = \frac{3\pi n^2}{k_B E_F} J_{sf}^2 (g_J - 1)^2 J(J + 1) \sum_{i \neq o} F(2k_F R_{Oi}) \cos(\vec{k}_O \cdot \vec{R}_{Oi}) \qquad (3.5)$$

The RKKY model shows also that for isostructural rare-earth compounds, the
ordering temperatures are expected to scale with $G = (g_J - 1)^2 J (J + 1)$
(de Gennes, 1962).

The critical review of the applications and modifications of the RKKY
model for the interpretation of the magnetic properties of different classes of
compounds has been published (Kirchmayer and Poldy, 1978). The RKKY
model usually has been applied for interpretation of the experimental data.
In a large amount of compounds hitherto studied (Skrabek and Wallace, 1963)
it is observed that the ordering temperatures generally follow the same universal curve when plotted against the de Gennes factor G (see Figure 3.2).
A different coupling scheme between the localized 4f moments was proposed
by Campbell (1972). In this model, the 5d electrons of the rare-earth component play an important role. The 5d electrons are far less localized than
are the 4f electrons, and a considerable overlap might occur between the 5d
wave functions of neighbor atoms. In compounds with sufficiently large rare-
earth concentration, one can expect, therefore, a direct d-d interaction. This
offers the possibility of an alternative form of interaction involving a positive
ordinary f-d exchange, combined with the positive direct d-d interaction mentioned above (Campbell, 1972). The overall indirect interaction between the

4f moments is, therefore, always ferromagnetic. In contrast to the RKKY interaction, it has a short range and treats d and s electrons separately.

The observed experiments in magnetic order in the rare earth sublattice result from the competition of two factors: the magnetic exchange interactions (H_{exch}) and the crystal electric field (CEF) (H_{cef}). The respective role of both factors is discussed below:

3.1.1.1. The Case $H_{exch} > H_{cef}$

This situation occurs when the exchange interaction is very large, or when the CEF splitting is very small. In that case, H_{cf} can be treated as a perturbation in comparison with H_{exch}. The anisotropy can be expressed by the classical formulation as

$$E_A = K_1 \sin^2 \Theta + K_2 \sin^4 \Theta + ... \qquad (3.6)$$

The magnetic moment reaches the maximum value $g_J J$. The preferred magnetization direction depends on the crystal structure via the CEF parameters B_n^m and the shape of the 4f electron charge cloud via the so-called Stevens coefficients. In the case of uniaxial crystal structures the lowest-order parameters are B_2^0 and α_J, respectively.

3.1.1.2. The Case $H_{exch} < H_{cef}$

This situation usually applies to ionic rare-earth compounds, but it also occurs very often in rare-earth intermetallics that have a low ordering temperature (Rossat-Mignot, 1983). Two parameters, δ which is the energy of the first excited CEF level and Δ which is the total CEF-splitting, must be introduced.

A simple and more common situation corresponds to the case that occurs when the energy of magnetic interactions is smaller than δ. Then, the Hamiltonian H_{exch} can be projected on the crystal-field ground level. In this case, the CEF anisotropy may lead to noncollinear magnetic structures. For rare-earth compounds in which the magnetic interactions are smaller than the energy δ of the first excited CEF level, the magnetic behavior is dominated by the CEF anisotropy.

3.1.1.3. The Case $H_{exch} \sim H_{cef}$

When $H_{exch} \sim H_{cef}$ no simple approximation can be made and the complete Hamiltonian $H_{exch} + H_{cef}$ must be diagonalized. The problem is then much more complex, but qualitative results can be obtained by also using a semi-classical description. By decreasing the temperature, the population of excited levels decreases, and a rotation moment may occur due to the competition between the CEF anisotropy and the entropy. Such a rotation moment has been observed, e.g., in $CeCu_2Ge_2$ (Knopp et al., 1989). In that case, it is not possible to formulate general trends; the magnetic properties depend very much on the strength of both CEF and exchange terms. The symmetry of the rare-earth site always remains an important parameter, but a large variety of anomalous magnetic behaviors can occur.

3.1.2. BINARY LANTHANIDE –3d– TRANSITION METAL COMPOUNDS

Although the magnetic properties of binary intermettalics with the general formula R_xT_y (R-lanthanide, T-3d transition metal) have been intensively investigated in the course of the last 25 years, the nature of magnetic interaction has not been well understood even now. The difficulties arise from the inherently different nature of magnetic moments — the R and T atoms and the interactions among them. Three main types of interactions can be distinguished: R–R, R–T and T–T. To explain their influence on the general magnetic behavior of R_mT_n binaries, the Néel molecular-field model is usually applied. The respective molecular-fields are written as:

$$H_R(T) = H + (n_{RR}\, \mu_R(T) + n_{RT}\, \mu_T(T))$$
$$H_T(T) = H + (n_{TT}\, \mu_T(T) + n_{RT}\, \mu_R(T)) \qquad (3.7)$$

H is applied external field, μ_R and μ_T are the magnetic moments on the R and T atoms, respectively. The molecular-field coefficients n_{RR}, n_{RT}, and n_{TT} are related to the R–R, R–T, and T–T magnetic interactions, respectively. It is assumed that the temperature dependence of each moment follows the Brillouin function

$$\mu_R(T) = \mu_R(0)B_J(\mu_B\mu_R(0)H_R(T)(kT)^{-1})$$
$$\mu_T(T) = \mu_T(0)Bs_{3d}(\mu_B\mu_T(0)\, H_T(T)(kT^{-1})) \qquad (3.8)$$

$\mu_R(0)$ and $\mu_T(0)$ are the zero-temperature moments, J is the angular momentum of the 4f shell of the R atom; S_T is the spin of the 3d atom, $B_J(x)$ is the Brillouin function.

The physical condition that $\mu_R(T)$ and $\mu_T(T)$ separately vanish when $T = T_c$ leads to a relation between T_c and the molecular-field coefficient n_{ij}

$$Tc = 1/2\{\alpha n_{T-T} + \beta_{R-R} + [(\alpha n_{T-T} - \beta n_{R-R})^2 + 4\alpha\beta n_{R-T}^2]\} \qquad (3.9)$$

where

$$\alpha = \mu_T^2(0)\left[\frac{\mu_B}{k}\right]\left[\frac{S_T + 1}{3S_T}\right]$$

$$\beta = \mu_R^2(0)\left[\frac{\mu_B}{k}\right]\left[\frac{J + 1}{J}\right] \qquad (3.10)$$

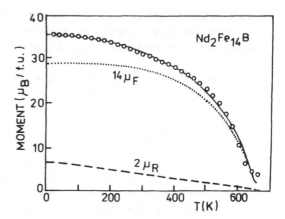

FIGURE 3.3. Molecular-field analysis of $Nd_2Fe_{14}B$. Open circles denote the measured moment per formula unit. The solid line is the calculated total moment, which is the sum of the iron (dotted line) and neodymium (dashed line) contributions (Fuerst et al., 1986).

The n_{ij} coefficients are determined by the numerical solution of Equation 3.8, subject to the condition, that the calculated magnetization

$$M(T) = N_R\mu_R(T) + N_T\mu_T(T) \tag{3.11}$$

agrees with the experiment. N_R is the number of lanthanide atoms in the chemical "molecule", N_T — the number of transition metal atoms. For example, a fair correspondence with the measured magnetization of $Nd_2Fe_{14}B$ has been obtained for $n_{R-R} = 5.9 \times 10^3$; $n_{R-T} = 2.2 \times 10^3$ and $n_{R-R} = 3.3 \times 10^2$ (see Figure 3.3) (Fuerst et al., 1986).

Exchange interaction energies J_{ij} are also related to the molecular-field coefficients n_{ij}. This follows from the equivalence of the molecular-field model with the two-sublattice Heisenberg model (the nearest neighbor self consistent-field approximation).

$$J_{T-T} = \frac{\mu_B}{Z_{T-T}} g_T^2 n_{T-T}$$

$$J_{R-T} = \frac{\mu_B}{Z_{R-T}} g_T \frac{g_J}{g_J - 1} n_{R-T}$$

$$J_{R-R} = \frac{\mu_B}{Z_{R-R}} \left[\frac{g_J}{g_J - 1}\right]^2 n_{R-R} \tag{3.12}$$

Z is the number of the respective nearest-neighbors. For $Nd_2Fe_{14}B$ the magnitudes of J have been determined to be: $J_{T-T} = -3.8$ meV (2.4×10^{-22} J); $J_{R-T} = -2.4$ meV (1.5×10^{-22} J) and $J_{R-R} = 0.2$ meV (0.12×10^{-22} J).

The above method is generally used to interpret the magnetic properties of lanthanide-3d metal binaries. As an example, Figure 3.4 shows the de-

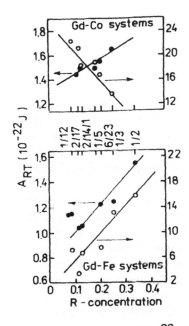

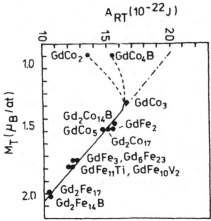

FIGURE 3.4. J_{R-3d} values (o) for Gd–Fe and Gd–Co compounds. For comparison, the values J_{TT} representing T–T interactions are also indicated (Duc et al., 1992).

FIGURE 3.5. The variations of J_{RT} as a function of the 3d magnetic moment in a number of Gd–T intermetallics (Duc, 1991).

pendence of J_{R-T} and J_{T-T} exchange energies on the composition of Gd_mT_n (T = Fe,Co) binaries. As the concentration of the 3d metal falls, an increase of J_{R-T} values is observed in both the Gd_mFe_n and Gd_mCo_n systems. The variation of J_{R-T} as a function of the 3d magnetic moment is displayed in Figure 3.5. This result suggests that the 3d-5d hybridization effects rise with the concentration of lanthanides.

3.2. THE CRYSTAL ELECTRIC FIELD (CEF) MODEL

3.2.1. INTRODUCTION

Quantum theory provides three theoretical approaches which describe the behavior of lanthanide elements in solids:

- The valence bond treatment
- The crystal electric field (CEF) model
- The molecular field theory

The latter two are in current use.

In contrast to the molecular field theory which is difficult to handle computationally, the CEF model is based on very simple and elegant concepts. Although it involves parameters that cannot be determined theoretically at present, the CEF model provides a powerful parameterization scheme for the description of spectroscopic, magnetic, and related properties of lanthanide ions in solids.

When describing the properties of lanthanide atoms in crystals using the CEF model, it is assumed that the atom is in a definite ionization state, and the interaction of the atom with its surroundings can be expressed in terms of a classical electrostatic potential. The ionization state of the atom is deduced from chemical and magnetic properties of elements, so that it can be treated quantum-mechanically in a straightforward way. The determination of the magnitude of the electrical potential can be done using the so-called point-charge model (Hutchings, 1964). In this approximation the electrostatic potentials of the most important neighbors are summed, under the assumption that their charge distributions can be approximated by point charges. Although this approximation is rarely justified, it is useful in defining the CEF parameters.

The typical CEF splitting in transition metals is 1 eV. In the case of lanthanide elements, the ions have, in general, a valence of $+3$ and the krypton $4d^{10}4f^n5s^25p^6$ electron configuration. As can be seen from Figure 3.6, which is typical for all R^{+3} lanthanide ions, the uncompensated 4f electrons are well shielded by the closed $5f^25p^6$ shells. Therefore, the CEF splitting is reduced typically to the order of 10 meV for ions when orbital momentum cannot be neglected. It is also possible to observe higher-order CEF effects on S-state ions (e.g., Gd^{+3}), which are of order 10^{-1} meV. These estimates turn out to be valid for insulators as well as for metals, and no significant effect due to conduction-electron shielding is observed.

In applying the CEF model, the parameters that are essentially products of the strength of the electrical potential and of the free ion properties are used. The number of these parameters depends strongly on point group symmetry and on the kind of electrons that are affected. For low-point group symmetries the number of parameters can be rather large, but is usually reduced to a few in the case of high point group symmetries. Large CEF effects cannot be observed in compounds with lanthanide ions having empty, completely filled, or half-filled 4f shells, like Sc, Y, La, Gd, and Lu. Also, in Ce, Sm, Eu, Tm, and Yb the $4f^n5d^x6s^2$ ($x = 0, 1$) and $4f^{n-1}5d^{x+1}6s^2$ configurations are near to each other in energy, which leads to nonsystematic behavior. Table 3.1 lists the most important R^{+3} free ion parameters.

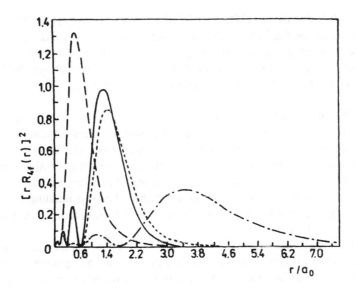

FIGURE 3.6. Radial distribution of wave functions of Gd obtained from nonrelativistic Hartree-Fock calculations (Freeman, 1972) (—, 4f; -, 5s; ——, 5p; –·–, 6s). The radius is measured in units of $a_0 = 0.529$ Å.

TABLE 3.1
Free-ion Data for R^{3+}

Ion	Number n of 4f electrons	L	S	J	g_J	Ground state	Δ(meV)
La^{3+}	0	0	0	0	—	$^1S_0$	—
Ce^{3+}	1	3	1/2	5/2	6/7	$^2F_{5/2}$	275.7
Pr^{3+}	2	5	1	4	4/5	$^3H_4$	267.1
Nd^{3+}	3	6	3/2	9/2	5/11	$^4I_{9/2}$	232.7
Pm^{3+}	4	6	2	4	3/5	$^5I_4$	275.7
Sm^{3+}	5	5	5/2	5/2	2/7	$^6H_{5/2}$	120.6
Eu^{3+}	6	3	3	0	—	$^7F_0$	43.1
Gd^{3+}	7	0	7/2	7/2	2	$^8S_{7/2}$	—
Tb^{3+}	8	3	3	6	3/2	$^7F_6$	249.9
Dy^{3+}	9	5	5/2	15/2	4/3	$^6H_{15/2}$	405.0
Ho^{3+}	10	6	2	8	5/4	$^5I_8$	646.3
Er^{3+}	11	6	3/2	15/2	6/5	$^4I_{15/2}$	810.0
Tm^{3+}	12	5	1	6	7/6	$^3H_6$	732.4
Yb^{3+}	13	3	1/2	7/2	8/7	$^2F_{7/2}$	1293.5
Lu^{3+}	14	0	0	0	—	$^1S_0$	—

Note: L, S, and J are the orbital, spin, and total angular-momentum quantum numbers; g_J is the Lande factor and Δ is the distance between the ground and the first excited multiplet.

3.2.2. THE INTERACTION OF THE FREE ION WITH CEF

Including the CEF as a perturbation on the free ion state, three different cases depending on the relative magnitude of the CEF with respect to the terms occurring in the Hamiltonian of the free ion can be distinguished. This is convenient in order to obtain a rapidly converging perturbation expansion.

1. The strong-field case. The CEF is of the order of the interelectronic repulsion. The kinetic energy together with Coulomb interaction of the electron with the nucleus has to be treated first. The zero-order wave functions of the perturbation calculations are thus classified by the different projections of the angular momentum of the kth electron l_k in the z direction. In this case, considerable chemical bonding is usually present, and often the problem has to be treated in terms of the molecular-orbital theory.

2. The intermediate-field case. The CEF is smaller than the kinetic energy plus the Coulomb energy V_{ck}, but larger than the spin-orbit coupling V_{LS}. V_{CF} define states with total angular momentum quantum number L as a good quantum number, but the zero-order wave functions are classified by the projection m of the total angular momentum in the z direction. This case often applies to the d transition metals.

3. The weak-field case. The CEF is smaller than the spin-orbit coupling, but larger than the spin-spin and hyperfine interactions. Assuming Russel-Saunders coupling to be valid, one obtains functions labeled by the total angular-momentum quantum number L, by the total spin S, by J resulting from the sum of angular momentum, and by M resulting from the projection of J in the z direction. The set of zero-order wave functions of this perturbation approach can then be defined by these quantum numbers. The lanthanide elements are usually treated within this scheme.

An example of the effect of CEF on the level scheme of an ion with L = 3 and S = 1 is shown in Figure 3.7. Its left-hand part belongs to Case 2 and the right-hand part to Case 3.

In each of the three cases, the unperturbed wave functions $|J'M'\rangle$ are eigenfunctions of the square of angular momentum operator J', which can be l_k, L, or J and, simultaneously, the eigenfunctions of M. In the absence of a magnetic field, the states with definite angular momentum are $2J'+1$ degenerate. To calculate the level splitting due to the CEF perturbation, one has to diagonalize a $(2J'+1) \times (2J'+1)$ matrix. To solve the equation, it is necessary to introduce the explicit form of V_{CEF}. Assuming that the CEF is produced by an array of point charges surrounding the central ion or spatially extended charges that do not overlap with the electrons of the free ion, V_{CF} must satisfy the Laplace equation:

$$\nabla^2 V_{CF} = 0 \qquad C \qquad (3.13)$$

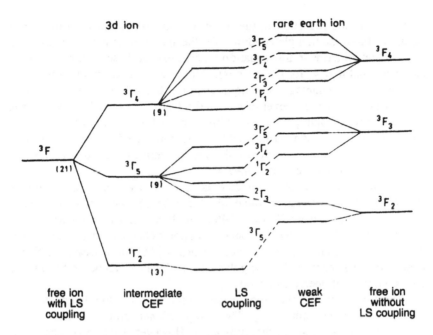

FIGURE 3.7. Interaction of the CEF with an ion with L = 3 and S = 1. Left: CEF of intermediate strength; right: weak CEF (schematic).

In the region around the central ion, the solution to the above equation can be expanded in terms of spherical harmonics Y_n^m as

$$V_{CEF} = \sum_{n=0}^{\infty} \sum_{m=-n}^{n} A_n^m r^n Y_n^m(\Theta,\varphi) \tag{3.14}$$

Here, the A_n^m are constants. If there is more than one electron, it becomes necessary to sum this equation over the number of electrons of the central ion.

Now the matrix elements of the form

$$\langle J'M_i'|A_n^m r^n Y_n^m(\Theta,\varphi)|J'M_k'\rangle = A_n^m \langle R|r^n|R\rangle \langle \phi_i|Y_n^m(\vartheta,\varphi)|\phi_k\rangle \tag{3.15}$$

should be found. To solve the above equation, the electron wave functions are presented as the product of a radial function $|R\rangle$ and a function $|\Phi_i\rangle$, depending only on the angle. $|R\rangle$ is obtained from nonrelativistic Hartree-Fock calculations (Freeman and Watson, 1962) and also from relativistic calculations (Lewis, 1971; Freeman and Desclaux, 1979) and is the same for all states of a given electron configuration ($3d^n4f^m$, etc). The crucial point in the CEF model is that the $|\Phi_i\rangle$ can be expanded in terms of spherical harmonics (for d electrons up to order n = 2; for f electrons up to order n = 3), which leads to a drastic reduction in the number of matrix elements in the above equation. From the triangle condition for the spherical harmonics integral, it follows that all terms in this equation vanish, except those with $n < 6$ (for f electrons) and $n < 4$ (for d electrons). All terms with odd n also vanish for configurations containing solely equivalent electrons. The term with $n = 0$ leads to a shift of the total level scheme and can be neglected when considering only the spectroscopic properties within one configuration. The point symmetry of the central ion further reduces the number of parameters, so that finally, in the systems with high point group symmetry, a small number of nonvanishing matrix elements appears.

To determine the matrix elements numerically, one has to obtain $\langle R|r^n|R \rangle A_n^m$. The expectation value of r^n can be obtained from the results published in Freeman and Watson (1962) and Freeman and Desclaux (1979), and the A_n^m can, in principle, be found from point-charge models. Since these models are quantitatively unjustified, the $\langle R|r^n|R \rangle A_n^m$ can be treated as fitting parameters. The nonvanishing angular matrix elements can be best evaluated by the equivalent-operator method.

The final step in perturbation calculation consists in the diagonalization which is performed numerically. The computational effort becomes considerably reduced when group theory is used. However, when magnetic fields that show effects of the order of magnitude of CEF are included, the perturbation matrix should be diagonalized in one step. The results of a solution of the perturbation problem in situations occurring in RT_2X_2 intermetallic compounds will be discussed later.

The calculations of matrix elements by direct integration is very tedious if one is dealing with more than one 4f electron, but if only the lowest multiplet J is considered, the method of Stevens equivalent operators (Stevens, 1952) may be applied. The main idea is to replace polymonials of the spatial coordinate operators x_i, y_i, and z_i with the corresponding products of the angular momentum operators $\hat{J}_x$, $\hat{J}_y$, and $\hat{J}_z$. Since the angular-momentum operators do not commute, care has to taken to write the corresponding products in a symmetrical form, i.e., xy $1/2(\hat{J}_x\hat{J}_y + \hat{J}_y\hat{J}_z)$. The operator equivalents act on the 4f shell as a whole, thus avoiding the necessity of going back to single-electron wave functions. For example, the equivalent of $\sum_i(x_i^4 - 6x_i^2 y_i^2 + y_i^4)$ is given by:

$$\sum_i (x_i^4 - 6x_i^2y_i^2 + y_i^4) = O_4\langle r^4 \rangle \frac{1}{2}(\hat{J}_+^4 + \hat{J}_-^4) = O_4\langle r^4 \rangle \hat{O}_4^4 \quad (3.16)$$

A complete list of Stevens operator equivalents O_n^m has been published (Hutchings, 1964), together with the values of the reduced matrix elements O_2, O_4, and O_6 (often called α_j, β_j, and γ_j). They are commonly called the Stevens coefficients. The CEF Hamiltonian can, therefore, be expressed in general form as:

$$H_{CEF} = \sum_{n,m} A_n^m \langle r^n \rangle 0_n \hat{O}_n^m \qquad (3.17)$$

The A_n^m are the CEF parameters which are to be determined in an experiment. In the case of 4f electrons, the summation can be limited to terms with $n < 6$. The point-group symmetry will further reduce the number of nonvanishing terms in this equation. For example, in the case of the cubic point group, the CEF Hamiltonian becomes reduced to:

$$H_{CEF} = B_4(O_4^0 + 5O_4^4) + B_6(O_6^0 - 21O_6^4) \qquad (3.18)$$

Here, the fourfold axis has been chosen as a quantization axis. The CEF parameters B_4 and B_6 are given by:

$$B_n = A_n^0 \langle r^n \rangle O_n \qquad (3.19)$$

The theoretical calculation of the CEF parameters $A_n^m \langle r^n \rangle$ is difficult. Often a simple point-charge model is used. The charge distribution (R) is replaced by point charges q_j located at the neighboring lattice sites:

$$\rho(R) = \sum_j q_j(R_j - R) \qquad (3.20)$$

Several theoretical attempts have been made to deal with the influence of the conduction electrons on CEF. In one model (Duthie and Heine, 1979), the electrons are regarded as free electrons. In a more realistic approach, the d and f characters of the conduction band have been allowed for. The contribution of the conduction electrons to CEF consists of two parts. The first is a direct Coulomb part, while the second is an exchange Coulomb part. The d and f electrons both contribute to the CEF parameters of the fourth order while the f electrons also exert influence on the six-order term. However, similar to the case of the point-charge model, the calculated CEF parameters do not fit the experimentally determined values.

3.2.3. EFFECT OF THE CEF ON PHYSICAL PROPERTIES
3.2.3.1. Magnetic Moment, Anisotropy and Susceptibility

For a free ion in a magnetic field the ground state has the maximum moment in the z direction. For the case of the lanthanides the moment has the value g_J (in Bohr magnetons). In the presence of CEF, the new eigenfunctions are linear combinations of functions with one J, but different M. For the ground state in the magnetic field, the admixture of wave functions with different M has the consequence that the expectation value $\mu_B g_J \langle M \rangle$ of the magnetic moment of the ground state in z-direction is reduced. However, when $\mu_B g_J JH \gg V_{CEF}$, all admixtures from $M < J$ can be neglected so that free ion magnetic moment is obtained.

The susceptibility of a free ion can be presented as:

$$\chi_0 = \frac{\mu_B^2 g_J^2 J(J + 1)}{3K_B T} \tag{3.21}$$

for $\mu_B g_J J H \ll k_B T$

For the ions implanted in solids, this relation will be modified by CEF and by exchange interactions. The latter will be considerably reduced when the concentration of magnetic ions will be very small. CEF will be thus left as the only interaction causing deviations from the above equation. The measured values of the Lande factors for lanthanides are generally smaller than 2 (see Table 3.1).

Therefore the details of the effect of CEF on the orbital contributions to the magnetic moment cannot be neglected. The energy of the spectroscopic levels of a lanthanide ion which experiences both the CEF and the magnetic field can be written as:

$$E_{nm} = E_n + E_{nm}^{(1)} H + E_{nm}^{(2)} H^2 + \ldots \tag{3.22}$$

Here, n is the CEF quantum number labeling the CEF levels in zero magnetic field, and m is the quantum number which labels the different levels due to the splitting of the n-th CEF level in the magnetic field. The magnetic moment is given by:

$$\mu_{nm} = -\frac{\partial}{\partial H} E_{nm} = -E_{nm}^{(1)} - 2E_{nm}^{(2)} H + \ldots \tag{3.23}$$

and the susceptibility per ion by:

$$\chi = \frac{\partial}{\partial H} \frac{\sum\limits_{n,m} \mu_{nm} \exp(-E_{nm}/k_B T)}{\sum\limits_{n,m} \exp(-E_{nm}/k_B T)} \tag{3.24}$$

For the restriction that $\mu_B g_J J H \ll kT$, which can always be satisfied by choosing H sufficiently small, one gets:

$$\chi = \frac{\sum\limits_{nm} [(E_{nm}^{(1)})^2/k_B T - 2E_{nm}^{(2)}] \exp(-E_n/k_B T)}{\sum\limits_{n} \exp(-E_n/k_B T)} \tag{3.25}$$

The values of $E_{nm}^{(1)}$ are conveniently obtainable by computer calculations. In order to determine the CEF parameters, the thermal dependence of inverse susceptibility along particular axes is usually analyzed.

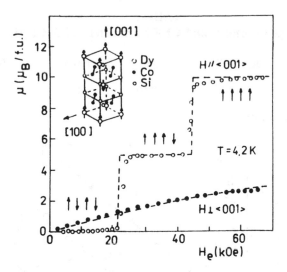

FIGURE 3.8. Plots of magnetization of $DyCo_2Si_2$ at 4.2 K . The broken lines represent the calculated magnetization. The inset represents the crystal structure and the magnetic structure. Possible moment alignment under a magnetic field along the (001) axis (Iwata et al., 1990).

Magnetic properties of $DyCo_2Si_2$ (Iwata et al., 1990) provide an illuminating example which shows the role of the CEF. The crystal structure of $DyCo_2Si_2$ is of tetragonal $ThCr_2Si_2$ type (space group I4/mmm). It is shown in the insert in Figure 3.8. In molecular field theory, the effective Hamiltonian of a Dy ion in the i-th plane in a field H applied along the {001} axis is as shown below:

$$\mathscr{H}_i = -g_J\mu_B J_{zi}(H_{mi} + H) + \frac{1}{2} g_J\mu_B\langle J_{zi}\rangle H_{mi}$$
$$+ B_2^0 O_2^0(i) + B_4^0 O_4^0(i) + \dots \quad (3.26)$$

Here B_n and O_n are the CEF parameters and Stevens operators respectively. The molecular field H_{mi} acting on the Dy ion in the i-th plane may be written as:

$$H_{ni} = \frac{(g_J - 1)^2}{g_J\mu_B} \sum_n J_n J_{z(i+n)} \quad (3.27)$$

where adjacent planes are coupled by exchange constants J_1, next nearest planes by J_2, etc. The coupling in the plane is denoted by J_0.

The observed large anisotropy due to the CEF effect leads to the description of magnetic interactions in $DyCo_2Si_2$ as an Ising linear-chain model. The determined exchange constants and CEF parameters are listed in Table 3.2. The values of magnetization and inverse susceptibility were reproduced by

TABLE 3.2
Exchange Constants and CEF Parameters of DyCo$_2$Si$_2$ (in K)[a]

J$_0$(K)	J$_1$ + J$_3$(K)	J$_2$(K)	B$_2$(K)	B$_4$(K)
1.98	−1.82	−0.60	−1.75	3.0 × 10^{-2}

[a] Iwata et al., 1990.

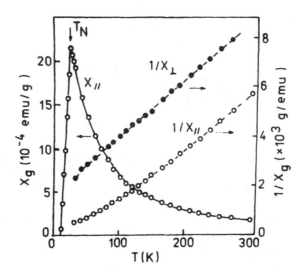

FIGURE 3.9. Plots of susceptibility χ_g along the (001)-axis and χ_g^{-1} along the (001) axis, and along the direction perpendicular to the axis. The broken lines represent calculated χ_g^{-1} (Iwata et al., 1990).

numerical calculations using the parameters listed in Table 3.2. The results are displayed in Figures 3.8 and 3.9 by broken lines. It can be seen that the agreement is fairly good.

The exchange-constant J$_o$ is positive, indicating ferromagnetic coupling, whereas J$_1$<0, J$_2$<0, and J$_3$<0 indicate antiferromagnetic interactions. The J$_1$ and J$_3$ appear in H$_{m1}$ in the form of (J$_1$ + J$_3$) for all of the three structures; hence they cannot be determined separately. An estimate of T$_N$ with the parameters given in Table 3.2 yields T$_N$ = 24 K, exactly as observed in the experiment.

3.2.3.2. Transport Properties

The energy splitting of the Hund ground state also has pronounced effects on transport properties, such as electrical resistivity thermoelectrical power, thermal conductivity, and the Hall effect.

The resistivity of magnetic lanthanide compounds can be expressed as the sum of the residual resistivity ρ_R, the lattice resistivity due to the phonons and the magnetic resistivity (Gratz and Zuckermann, 1982):

$$\rho(T) = \rho_R + \rho_{ph}(T) + \rho_{mag}(T) \qquad (3.28)$$

The magnetic part of the electrical resistivity is due to the 4f electrons reflecting the state of the 4f shell. In the first Born approximation the magnetic part of the resistivity in the single ion model is expressed as:

$$\rho_{mag}(T) = \rho_0 tr(PQ) \qquad (3.29)$$

where P and Q are matrices with elements P_{ij} and Q_{ij} defined as:

$$P_{ij} = \frac{e^{-E_i/k_BT}}{Z} \frac{(E_i - E_j)/k_BT}{1 - e^{(E_i - E_j)/k_BT}}$$

$$Q_{ij} = |\langle i|J_z|j\rangle|^2 + \frac{1}{2}|\langle i|J_+|j\rangle|^2 + \frac{1}{2}|\langle i|J_-|j\rangle|^2 \qquad (3.30)$$

From the knowledge of eigenvalues E_i and the eigenfunctions $|i\rangle$ of the Hamiltonian, the temperature dependence of the magnetic part of the electrical resistivity can be calculated; however, the dependence is generally not strong enough to allow accurate determination of the CEF parameters. Moreover, additional effects such as quadrupolar scattering should be allowed for in order to obtain a satisfactory theoretical description.

Other transport properties, such as the thermal conductivity, the thermal electric power, and the Hall effect exhibit no systematic correlation with magnetic properties, but reflect in a complex manner the band structure of the studied material.

3.2.3.3. Magnetic Part of the Molar Capacity

The energy splitting due to the CEF and the magnetic interactions strongly influences the molar heat capacity if the splitting is comparable to the thermal energy. For this reason, molar capacity may be used to determine the CEF and exchange parameters. The relevant relation can be written as:

$$C_P(T,B) = C_{ce}(T) + C_{ph}(T) + C_n(T) + C_{mag}(T,B) \qquad (3.31)$$

In this formula, $C_{ce}(T)$ stands for the contribution due to the conduction electrons; C_{ph} is the lattice contribution; $C_n(T)$ is the nuclear part; and $C_{mag}(T)$ is the magnetic part, which is due to the 4f electrons and can be calculated. When a free ion has a first exited state sufficiently far from the ground state, the latter does not contribute to the specific heat. In the case where the ground state is split by the CEF into N levels (n = 1,2, . . . ,N) with energy difference Δ_n to the ground state ($\Delta_i = 0$) and the degeneracy g_n, the internal energy E increases with temperature. Consequently:

$$E = \frac{1}{Z} \sum_{n=1}^{N} \Delta_n g_n \exp(-\Delta_n/k_BT) \quad \text{with} \quad Z = \sum_{n=1}^{N} g_n \exp(-\Delta_n/k_BT) \qquad (3.32)$$

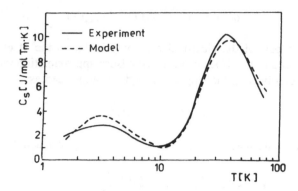

FIGURE 3.10. Schottky specific heat C_s (full line) of $Tm_{1-x}Lu_xCu_2Si_2$ vs T. The dashed line is a fit of the Schottky formula to the experimental results (Kozłowski et al., 1987).

The corresponding Schottky specific heat is expressed as:

$$C_{Sch} = \frac{1}{Zk_BT^2}\left[\sum_{n=1}^{N} g_n\Delta_n^2\exp(-\Delta_n/k_BT)\right.$$

$$\left. -\frac{1}{Z}\left(\sum_{n=1}^{N} g_n\Delta_n \exp(-\Delta_n/k_BT)\right)^2\right] \quad (3.33)$$

The entropy S of the ion at high temperature is

$$S = k_B\ln\left(\sum_{n=1}^{N} g_n\right) \quad (3.34)$$

and the gain in entropy when heating from low to high temperatures

$$S = S - k_B\ln g_i = \int_0^\infty \frac{1}{T} C_{Sch} \, dT$$

ΔS can be determined by appropriate integration of the measured specific heat and is often a valuable criterion for the degeneracy of the ground state. Figure 3.10 shows the Schottky specific heat vs. temperature curve determined for $TmCu_2Si_2$ (Kozłowski et al., 1987). The deduced energy level scheme is given in Table 3.3.

3.2.3.4. Hyperfine Interactions

The operator of magnetic hyperfine interactions V_M depends on the expectation values of I_n and S_n, which are determined by CEF. The resulting hyperfine splitting can be written as:

$$V_M = \mu_N g_n H_{hyp} I_z \quad (3.35)$$

TABLE 3.3
Eigen States and Eigen Values of CEF Hamiltonian of TmCu$_2$Si$_2$ as Obtained from the Fit of the Schottky Specific Heat to the Experimental Values[a]

Mixed states of J$_z$	Multiplicity	Energy (K)
$\|G\rangle\|6\rangle, \|2\rangle, \|-2\rangle, \|-6\rangle$	1	0
$\|E\rangle\|6\rangle, \|2\rangle, \|-2\rangle\|-6\rangle$	1	7.2
$\|6\rangle, \|2\rangle, \|-2\rangle, \|-6\rangle$	1	91.5
$\|5\rangle, \|3\rangle, \|1\rangle, \|-1\rangle$		
$\|-3\rangle, \|-5\rangle$	2	97.5

Note: The CEF parameters are w = 4.42 K, x_1 = 0.24, x_2 = −0.24, x_3 = 0.04, x_4 = 0.44, x_5 = 0.04.

[a] Kozłowski et al., 1987.

where H$_{hyp}$ is the magnetic hyperfine field and I$_z$ is the nuclear angular momentum in the z direction. H$_{hyp}$ consists of different parts which are due to the orbital momentum of the unfilled shells, the spin momentum of the atom, and the net spin magnetization of the conduction electrons. In lanthanide ions, the orbital momentum part plays the dominant role in H$_{hyp}$ so that:

$$H_{orb} = 2 \ \mu_B \langle J\|N\|J\rangle \langle r^{-3}\rangle (1 - R_m)\langle M\rangle \tag{3.36}$$

Here $\langle J\|N\|J\rangle$ is the reduced matrix element of the magnetic dipole interaction; R_m, which is of the order of 10^{-10}, represents the diamagnetic shielding due to the filled shells of the atom. H$_{orb}$ is proportional to $\langle M\rangle$ and, therefore, proportional to the expectation value of the magnetic moment.

The electrical hyperfine interaction is due to the electric field gradient produced by the surrounding ions at the site of the nucleus plus the electric field gradient on the ion itself, i.e., by the uncompensated electrons which lead to the formation of the orbital magnetic moment. This field gradient interacts with the nuclear quadrupole moment. Since the former contribution vanishes in cubic point symmetry, it can be neglected. In the case of axial symmetry or, when the magnetic hyperfine splitting is much larger than the electrical hyperfine splitting, the orbital part can be expressed as:

$$V_E = \frac{eQV_{zz}}{4I(2I - 1)} \langle 3I_z^2 - I(I + 1)\rangle \tag{3.37}$$

V_{zz} is the electrical field gradient. Explicitly, for 4f electrons it is as follows:

$$V = -e\langle J|\alpha|J\rangle\langle r^{-3}\rangle(1 - R_0)\langle 3M^2 - J(J + 1)\rangle \tag{3.38}$$

where R_o (of order 0.3) describes electric shielding and $\langle J|\alpha|J \rangle$ is the matrix element of the electrical quadrupole interaction. V_{zz} is proportional to $\langle 3M^2 - J(J+1) \rangle$ which, in turn, reflects the effect of the CEF on the wave functions.

The V_{zz} component of the EFG tensor induced by the crystal lattice, which can be directly deduced from the quadrupole interaction constant, is proportional to the B_2^0 term of the CEF Hamiltonian. The relationship between B_2^0 and V_{zz} has the following form:

$$V_{zz} = -4A_2^0(1 - \gamma_\infty)/e \qquad (3.39)$$

and

$$B_2^0 = A_2^0 \langle r_{4f}^2 \rangle (1 - \delta_2) \qquad (3.40)$$

γ_∞ and σ_2 denote the screening of the atomic nucleus and electronic f shell against external electrostatic potential. $\langle r^2 \rangle_{4f}$ is the second moment of the radial function of 4f electrons and represents a numerical Stevens factor. The quantity A_2^0 is a universal factor applicable to all isostructural lanthanide compounds, provided their structural parameters are constant. Values for A_2^0 can be derived from the observed quadrupole splittings E_q, using the relation:

$$A_2^0 = -\Delta E \cdot E_\gamma/[(1 - \gamma_\infty)Qc] \qquad (3.41)$$

This formalism has been used to determine the values of B_2^0 parameters for the RT_2Si_2 compounds (R-heavy lathanide) (Łątka, 1989). The obtained results are listed in Table 3.4.

3.2.3.5. Neutron Scattering by CEF Transitions

When neutrons are magnetically scattered by CEF split ions, they may induce inelastic transition from a CEF level Γ_{nm} to a level $\Gamma_{n'm'}$ or quasielastic transition from state Γ_{nm} to $\Gamma_{nm'}$. For these transitions, it can be shown that for only magnetic dipole transitions are important (Balcar and Lovesey, 1970). The double differential cross section can be expressed as:

$$\frac{\partial^2 \sigma}{\partial \Omega \partial \omega} = N(\gamma e^2/2mc^2)F^2(\varkappa) \frac{k_f}{k_i} \sum_{\substack{nm \\ n'm'}} (\exp(-E_{nm}/k_BT)/\sum_{n''m''} \exp$$

$$(-E_{n''m''}/k_BT)\langle|\Gamma n'm'|\hat{J}_\perp|\Gamma nm\rangle|^2 \delta(E_{nm} - E_{n'm'} - n\omega) \qquad (3.42)$$

In this formula, k_i and k_f are the moments of the in- and out-going neutrons, Γ_{nm} is the label of the CEF level where n indicates the CEF level and m runs over all its degenerate states. $\hat{J}_\perp$ is the total angular momentum operator

TABLE 3.4
Crystalline Electric Field Parameters B_2^0 for RT_2Si_2 Intermetallic Compounds[a]

T/R	Tb^{3+}	Dy^{3+}	Ho^{3+}	Er^{3+}	Tm^{3+}	Yb^{3+}
Fe	-4.07	-2.41	-0.80	0.87	3.37	10.12
Co	-2.24	-1.33	-0.44	0.48	1.86	5.58
Ni	-0.66	-0.39	-0.13	0.14	0.55	1.65
Cu	1.30	0.77	0.26	-0.28	-1.07	-3.23
Ru	-8.33	-4.94	-1.64	1.78	6.89	20.74
Rh	-3.26	-1.93	-0.64	0.70	2.69	8.11
Pd	-0.18	-0.11	0.04	0.04	0.15	0.45
Ag	1.77	1.05	0.35	-0.38	-1.46	-4.41
Os	-8.32	-4.93	-1.64	1.78	6.88	20.71
Ir	-3.91	-2.32	-0.77	0.84	3.24	9.75
Pt	2.48	1.47	0.49	-0.53	-2.06	-6.19
Au	-0.25	-0.15	-0.05	0.05	0.21	0.62

[a] Łątka, 1989.

normal to x. The dipole matrix elements can be calculated in a straightforward manner. For lanthanide ions in cubic CEF, they have been fully tabulated (Birgenau, 1972). The inelastic cross section is typically of the order of 0.1 barn/sterad per ion which is somewhat larger than the phonon cross section. Furthermore, the energy transfer is of the order of 10 meV, which is the range well covered by thermal neutrons. Consequently, neutron scattering is a direct method to study the level structure of the CEF split ion. In particular, this method is suitable for metals, where large absorption makes it difficult to measure the CEF level structure by optical methods.

In the neutron experiment, one has to be sure that the exchange interactions among the lanthanide ions are at least smaller than the level splitting. This can be achieved by partially replacing the magnetic active ion by La or Y. In general, neither the structure nor the phonon and CEF scattering are expected to change drastically. In the next step one has to separate the CEF scattering from other scattering contributions, in particular from phonons. In principle, this can be done by increasing x, since the cross section of CEF scattering falls off with $F^2(x)$, whereas for the phonons it should increase with η^2. However, this test is usually not used because of calibration difficulties. Instead, one compares the calculated intensities with those obtained from the experiment. One varies the temperature to excite consecutively the CEF levels and then observes the corresponding peaks in neutron gain configuration. Finally, one can measure the material after replacing all magnetic lanthanide ions by La or Y. This gives the check of phonon and apparatus background. As an example, inelastic neutron scattering curves obtained for tetragonal $CeCu_2Si_2$ and $LaCu_2Si_2$ are shown in Figure 3.11. The CEF level scheme deduced from these data is displayed in Figure 3.12. Transition energies and matrix elements are collected in Table 3.5 (Horn et al., 1981).

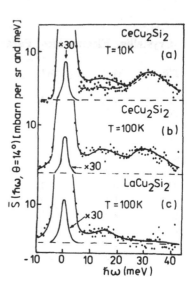

FIGURE 3.11. Scattering law for average scattering angle $\theta = 14°$ as a function of energy transfer h as obtained for $CeCu_2Si_2$ (a) and (b) and $LaCu_2Si_2$ (c). Solid lines represent a fit to the data points, as described in the text. The difference spectrum (triangles) $CeCu_2Si_2$ (T = 10 K) minus (0.72) $LaCu_2Si_2$ (T = 100 K) shows the existence of the inelastic magnetic line around h = 12.5 meV, which is masked by phonon scattering in the original spectrum. The dashed line represents the difference between the fit spectrum of $CeCu_2Si_2$ (T = 10 K) and 0.72 times the fit spectrum of $LaCu_2Si_2$ (T = 100 K). The factor 0.72 is estimated from the different nuclear coherent scattering lengths of $CeCu_2Si_2$ and $LaCu_2Si_2$ and the different phonon population at 10 and 100 K. (Horn et al., 1981).

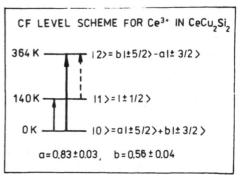

FIGURE 3.12. Crystal-field level scheme of Ce^{3+} in the tetragonal compound $CeCu_2Si_2$ as obtained from the measured spectra of Figure 3.11 and by diagonalization of Equation 3.1, yielding $B_2^0 = -3.0 \pm 1.0$, $B_4^0 = -0.4 \pm 0.1$, $B_4^4 = -6.25 \pm 0.05$ (Horn et al., 1981).

TABLE 3.5

Matrix Elements $M_{nm} = |\langle n|J\perp|m\rangle|^2$ and Transition Energies $\hbar\omega_{nm}$ Calculated and Measured for $CeCu_2Si_2$[a]

Transition	M_{nm}^{calc}	M_{nm}^{meas}	$\dfrac{\hbar\omega_{nm}^{calc}}{k_B}$ (K)	$\dfrac{\hbar\omega_{nm}^{meas}}{k_B}$ (K)
n = 0 → m = 1	1.61	1.65	139	175 + 15
n = 0 → m = 2	5.1	5.6	364	360 + 20
n = 1 → m = 2	3.7		225	

[a] Horn et al., 1981.

Chapter 4

MAGNETIC PROPERTIES OF THE INTERMETALLIC $R_xT_yX_z$ COMPOUNDS WITH y/x ≤ z

4.1. RTX PHASES

A large number of equiatomic ternary rare-earth intermetallic compounds with the general formula RTX (R = rare earth, T = transition element, and X = metalloid) are known to exist (Dwight et al., 1968; Hovestreyt et al., 1982; Bażela, 1987). They crystallize in several different types of structure, such as $MgCu_2$, MgAgAs, ZrOS, Fe_2P, AlB_2, Ni_2In, LaPtSi, PbFCl, $MgZn_2$, TiNiSi, and $CeCu_2$.

4.1.1. COMPOUNDS WITH THE $MgCu_2$-TYPE STRUCTURE

The ternary RMnGa (R = Ce or Ho) compounds show the crystal structure which belongs to the cubic Laves phase type ($MgCu_2$, space group Fd3m).

RMnGa remain paramagnetic at 5 K. At high temperatures, Curie-Weiss behavior with effective moments higher than expected for free trivalent R ions is observed. This can be considered as an indication of the existence of a moment on the manganese atoms (Brabers et al., 1992). In the compounds with R = Ce, Pr, and Nd, the temperature dependence of the electrical resistivity and the magnetic susceptibility imply spin glass-like behavior in the temperature range from 4.2 to 300 K (Tagawa et al., 1988).

Neutron diffraction, electric resistivity and magnetic measurements carried out for the DyMnGa show a spin glass state with a spin glass temperature T_{sg} = 40 K (Sakurai et al., 1988).

TbMnAl and ErMnAl are antiferromagnets with a Néel temperature T_N = 34 K and 15 K, respectively (Oesterreicher, 1972). Neutron diffraction data available for TbMnAl suggest a modulated magnetic structure similar to that found in $TbMn_2$ (Corliss and Hastings, 1964).

4.1.2. COMPOUNDS WITH THE MgAgAs-TYPE STRUCTURE

Only a small number of RTX compounds crystallize in the cubic MgAgAs-type structure (space group F$\bar{4}$3m). In RNiSb compounds in which R = Ho, Er, Tm or Y, no magnetic ordering is observed above 5 K (Aliev et al., 1988).

The magnetic susceptibility of YbPdX (X = Sb or Bi) satisfies the Curie-Weiss law in the temperature range 4.2 to 300 K (Dhar et al., 1988). GdPtSn is a paramagnet with a paramagnetic Curie temperature Θ_p = 24 K and an effective magnetic moment of 8.28 μ_B (De Vries et al., 1985).

4.1.3. COMPOUNDS WITH THE LaIrSi (ZrOS)-TYPE STRUCTURE

The RTSi compounds, in which R is a light rare-earth atom (La–Eu) and T represents Rh or Ir, crystallize in a primitive cubic structure (space group $P2_13$).

LaRhSi and LaIrSi exhibit a superconducting transition at 4.35 and 2.3 K, respectively. Above it, the magnetic susceptibility has been found to be positive and almost temperature independent (Chevalier et al., 1982a).

NdIrSi shows a spontaneous magnetization below the Curie temperature at $T_c = 10$ K. The fact that the magnetic saturation is not reached up to 20 kOe suggests that a noncolinear magnetic ordering occurs below T_c. A hysteresis loop was obtained at 4.2 K with a coercive field of 0.5 kOe. Above T_c, the magnetic susceptibility obeys the Curie-Weiss law, with a positive value of the paramagnetic Curie temperature, $\Theta_p = 12$ K. The paramagnetic moment is equal to 3.62 μ_B (Chevalier et al., 1982a).

EuPtSi and EuPdSi are also isostructural with the LaIrSi. The magnetic susceptibility for both compounds obeys the Curie-Weiss law between 10 and 300 K, with an effective paramagnetic moment close to the free Eu^{2+} ion value. At 4.2 K, a symmetric, unresolved, hyperfine split Mössbauer spectrum is observed in EuPtSi, indicating the onset of magnetic ordering. Only a single Mössbauer line is observed at $T = 4.2$ K for EuPdSi (Adroja et al., 1988b).

4.1.4. COMPOUNDS WITH THE Fe₂P (ZrNiAl)-TYPE STRUCTURE

The hexagonal structure of the Fe_2P type belongs to the space group $P\bar{6}m2$. In the case of the ternary RTX compounds, the T atoms occupy the phosphorus sites, and the R and X atoms are situated in the two nonequivalent iron sublattice sites. RNiAl and RCuAl crystallize in this structure type (Dwight et al., 1968). All are ferromagnets at low temperatures. Their magnetic parameters are summarized in Table 4.1 (Buschow, 1980). GdTAl and GdTSn compounds were also found to be ferromagnets (Buschow, 1971, 1973). A rather unusual variation of the paramagnetic Curie temperature Θ_p was observed in the $Gd_{1-x}Th_xCuAl$ solid solution: they pass through a maximum for x = 0.3. In the $Gd_{1-x}Th_xPdIn$ series, a change in the sign of Θ_p from positive to negative was observed at about the same concentration. Taking into account the ^{27}Al NMR data obtained for GdCuAl, attempts have been made to explain such behavior in terms of the RKKY model (Buschow et al., 1971, 1973).

[155]Gd Mössbauer spectra obtained for some of GdTX compounds show magnetic ordering at 4.2 K. Their analysis indicates that in GdCuAl, GdNiIn, and GdPdIn, the magnetic moments of the Gd ions are oriented parallel to the c axis, while in GdPdSn and GdPdAl they make an angle of $\phi = 47°$ with the c axis (De Vries et al., 1985).

CeTIn (T = Ni, Pd, Pt, or Rh) also crystallize in this structure type. The temperature dependence of the inverse susceptibility χ_g^{-1} in CePdIn and CePtIn

TABLE 4.1
Magnetic Data for RTX Compounds

Compound	Crystal structure	Type of magnetic ordering	$T_{C,N}(K)$	$\Theta_p(K)$	$\mu_{eff}(\mu_B)$	$\mu_R(\mu_B)$	Ref.
NdMnGa	MgCu$_2$	Spin glass	10	−11			1
DyMnGa	MgCu$_2$	Spin glass	40	18	10.6		2
TbMnAl	MgCu$_2$	AF	34				3
ErMnAl	MgCu$_2$	AF	15				3
HoPdSb	MgAgAs			−7	10.63		4
ErPdSb	MgAgAs			−1	9.77		4
TmPdSb	MgAgAs			−3	8.06		4
YbPdSb	MgAgAs			−9	4.39		5
YbPdBi	MgAgAs			−9	4.04		5
GdPtSn	MgAgAs			24	8.28		6
NdIrSi	ZrOS	F	10	12	3.62	1.5	7
PrNiAl	Fe$_2$P			−10	3.73		8, 9
NdNiAl	Fe$_2$P	F	15–17	5	3.84	1.6	8, 9
GdNiAl	Fe$_2$P	F	61–70	53–70	8.5–8.9	7.38–7.42	8–11
TbNiAl	Fe$_2$P	F	57–65	45–52	10.1–10.2	7.48–8.01	8–11
DyNiAl	Fe$_2$P	F	39–47	30	11.0–11.1	7.38–7.82	8–11
HoNiAl	Fe$_2$P	F	25–27	11–12	10.6–10.8	7.25–8.86	8–11
ErNiAl	Fe$_2$P	F	15–16	−1	9.8–9.85	7.39–7.4	8–11
TmNiAl	Fe$_2$P	F	4.2,12	−11	7.8	4.72	8, 11
LuNiAl	Fe$_2$P	Pauli paramagnetic					8, 11
PrCuAl	Fe$_2$P	F	36			1.7	8, 11
NdCuAl	Fe$_2$P	F	25			1.8	8, 11
GdCuAl	Fe$_2$P	F	67–90	55–90	8.2	7.0	8–11
TbCuAl	Fe$_2$P	F	52	42	10.1	7.41	8, 11
DyCuAl	Fe$_2$P	F	35	29	11.0	8.66	8, 11
HoCuAl	Fe$_2$P	F	23	13	10.9	8.59	8, 11
ErCuAl	Fe$_2$P	F	17	3	10.0	7.27	8, 11
TmCuAl	Fe$_2$P	F	13	−8	7.6	4.71	8, 11
YbCuAl	Fe$_2$P			−34	4.35		12
LuCuAl	Fe$_2$P	Pauli paramagnetic					8, 11
CeNiIn	Fe$_2$P	Pauli paramagnetic					13
NdNiIn	Fe$_2$P	F				1.6	13
GdNiIn	Fe$_2$P	F	83	80	7.28		6, 10
CePdIn	Fe$_2$P	AF	1.8	−43	2.61		13
NdPdIn	Fe$_2$P	F				1.8	13
GdPdIn	Fe$_2$P	F	102	103	7.73		6, 10
CePtIn	Fe$_2$P			−73	2.58		14
CeAuIn	Fe$_2$P	AF	5.7	−10	2.1		14
CeRhSn	Fe$_2$P				2.56		15
PrRhSn	Fe$_2$P			10	3.62		15
NdRhSn	Fe$_2$P			12	3.68		15
GdRhSn	Fe$_2$P	AF	14.8	18	8.0		16
TbRhSn	Fe$_2$P	AF	18.4	4	9.9		16
DyRhSn	Fe$_2$P	AF	8	−12	10.85		16
HoRhSn	Fe$_2$P			6	10.7		16
ErRhSn	Fe$_2$P			−4	9.7		16
CePtSn	Fe$_2$P	AF	5.5	−28	2.5		16
PrPtSn	Fe$_2$P			−8	3.78		16
NdPtSn	Fe$_2$P			−10	4.05		16
GdPtSn	Fe$_2$P			32	8.09		16
TbPtSn	Fe$_2$P	AF	12	−6	10.2		16
DyPtSn	Fe$_2$P	AF	8	−4	10.77		16

TABLE 4.1 (continued)
Magnetic Data for RTX Compounds

Compound	Crystal structure	Type of magnetic ordering	$T_{C,N}(K)$	$\Theta_p(K)$	$\mu_{eff}(\mu_B)$	$\mu_R(\mu_B)$	Ref.
HoPtSn	Fe_2P			4	10.7		16
ErPtSn	Fe_2P			10	10.0		16
TmPtSn	Fe_2P			22	7.93		16
YFeAl	$MgZn_2$	F	38			0.1	17
GdFeAl	$MgZn_2$	F	260			5.81	17
TbFeAl	$MgZn_2$	F	195			6.44	17
DyFeAl	$MgZn_2$	F	125–144.5			7.12–7.6	17–19
HoFeAl	$MgZn_2$	F	92			8.11	17
ErFeAl	$MgZn_2$	F	56			6.32	17
TmFeAl	$MgZn_2$	F	38			2.93	17
LuFeAl	$MgZn_2$	F	39			0.1	17
TbCoAl	$MgZn_2$	F	48			6.42	20
DyCoAl	$MgZn_2$	F	47	36	10.9	6.55	21, 22
HoCoAl	$MgZn_2$	F	34			8.54	20
ErCoAl	$MgZn_2$	F	25		10.4	8.3	23
CeCuSi	AlB_2	F	15.5	−30	3.3	1.25	24, 25
PrCuSi	AlB_2	F	14	8	3.39	2.02	26
NdCuSi	AlB_2			−45	4.2		24
GdCuSi	AlB_2	F	49	30–58	7.0–7.32	6.9	24, 26
TbCuSi	AlB_2	F	47	52	9.62	7.3	26
	Ni_2In	AF	16			8.7	27
HoCuSi	AlB_2			30	10.2		24
TmCuSi	Ni_2In	F	9			6.1	28
CeZnSi	AlB_2			12	2.54		29
NdZnSi	AlB_2			30	3.62		29
GdZnSi	AlB_2			63	7.94		29
TbZnSi	AlB_2			40	9.72		29
HoZnSi	AlB_2			50	10.61		29
GdCuGe	AlB_2	AF	16				30
NdAgSi	AlB_2	F	20	17	3.62	0.2	31
$EuAg_{0.67}Si_{1.33}$	AlB_2	F	34	21	7.94	0.2	31
$NdNi_{0.67}Si_{1.33}$	AlB_2	F	23	12	3.68	1.0	31
$CeCo_{0.4}Si_{1.6}$	AlB_2			0	4.9		31
$SmFe_{0.4}Si_{1.6}$	AlB_2			0	5.75		31
$NdNi_{0.4}Si_{1.6}$	AlB_2			0	4.9		31
$GdCo_{0.4}Si_{1.6}$	AlB_2			0	9.0		32
$GdFe_{0.4}Si_{1.6}$	AlB_2			−30	9.06		32
$SmGe_{0.67}Ge_{1.33}$	AlB_2	AF	26	33	0.07		32
NdAlGa	AlB_2	AF	2.5				33
TbAlGa	AlB_2	AF	47, 23			6.7	34
DyAlGa	AlB_2	AF	51.5, 17	1	10.8	6.8	35
HoAlGa	AlB_2	AF	30, 17.8			8.2	34
ErAlGa	AlB_2	AF	2.8			4.9	33
CeCuSn	$CaIn_2$	AF	4.2	5	2.59		36
GdCuSn	$CaIn_2$	AF	24	−32		2.3	37
CeAgSn	$CaIn_2$	AF	5.5				38
PrAgSn	$CaIn_2$	AF	9				38
NdAgSn	$CaIn_2$	AF	7				38
SmAgSn	$CaIn_2$	AF	28				39
TbAgSn	$CaIn_2$	AF	33			8.5	40
HoAgSn	$CaIn_2$	AF	15			8.8	39, 40

TABLE 4.1 (continued)
Magnetic Data for RTX Compounds

Compound	Crystal structure	Type of magnetic ordering	$T_{C,N}(K)$	$\Theta_p(K)$	$\mu_{eff}(\mu_B)$	$\mu_R(\mu_B)$	Ref.
GdAuSn	CaIn$_2$	AF	35	−10		1.6	37
CePtSi	LaPtSi			−47	2.56		41
NdPtSi	LaPtSi	AF	15				42
SmPtSi	LaPtSi	AF	4				42
YMnSi	PbFCl	F, AF	275, 130	290		1.3	43
LaMnSi	PbFCl	F	295		2.0	0.24	38
GdMnSi	PbFCl	F	314–320	220–314	7.8	5.37	44, 45
DyMnSi	PbFCl	AF	30	30	10.6	6.7	44
HoMnSi	PbFCl	AF	36	−10	11.7	7.35	44
CeMnSi	PbFCl	AF	228				46
PrMnSi	PbFCl	AF	250				46
NdMnSi	PbFCl	AF	280				46
LaFeSi	PbFCl	Pauli paramagnetic					47
CeFeSi	PbFCl				2.54		47
PrFeSi	PbFCl			35	3.58		47
NdFeSi	PbFCl	F	25	20	3.90	1.4	47
SmFeSi	PbFCl	F	40			0.3	47
GdFeSi	PbFCl	F	135	165	8.09	7.1	47
TbFeSi	PbFCl	F	125	110	9.62	4.1	47
DyFeSi	PbFCl	F	110	75	10.57	5.2	47
YNiSi	TiNiSi	Pauli paramagnetic					48
LaNiSi	TiNiSi	Pauli paramagnetic					48
CeNiSi	TiNiSi			−57	2.86		48
PrNiSi	TiNiSi			17	3.56		48
NdNiSi	TiNiSi			−15	3.50		48
SmNiSi	TiNiSi	Pauli paramagnetic					48
GdNiSi	TiNiSi			0	8.12		48
TbNiSi	TiNiSi			−2	9.83		48
DyNiSi	TiNiSi			0	10.4		48
HoNiSi	TiNiSi			0	10.4		48
ErNiSi	TiNiSi			5	9.53		48
TmNiSi	TiNiSi			8	7.58		48
YbNiSi	TiNiSi			−65	4.57		48
LuNiSi	TiNiSi	Pauli paramagnetic					48
GdRhSi	TiNiSi	F	100	90	7.55, 7.95	2.2	49, 50
TbRhSi	TiNiSi	F	55	48	9.92		49
		AF	29, 13			8.1	51, 52
DyRhSi	TiNiSi	F	25	11.5	10.31		49
HoRhSi	TiNiSi	AF	11, 8	10.5	10.71	8.7, 9.1	49, 51, 52
ErRhSi	TiNiSi	AF	12, 7.5	−3	9.54	6.6	49, 51
GdNiGe	TiNiSi	AF	11	−10	7.85		53
TbNiGe	TiNiSi	AF	18.5	4.1	8.97	9.1	54
DyNiGe	TiNiSi	AF	4.7	−8.2	10.42	7.54	54
HoNiGe	TiNiSi	AF	5.0	−2.6	11.2	9.98	55
ErNiGe	TiNiSi	AF	2.9	−1.4	9.19	9.05	55
CeRhGe	TiNiSi	AF	9.3	−56	2.3		56
NdRhGe	TiNiSi	AF	14	−10	3.73		57, 58
TbRhGe	TiNiSi	AF	15			9.26	59
CeIrGe	TiNiSi			−10	0.27		56
NdIrGe	TiNiSi	AF	12.5				57
TbCoSn	TiNiSi	AF	20.5	15	9.81	5.5	60

TABLE 4.1 (continued)
Magnetic Data for RTX Compounds

Compound	Crystal structure	Type of magnetic ordering	$T_{C,N}(K)$	$\Theta_p(K)$	$\mu_{eff}(\mu_B)$	$\mu_R(\mu_B)$	Ref.
DyCoSn	TiNiSi	AF	10	9	10.49	5.6	60
HoCoSn	TiNiSi	AF	7.8	6.5	10.44	6.0	60
ErCoSn	TiNiSi			0	9.55		60
TmCoSn	TiNiSi			5	7.75		61
LuCoSn	TiNiSi			31	0.79		61
CeNiSn	TiNiSi	Pauli paramagnetic					15
PrNiSn	TiNiSi			−10	3.67		15
NdNiSn	TiNiSi			−4	4.2		15
SmNiSn	TiNiSi	AF	9	−40	1.37		15
GdNiSn	TiNiSi	AF	10.5	−3	8.8		62
TbNiSn	TiNiSi	AF	7.2	6	11.27		62
DyNiSn	TiNiSi	AF	8.2	−1	11.06		62
HoNiSn	TiNiSi			−2	10.75		62
ErNiSn	TiNiSi			6	9.85		62
TmNiSn	TiNiSi			−4	7.65		62
CePdSn	TiNiSi	AF	7.5	−68	2.67		63
PrPdSn	TiNiSi			−2	3.60		63
NdPdSn	TiNiSi			−8	4.93		63
SmPdSn	TiNiSi	AF	11				63
EuPdSn	TiNiSi	AF	13	5	8.27		63
GdPdSn	TiNiSi	AF	14.5	−27	8.16		63
TbPdSn	TiNiSi	AF	23.5	−16	10.17	7.6	63, 64
DyPdSn	TiNiSi	AF	11.4	−2	11.1		63
HoPdSn	TiNiSi	AF	3.5	−7	11.07		63, 58
ErPdSn	TiNiSi	AF	5.6	−0.3	9.51		63
YbPdSn	TiNiSi			−5	1.45		63
CePtGa	TiNiSi	AF	3.2				66
CePdGa	TiNiSi	AF	1.7				66
GdAuGa	TiNiSi	AF	6	−8.5	8.06		67
TbAuGa	TiNiSi			−10	9.7		67
DyAuGa	TiNiSi			−4.5	10.63		67
HoAuGa	TiNiSi			3.5	10.58		67
ErAuGa	TiNiSi			1.5	9.6		67
TmAuGa	TiNiSi			−2.0	7.59		67
EuPdSb	TiNiSi	AF	13	−35	8.19		4
CePdGe	CeCu$_2$	AF	3.4	−37	2.55		56
CePtGe	CeCu$_2$	AF	3.4	−82	2.54		56
TbNiGa	CeCu$_2$	AF	23			6.8	68
HoNiGa	CeCu$_2$	AF	12			7.9	65
PrAgGa	CeCu$_2$			31	3.18		69
NdAgGa	CeCu$_2$			4	3.65		69
GdAgGa	CeCu$_2$	AF	27	52	7.95		69
TbAgGa	CeCu$_2$	AF	18	20	10.03		69
DyAgGa	CeCu$_2$			17	10.0		69
HoAgGa	CeCu$_2$	AF	4.7	14	10.43		69
ErAgGa	CeCu$_2$	AF	3	12	9.43		69
TmAgGa	CeCu$_2$			9	7.38		69
EuCuGa	CeCu$_2$	AF	10				70
LaPdSb	CaIn$_2$	Pauli paramagnetic					4
CePdSb	CaIn$_2$	F	17	10	2.6	1.2	4
PrPdSb	CaIn$_2$			−2	3.63		4
NdPdSb	CaIn$_2$	AF	11	7	3.71		4

TABLE 4.1 (continued)
Magnetic Data for RTX Compounds

Compound	Crystal structure	Type of magnetic ordering	$T_{C,N}(K)$	$\Theta_p(K)$	$\mu_{eff}(\mu_B)$	$\mu_R(\mu_B)$	Ref.
SmPdSb	CaIn$_2$	AF	18				4
GdPdSb	CaIn$_2$	AF	17	-14.5	8.10		4
TbPdSb	CaIn$_2$			-7	10.2		4

References: 1. Tagawa et al., 1988; 2. Sakurai et al., 1988; 3. Oesterreicher, 1972; 4. Malik and Adroja, 1991; 5. De Vries et al., 1985; 6. Dhar et al., 1988; 7. Chevalier et al., 1982a; 8. Oesterreicher, 1973; 9. Leon and Wallace, 1970; 10. Buschow, 1975; 11. Buschow, 1980; 12. Mattens et al., 1980; 13. Fujii et al., 1992; 14. Fujii et al., 1987; 15. Routsi et al., 1992a; 16. Routsi et al., 1992b; 17. Oesterreicher, 1977b; 18. Sima et al., 1983; 19. Bara et al., 1982; 20. Oesterreicher, 1973; 21. Oesterreicher, 1977a; 22. Slebarski, 1980; 23. Oesterreicher et al., 1970; 24. Kido et al., 1983b; 25. Gignoux et al., 1984; 26. Oesterreicher, 1976; 27. Bażela et al., 1985a; 28. Allain et al., 1988; 29. Kido et al., 1983a; 30. Oesterreicher, 1977c; 31. Felner and Schieber, 1973; 32. Felner et al., 1972; 33. Martin et al., 1983; 34. Girgis and Fischer, 1979; 35. Doukouré et al., 1986; 36. Adroja et al., 1988b; 37. Oesterreicher, 1977a; 38. Adam et al., 1990; 39. Sakurai et al., 1992; 40. Bażela et al., 1992; 41. Lee and Shelton, 1987; 42. Braun, 1984; 43. Kido et al., 1985c; 44. Nikitin et al., 1987; 45. Kido et al., 1982; 46. Welter et al., 1991; 47. Welter et al., 1992; 48. Skolozdra et al., 1984; 49. Chevalier et al., 1982b; 50. Szytuła, 1989; 51. Bażela et al., 1985b; 52. Quezel et al., 1985; 53. Kotsanidis et al., 1990; 54. André et al., 1992; 55. André et al., 1993a; 56. Rogl et al., 1989; 57. Chevalier et al., 1991; 58. Szytuła, 1992; 59. Szytuła et al., 1988a; 60. Bażela et al., 1990; 61. Skolozdra et al., 1982; 62. Routsi et al., 1991; 63. Adroja and Malik, 1992; 64. André et al., 1993b; 65. Kotsanidis et al., 1991; 66. Malik et al., 1988a; 67. Sill and Hitzman, 1981; 68. Kotsanidis and Yakinthos, 1989; 69. Sill and Esau, 1984; 70. Malik et al., 1987.

follows the Curie-Weiss law, with effective moments which agree well with the theoretical free-ion value for the Ce^{3+} ion. At low temperatures, CePdIn exhibits antiferromagnetic ordering below $T_N = 1.8$ K, whereas CePtIn has been discovered to be a heavy fermion compound (Fujii et al., 1987). The temperature dependence of the magnetic susceptibility and the electrical resistivity suggests that CeNiIn is an intermediate-valence compound (Fujii et al., 1987). The temperature dependence of the magnetic susceptibility of CeRhIn also indicates the mixed-valence behavior (Adroja et al., 1989).

The temperature dependence of the magnetic susceptibility and the specific heat of CeAuIn both indicate antiferromagnetic ordering below $T_N = 5.7$ K. Above it, the magnetic susceptibility obeys the Curie-Weiss law, with an effective moment that appears to be reduced with respect to that expected for the 4f^1 configuration of Ce^{3+} (Pleger et al., 1987).

GdRhSn, TbRhSn and DyRhSn compounds order antiferromagnetically with Néel temperatures of 14.8, 18.4, and 8 K, respectively. The compounds with R = Ho or Er are paramagnets at 4.2 K (Routsi et al., 1992b).

4.1.5. COMPOUNDS WITH THE MgZn$_2$-TYPE STRUCTURE

A group of intermetallic compounds RTAl with T = Fe or Co exhibits the hexagonal MgZn$_2$-type structure (space group P6$_3$/mmc). The magnetic measurements of RFeAl compounds in which R is a heavy lanthanide atom show that they are ferromagnets with high Curie temperatures (Oesterreicher, 1977c).

Systematic studies were only performed on DyFeAl. Neutron diffraction data indicate a ferromagnetic structure in which the magnetic moments of the $7.6(1)\mu_B$/Dy atom order ferromagnetically. The Fe sublattice orders ferromagnetically with Fe moments equal to $\mu(2a) = 0.8(4)\ \mu_B$ and $\mu(6h) = 0.5(2)$ μ_B, The Fe sublattice is coupled antiferromagnetically to the Dy sublattice. There is a strong reduction of the Dy moment compared to the free-ion value. The magnetic moment lies in the basal plane (Šima et al., 1983). The L_{III} emission spectra of iron in DyFeAl give evidence for a charge transfer between the 3dFe and 5dDy bands (Ślebarski and Zachorowski, 1984). The results of Mössbauer investigations indicate a more complete transfer of Dy 5d electrons to the 3d band and of Al 3p electrons to the iron sites. This effect is in accordance with the observed increase in the intensity of the L_{III} spectra. The total magnetic moment localized on the Dy atoms is reduced at T = 4.2 K, as a consequence of an opposite polarization of the 5d and 4f bands (Bara et al., 1982, Ślebarski, 1987).

In the case of the RCoAl compounds (R = Tb–Er), a ferromagnetic ordering is observed at low temperatures. The relevant magnetic data are listed in Table 4.1 (Oesterreicher, 1973, 1977a). Neutron diffraction study of ErCoAl indicates a ferromagnetic ordering at T = 4.2 K, with the magnetic moments per Er atom equal to μ = 7.0 μ_B parallel to the c axis. No moment is observed on Co atoms (Oesterreicher et al., 1970).

4.1.6. COMPOUNDS WITH THE AlB$_2$ AND Ni$_2$In-TYPE STRUCTURE

Many ternary equiatomic compounds crystallize in two closely related hexagonal structures represented by the AlB$_2$ type (space group P6/mmm) (Rieger and Parthé, 1969) and the Ni$_2$In type (space group P6$_3$/mmc) (Iandelli, 1983). These two structural types are shown in Figure 2.8. On the basis of neutron diffraction data, Mugnoli et al. (1984) concluded that LaCuSi exists in two polymorphic modifications: a low-temperature Ni$_2$In type and a high-temperature AlB$_2$ type.

The magnetic properties of RCoSi (R = Y, Ce, Nd, Sm, Gd, or Ho) were investigated by Kido et al. (1983b). The magnetic susceptibilities of YCuSi and SmCuSi are 10^2 times smaller than those of the other compounds. In addition, they show no temperature dependence. In the other compounds, the magnetic susceptibility obeys the Curie-Weiss law, with effective magnetic moments close to the free-ion values (see Table 4.1). The magnetic properties of RCuSi with R = Pr, Gd, or Tb were investigated from 4.2 to 150 K in magnetic fields up to 5 T. As may be seen in Table 4.1, all of these compounds order ferromagnetically (Oesterreicher, 1976).

The magnetic properties of CeCuSi were studied by neutron diffraction and magnetization measurements. CeCuSi shows ferromagnetic ordering below T = 15.5 K, with a magnetic moment of 1.25 μ_B at T = 2.5 K, perpendicular to the c axis (Gignoux et al., 1986a).

Neutron diffraction studies of TbCuSi indicate a cosinusoidally modulated transverse spin structure below T_N = 16 ± 2 K, while DyCuSi and HoCuSi remain paramagnetic down to T = 4.2 K (Bażela et al., 1985b). TmCuSi is a colinear ferromagnet with T_c = 9 K and a magnetic moment μ = 6.1(2) μ_B at T = 2.1 K oriented parallel to the c axis (Allain et al., 1988).

GdCuGe is an antiferromagnet with T_N = 17 K (Oesterreicher, 1977c). The RZnSi compounds (R = Ce, Nd, Sm, Gd, Tb, or Ho) are paramagnetic in the temperature range between 77 and 300 K. Their effective magnetic moments are in good agreement with the corresponding free-ion values. YZnSi is a Pauli paramagnet (Kido et al., 1983a).

Studies of the magnetic properties of pseudoternary $RCu_{1-x}Zn_xSi$ (0 ≤ x ≤ 1) systems, for R = Gd (Kido et al., 1984b), R = Tb (Bażela and Szytuła, 1989), and R = Ho (Kido et al., 1985c), show that in all of these systems, the paramagnetic Curie temperatures, and for $GdCu_{1-x}Zn_xSi$ the Curie temperatures, have a maximum at the concentration x = 0.4. The values of μ_{eff} are about equal to the free R^{3+} ion value in the whole composition range.

The magnetic susceptibility of GdTiSi shows Curie-Weiss behavior, with μ_{eff} = 7.07μ_B and Θ_p = 10.4 K (Kido et al., 1984a).

The RAlGa compounds (R = Nd, Tb, Dy, Ho, or Er) crystallize in the hexagonal AlB_2-type structure (Martin and Girgis, 1983). NdAlGa is an antiferromagnet with T_N = 2.5(3) K and a magnetic structure incommensurate with the crystal lattice in the basal plane. The cycloidal spin structure of the Nd moments rotates in the basal plane. All magnetic moments are perpendicular to the c axis. The rotation angles are 62.7° (Martin et al., 1983).

In TbAlGa and HoAlGa, two phase transitions are observed (see Table 4.1). The magnetic structure is described by the propagation vector k = (1/3, 1/3, k_z). At low temperatures, k_z = 1/2, which leads to a commensurate structure. In the temperature range T_1 < T < T_N, incommensurate configurations were found (Girgis and Fischer, 1979).

The thermal variation of the susceptibility of DyAlGa in the 4 to 100 K temperature range exhibits two transitions: at T = 17 and at T_N = 51.5 K, the latter being the Néel temperature. Neutron diffraction data at T = 25 K indicate magnetic ordering with a propagation vector K = (1/3, 1/3, 1/2) corresponding to a hexagonal magnetic cell six times larger than the crystallographic cell (a_m = a$\sqrt{3}$, c_m = 2c). In each Dy layer, the magnetic alignment is triangular (see Figure 4.1a). At T = 4.2 K, the ordering is colinear and the Dy moments are parallel to the c axis (Doukouré et al., 1986). The magnetic cell of ErAlGa is incommensurate with the crystal lattice in the c direction, the magnetic structure exhibiting a trigonal spin structure in the basal plane. The spin vectors are perpendicular to the c axis. All magnetic moments rotate in the basal plane with angles of 120°. The rotation angle is 170° from plane to plane in the c direction (Martin et al., 1983).

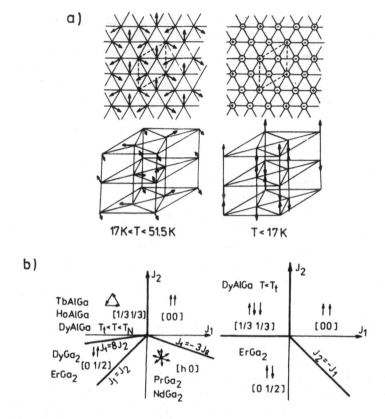

FIGURE 4.1. (a) Magnetic structures of DyAlGa at 25 and 4.2 K (Doukouré et al., 1986). (b) Stability diagrams of possible magnetic structures in a two-dimensional triangular (hexagonal) lattice when the moments lie in the basal plane and along the c axis. Numbers in brackets are the propagation vectors in the plane (Doukouré et al., 1986).

Bertaut (1961) discussed the stability of the possible magnetic structures in a two-dimensional hexagonal (or triangular) lattice. Putting the exchange energy as

$$E \sim \sum_{i \neq j} J_{ij} \vec{S}_i \cdot \vec{S}_j$$

he took into account the exchange integrals between the first (J_1) and the second (J_2) nearest-neighbors. The magnetic models obtained are represented in Figure 4.1a. An analogous study can be done with an Ising-like model in which the strong anisotropy forces the magnetic moment to be parallel to the c axis (Doukouré and Gignoux, 1982). The predicted magnetic structures are shown in Figure 4.1b.

$RT_{2-x}Si_x$ and $RT_{2-x}Ge_x$, where R = La, Ce, Nd, Sm, Eu, or Gd and T = Fe, Co, Ni, or Ag, crystallize in the AlB_2-type structure. The magnetic

TABLE 4.2
Magnetic Data of the R_2TSi_3 Compounds

Compound	Type of magnetic ordering	$T_{C,N}(K)$	$\Theta_p(K)$	$\mu_{eff}(\mu_B)$	$\mu_R(\mu_B)$	Ref.
Y_2RhSi_3	Diamagnetic					1
La_2RhSi_3	Diamagnetic					1
Ce_2RhSi_3	AF	6	-83	2.40		1
Nd_2RhSi_3	AF	15	7	3.76	3.2	1, 2
Gd_2RhSi_3	AF	14	2	7.61		1
Tb_2RhSi_3	AF	11	6	9.7	7.0	1, 2
Dy_2RhSi_3			-7	10.67		1
Ho_2RhSi_3			1	10.66		1
Er_2RhSi_3			6	9.36		1
Pr_2PdSi_3			8	3.47		3
Nd_2PdSi_3	F	16	18	3.54		3
Gd_2PdSi_3	AF	21	33	7.97		3
Tb_2PdSi_3	AF	19	28	9.70		3
Dy_2PdSi_3	AF	7	4	10.43		3
Ho_2PdSi_3	AF	6	5	10.58		3
Er_2PdSi_3	AF	8	6	9.50		3
Tm_2PdSi_3			-4	7.30		3

References: 1. Chevalier et al., 1984; 2. Szytuła et al., 1992; 3. Kotsanidis et al., 1990.

data are summarized in Table 4.1. NdAgSi, $NdFe_{0.67}Si_{1.33}$, $NdNi_{0.67}Si_{1.33}$, and $EuAg_{0.67}Si_{1.33}$ are ferromagnets, while $EuAg_{0.67}Si_{1.33}$ is an antiferromagnet. Other compounds are paramagnets at 4.2 K. In these compounds, the R ions are magnetically ordered, whereas the T atoms are nonmagnetic (Felner et al., 1972; Felner and Schieber, 1973).

4.1.7. R_2RhSi_3 COMPOUNDS

Ternary silicides R_2RhSi_3 (R = Y, La, Ce, Nd, Sm, or Gd–Er) crystallize in a hexagonal structure which is derived from the AlB_2 type.

Y_2RhSi_3 and La_2RhSi_3 are diamagnetic down to 1.6 K. Nd_2RhSi_3 is ferromagnetic below $T_c = 15$ K. Gd_2RhSi_3, Ce_2RhSi_3, and Tb_2RhSi_3 are antiferromagnets with metamagnetic phase transitions. Other silicides are paramagnets at 1.6 K. The results are summarized in Table 4.2. The effective moments are in fair agreement with the values calculated for the R^{3+} free ion. The saturation magnetization obtained for Nd_2RhSi_3 is smaller than the theoretical value expected for the Nd^{3+} free ion (Chevalier et al., 1984).

Neutron diffraction study of Nd_2RhSi_3 at 4.2 K shows ferromagnetic spiral magnetic ordering described by the wave vector k = (0.523, 0, 0.0174). The ordered moment amounts to 3.2(1) μ_B at 4.2 K and is almost equal to the free-ion value for the Nd^{3+} ion (gJ = 3.27 μ_B).

Tb_2RhSi_3 was found to exhibit colinear antiferromagnetic ordering with the wave vector k = (0, 0, 1/2). The magnetic moment localized on the Tb^{3+}

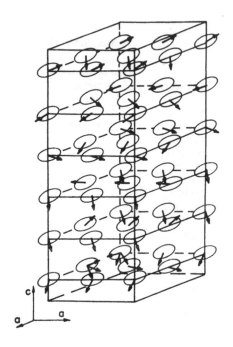

FIGURE 4.2. Magnetic structure of Nd$_2$RhSi$_3$
(Szytuła et al., 1992).

ions is 7.0(1) μ_B at 4.2 K and is smaller than the free-ion value (gJ = 9 μ_B). This effect may be ascribed to the crystal field interaction (Szytuła et al., 1992).

The magnetic properties of R$_2$PdSi$_3$ compounds, where R is Pr, Nd, Gd, Tb, Dy, Ho, Er, Tm, and Y, have been determined. Pr$_2$PdSi$_3$ does not show any magnetic ordering down to 4.2 K. Nd$_2$PdSi$_3$ (Figure 4.2) is ferromagnetic, with a Curie temperature of 16 K. The heavy rare-earth compounds show magnetic ordering temperatures in the region between 6 K for R = Ho and 21 K for R = Gd, except the compound with Tm, for which the ordering temperature is probably smaller than 4.2 K. Y$_2$PdSi$_3$ is a nonmagnetic compound (Kotsanidis et al., 1990a).

4.1.8. COMPOUNDS WITH THE CaIn$_2$-TYPE STRUCTURE

The RCuSn compounds crystallize in the hexagonal CaIn$_2$-type structure (space group P6$_3$/mmc).

The magnetic susceptibility of CeCuSn obeys the Curie-Weiss law between 40 and 300 K, with μ_{eff} = 2.59 μ_B and Θ_p = + 5 K. The susceptibility shows a deviation from the Curie-Weiss law below 40 K and a rapid rise below 10 K. The magnetic hyperfine splitting in the [119]Sn Mössbauer spectrum is observed at T = 4.2 K, confirming that this compound is ordered magnetically (Adroja et al., 1988a).

GdCuSn and GdAuSn are antiferromagnets with Néel temperatures of 24 and 35 K, respectively (Oesterreicher, 1977c).

Magnetic susceptibility, electrical resistivity, thermoelectric power, and magnetoresistance measurements indicate that in the family of RAgSn compounds, long-range magnetic ordering is observed for R = Ce, Sm, Gd, Tb, and Ho. The Néel temperatures range from 5.5 K in CeAgSn to 31 K in TbAgSn (Adam et al., 1990; Sakurai et al., 1992). A neutron diffraction study of TbAgSn and HoAgSn shows that they order antiferromagnetically below 33 and 15 K with magnetic moments at 4.2 K of 8.5(1) and 8.8(1) μ_B, respectively. The moments are aligned along the hexagonal axis of the magnetic cell with the dimensions $a_m = a_c$, $b_m = a_c\sqrt{3}$, and $c_m = c_c$, where a_c and c_c are parameters of the crystallographic unit cell (Bażela et al., 1992).

RPdSb (R = La–Dy) compounds crystallize in a CaIn$_2$-type crystal structure. CePdSb orders ferromagnetically with an ordering temperature of about 17 K, while NdPdSb, SmPdSb, EuPdSb, and GdPdSb order antiferromagnetically with Néel temperatures T_N of 11, 18, 13, and 17 K, respectively. Other compounds are paramagnetic at 4.2 K (Malik and Adroja, 1991).

4.1.9. COMPOUNDS WITH THE LaPtSi-TYPE STRUCTURE

The equiatomic RPtSi, RPtGe and RIrGe compounds with light rare earths (R = La–Gd) crystallize in the tetragonal LaPtSi-type structure (space group I4$_1$md) (Klepp and Parthé, 1982) which is a variant of the ThSi$_2$-type structure. The compounds with R = La are superconductors (Braun, 1984). The reciprocal magnetic susceptibility of CePtSi obeys the Curie-Weiss law, with an effective magnetic moment $\mu_{eff} = 2.56(5)$ μ_B. The temperature dependence of the magnetic susceptibility and the specific heat at low temperatures suggest that the compound belongs to the class of heavy fermions (Lee and Shelton, 1987; Rebelsky et al., 1988). RPtSi compounds with neodymium ($T_N = 15$ K) and samarium ($T_N = 4$ K) show magnetic ordering, while those with cerium and praseodymium remain paramagnetic down to 2 K (Braun, 1984).

Heat capacity measurements show that CeSiAl and CeGeAl order magnetically at 7.1 and 4 K, respectively. Magnetization measurements reveal that CeGeAl orders ferromagnetically, while CeSiAl orders antiferromagnetically (Dhar et al., 1992).

4.1.10. COMPOUNDS WITH THE PbFCl-TYPE STRUCTURE

The tetragonal structure of the PbFCl type belongs to the space group P4/nmm.

LaMnSi and GdMnSi are ferromagnets with Curie temperatures T_c of about 295 K, while DyMnSi and HoMnSi are antiferromagnets with a T_N of about 30 K (Nikitin et al., 1987).

In the YMnSi sample synthesized at 1000° C, ferromagnetic ordering exists below $T_c = 275$ K (Johnson, 1974/1976). Different properties are observed in the sample obtained at 1300° C and under a pressure of 1.0 GPa. At low temperatures, the sample is an antiferromagnet. With a rise of temperature, the transition to the ferromagnetic state is observed at 150 to 170

K. From the σ^2 temperature dependence, the Curie temperature was determined to be 282 K. In the paramagnetic region, the temperature dependence of χ_M^{-1} for YMnSi obeys the Curie-Weiss law, with an effective magnetic moment and a paramagnetic Curie temperature of 2.3 μ_B and 280 K, respectively (Kido et al., 1985b). GdCoSi is also a ferromagnet, with T = 250 K (Kido et al., 1982).

Among ternary silicides RFeSi (R = La–Sm, Gd–Dy), LaFeSi and CeFeSi are Pauli paramagnets, whereas PrFeSi is a Curie-Weiss paramagnet down to 2 K. The other compounds, with R = Nd, Sm, Gd, Tb, and Dy, are ferromagnetic below T_c = 25, 40, 135, 125, and 110 K, respectively. Neutron diffraction study of RFeSi (R = Nd, Tb, or Dy) compounds indicates a ferromagnetic colinear structure with magnetic moments localized on R^{3+} ions only and aligned along the c axis (they are all close to the theoretical free-ion values) (Welter et al., 1992).

4.1.11. COMPOUNDS WITH THE TiNiSi-TYPE STRUCTURE

The TiNiSi-type structure belongs to the Pnma space group.

In RNiSi compounds (R = Ce–Nd and Tb–Yb), the magnetic susceptibility obeys the Curie-Weiss law. The compounds with R = Y, La, Sm, or Lu are all Pauli paramagnets (Gladyshevskii et al., 1977).

The ternary silicides RRhSi, (R = Y, Gd, Tb, Dy, Ho, or Er) crystallize in the TiNiSi type of crystal structure (Chevalier et al., 1982b). The magnetometric data indicate that the Gd, Tb, and Dy compounds have a spontaneous magnetization, and their magnetic ordering probably arises from a noncolinear arrangement of the moments. HoRhSi and ErRhSi order antiferromagnetically at T_N = 8 and 7.5 K, respectively. At 4.2 K, both compounds undergo a metamagnetic transition at H_c = 6 kOe in HoRhSi, and at 12 kOe for ErRhSi (Chevalier et al., 1982b). Neutron diffraction data indicate that the magnetic structure of TbRhSi is a double flat spiral below T_N = 13 K. The magnetic structure of HoRhSi below T_N = 11 K is colinear, with a C(+ + − −) configuration. The magnetic moments of the holmium atoms are parallel to the b axis. ErRhSi also has a magnetic structure consisting of a double flat spiral below T_N = 12 K (Bażela et al., 1985a; Quezel et al., 1985).

Below T_N = 15(1) K, the magnetic structure of TbRhGe is incommensurate. It can be described in terms of a modulated transverse spin wave with a propagation vector k = (0, 0.388, 0.236) (Szytuła et al., 1988a).

In the ternary RNiGe (R = Gd–Tm, Y) systems, those with Gd, Tb, Dy, Ho, and Er are antiferromagnets with Néel temperatures of 11, 18, 6, 2.75, and 2.9 K, respectively (Kotsanidis et al., 1990; André et al., 1990a, 1992).

In TbNiGe and DyNiGe, a square modulated structure with the wave vector k = (2/3, 1/3, 0) was found at low temperatures. As the temperature rises, a change to a sinusoidally modulated structure in TbNiGe and cycloidal spiral structure in DyNiGe has been observed. In both structures and both phases, the magnetic moments are parallel to the c axis. In the case of TbNiGe, the coexistence of two magnetic structures over a wide temperature range has been observed (André et al., 1992). At low temperatures, the magnetic structures with the wave vectors $k_1 = (1/2, 0, 1/2)$ in HoNiGe and $k_2 = (0, 1/2, 0)$ in ErNiGe were discovered. With an increase of temperature, a change to a modulated magnetic ordering with the wave vector $k_1 = (0.48, 0.23, 0.42)$ in HoNiGe and $k_2 = (0, 0.5, 0.0837)$ in ErNiGe was detected (André et al., 1993a). Magnetic measurements performed for NdRhGe and NdIrGe gave Néel temperatures of 14 and 12.5 K, respectively. Similar magnetic measurements performed on PrRhGe and PrIrGe indicate a paramagnetic behavior at 4.2 K (Chevalier et al., 1991). A neutron diffraction experiment showed that both CeRhGe and NdRhGe exhibit a simple noncolinear antiferromagnetic structure (Bażela et al., 1993).

The magnetic susceptibility of RCoSn, where R = Tb–Lu, obeys the Curie-Weiss law in the temperature range 78 to 300 K. The YCoSn compound is a Pauli paramagnet (Skolozdra et al., 1982). Magnetization and magnetic susceptibility data obtained for RCoSn (R = Tb–Er) compounds indicate that in TbCoSn, DyCoSn, and HoCoSn, the magnetic moments of rare-earth atoms order antiferromagnetically below 20.5, 10, and 7.8 K, respectively, while ErCoSn remains paramagnetic at 4.2 K. In TbCoSn below T_N, an additional magnetic phase transition is observed at $T = 11.6$ K. Magnetization curves vs. H taken at 4.2 K reveal that all samples transform to a ferromagnetic state, but even at $H = 5$ T, the magnitudes of magnetic moments on R^{3+} ions do not reach free-ion values. Neutron diffraction data obtained at 4.2 K show the presence of a complex antiferromagnetic order in TbCoSn and HoCoSn characterized by propagation vectors $k = (0, 1/4, 0)$ and $k = (0, 1/3, 0)$, respectively. At 4.2 K, the values of magnetic moments localized on Tb^{3+} and Ho^{3+} ions are 4.9(1) and 5.1(1) μ_B, respectively, indicating strong magnetocrystalline anizotropy (Bażela et al., 1993).

The RNiSn compounds with Gd, Tb, and Dy are antiferromagnetic at low temperatures. No ordering has been observed at 4.2 K in the compounds with Ho, Er, and Tm. TbNiSn is paramagnetic above 20 K. In the temperature range $2 K < T < 8 K$ and $8 K < T < 20 K$, two different magnetic structures have been observed. In the range $8 K < T < 20 K$, a simple commensurate antiferromagnetic structure has been found. Below 8 K, a new incommensurate one has been detected (Routsi et al., 1991).

Among RPtSn compounds, antiferromagnetic ordering at low temperatures has been observed only for compounds with R = Ce ($T_N = 5.5$ K), Tb ($T_N = 12$ K), and Dy ($T_N = 8$ K). In other compounds, magnetic susceptibility obeys the Curie-Weiss law, even at $T = 4.2$ K. The determined values of the effective magnetic moments are in fair agreement with the corresponding free R^{3+} ion values (Routsi et al., 1992b).

CeRhGe orders antiferromagnetically below $T_N = 9.3$ K, whereas in CeIrGe, cerium is in the intermediate-valence state, and no transition to a magnetic ordered state is observed down to 1.5 K (Rogl et al., 1989).

In RAuGa compounds, in which R is Sm or one of the heavy rare-earth elements, the reciprocal susceptibility of all compounds, except for Sm, follow the Curie-Weiss law at high temperatures. The effective magnetic moments are in good agreement with those calculated for free R^{3+} ions. The paramagnetic Curie temperatures of the heavy rare-earth compounds do not vary linearly with the De Gennes function. At low temperatures, only the GdAuGa compound orders antiferromagnetically, with a Néel temperature of 6 K. It also exhibits metamagnetic behavior (Sill and Hitzman, 1981).

Magnetic susceptibility measurements reveal that CePdSn orders antiferromagnetically with $T_N = 7.5$ K. In the paramagnetic state, the susceptibility follows the Curie-Weiss law between 50 and 300 K, with $\Theta_p = -67$ K and $\mu_{eff} = 2.67 \, \mu_B$ (Adroja et al., 1988a). $T_N = 14.6$ K, as derived from magnetic susceptibility measurements of GdPdSn (Adroja et al., 1988a).

Magnetic measurements reveal that the RPdSn compounds with Ce, Sm, Eu, Gd, Tb, Dy, Ho, and Er order antiferromagnetically, with Néel temperatures (T_N) of 7.5, 11, 13, 14.5, 23.5, 11.4, 3.0, and 5.6 K, respectively. The compounds with R = Pr, Nd, and Tm remain paramagnetic at 4.2 K (Sakurai et al., 1990; Adroja and Malik, 1992; André et al., 1993). Electrical resistivity measurements carried out for SmPdSn and DyPdSn show a secondary phase transition at a temperature below T_N (Sakurai et al., 1992). Neutron diffraction data indicate that, in TbPdSn, the magnetic moments of Tb atoms order antiferromagnetically at $T_N = 19$ K, with a sine modulated spin arrangement and wave vector k = (0, 1/4, 0). Below $T_s = 10$ K, the low-temperature magnetic structure is characterized by the propagation vector K = (0, 0.25, 0.075). The magnetic moments form an antiferromagnetic cone spiral (André et al. 1993b).

The magnetic susceptibilities of CePdGa and CePtGa follow the Curie-Weiss law. The observed effective paramagnetic moments are close to those of the free Ce^{3+} ion in each case. At low temperatures, the susceptibility of CePtGa shows considerable deviation from the Curie-Weiss behavior, which may be due either to the effect of a crystalline electric field acting on cerium ions, or to a hybridization between the Ce 4f electrons and the conduction electrons. The low-temperature, specific-heat measurements reveal a peak in both compounds, implying magnetic ordering. The ordering temperatures are 1.7 K for CePdGa and 3.2 K for CePtGa (Malik et al., 1988).

4.1.12. COMPOUNDS WITH THE CeCu₂-TYPE STRUCTURE

The $CeCu_2$-type structure is orthorhombic and belongs to the space group Imma. The magnetic resistivity and specific-heat measurements reveal a magnetic transition near 3.4 K in CePdGe and CePtGe (Rogl et al., 1989).

A neutron diffraction study of TbNiGa revealed antiferromagnetic ordering with the Néel point at 23 K. The magnetic propagation vector k = (1/2, 0, 0) is along the a axis. The terbium magnetic moment is parallel to the b axis, with a magnetic moment value of 6.8(4) μ_B (Kotsanidis and Yakinthos, 1989).

Neutron diffraction data obtained for HoNiGa below the Néel temperature give a magnetic spin alignment of a linear transverse wave mode with k = (0.422, 0, 0) (Kotsanidis et al., 1991).

CeAgGa is a ferromagnet with a Curie temperature of 5.5 K and magnetic moments of 1.12 μ_B per formula (Malik, 1992).

The magnetic susceptibility of CeRhSb is weakly temperature dependent and exhibits a broad maximum at about 113 K, characteristic for valence-fluctuating Ce compounds (Malik and Adroja, 1991).

The magnetic parameters of the series of RAgGa compounds (where R is Pr, Nd, Gd, Tb, Dy, Ho, Er, or Tm) were determined in an applied field up to 26 kOe and at temperatures ranging from 3 to 300 K (Sill and Esau, 1984). At high temperatures, the reciprocal susceptibilities follow the Curie-Weiss law. The effective paramagnetic moments are in reasonable agreement with those calculated for R^{3+} free ions. The asymptotic Curie temperatures are relatively small and positive. At low temperatures, the Gd, Ho, and Er compounds order ferromagnetically. The Tb compound orders antiferromagnetically, while DyAgGa is metamagnetic.

^{151}Eu Mössbauer studies indicate Eu to be divalent in EuTGa. The magnetic susceptibilities of EuTGa compounds (T = Cu, Ag, or Au) follow the Curie-Weiss law, with an effective moment close to that of Eu^{2+}. Below 10 K, EuCuGa orders antiferromagnetically (Malik et al., 1988).

Magnetic susceptibility measurements were carried out on YbTGa (T = Cu, Ag, or Au) to study the Yb valence. In the susceptibility curve of YbCuGa, a broad maximum, typical of mixed-valence systems, is observed at about 210 K. The susceptibility of YbAgGa and YbAuGa varies slowly with temperatures between 50 and 300 K, suggesting the divalent state of the Yb ion (Malik et al., 1987).

4.2. RTX$_2$ PHASES

The RTX$_2$ compounds crystallize in different orthorhombic structures (see Chapter 2).

The RMnSi$_2$ compounds (R = La–Sm) crystallize in the TbFeSi$_2$-type structure (Yarovetz and Gorelenko, 1981). The manganese sublattice of the RMnSi$_2$ compounds orders ferromagnetically up to fairly high temperatures. The Curie temperatures of the RMnSi$_2$ compounds with R from La to Sm increase from 386 to 464 K. At low temperatures, PrMnSi$_2$ and NdMnSi$_2$ show an additional magnetic transition which corresponds to the ordering of the rare-earth sublattice. PrMnSi$_2$ becomes antiferromagnetic (T_N = 35 K), while NdMnSi$_2$ is ferromagnetic (T_c = 40 K) (Venturini et al., 1986). Neutron diffraction data taken at 4.2 K indicate that the magnetic structure of PrMnSi$_2$ is characterized by ferromagnetic layers of Pr atoms piled up along the b axis in the sequence + + − − (see Figure 4.3a). The Mn layers are ferromagnetically coupled with the adjacent Pr layers. The magnetic moment direction of Mn is in the (010) plane (Malaman et al., 1985). The moments of the Mn

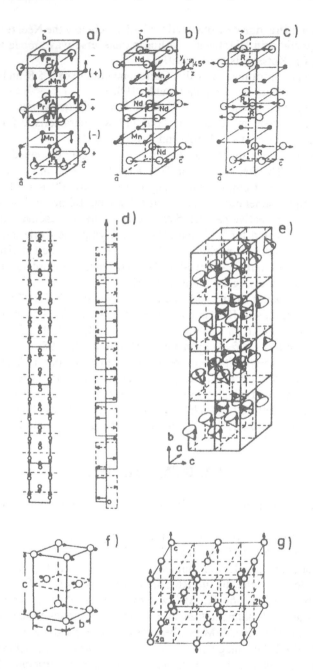

FIGURE 4.3. Magnetic structures of (a) PrMnSi$_2$, (b) NdMnSi$_2$, (c) RNiX$_2$, (d) NdFeSi$_2$, (e) TbCoSi$_2$, (f) TmNiC$_2$, and (g) TbNiC$_2$.

sublattices in the other $RMnSi_2$ compounds are parallel to the b axis. In $NdMnSi_2$, the moment of the Nd sublattice is parallel to the c axis, while the moments constituting the Mn sublattice are tilted 45° from the c direction (see Figure 4.3b). Table 4.3 shows that both compounds evidence large differences between the R moments and the free-ion values. The moments of the Mn sublattices in the other $RMnSi_2$ compounds (R = La, Ce, or Sm) are parallel to the b axis (Malaman et al., 1988).

RMn_xGe_2 compounds crystallize in the defected $CeNiSi_2$-type crystal structure. Magnetic susceptibility data indicate that at low temperatures, compounds with R = Nd, Gd, Tb, Dy, and Ho are antiferromagnets, whereas those with R = Sm and Er remain paramagnetic at 4.2 K (see Table 4.3) (Gil et al., 1993b). Neutron diffraction studies revealed antiferromagnetic structures (G-mode ordering) (see Figure 4.3c) in Nd, Tb, and Dy compounds. At low temperatures, only the magnetic moments of the rare earths are ordered. The magnitudes of magnetic moments at 4.2 K amount to 1.55(15) μ_B in $NdMn_{0.42}Ge_2$, 7.0(1) μ_B in $TbMn_{0.33}Ge_2$, and 7.3(1) μ_B in $HoMn_{0.33}Ge_2$. The moments are smaller than the free-ion values. This effect suggests the occurrence of strong crystal field anisotropy (Gil et al., 1993b).

The magnetic susceptibility of $LaFeSi_2$ and $CeFeSi_2$ is temperature independent. Moreover, $CeFeSi_2$ has been found to be a mixed-valence system. $PrFeSi_2$ is a colinear ferromagnet (T_c = 26 K) with the direction of the moment parallel to the (010) direction, while $NdFeSi_2$ (T_N = 6.5 K) exhibits an amplitude-modulated antiferromagnetic structure in which the (010) axis corresponds to the direction of both the modulation and the magnetic moment (see Figure 4.3d). In $RFeSi_2$ compounds, Fe atoms do not exhibit a magnetic moment. [57]Fe Mössbauer spectra of $PrFeSi_2$ and $NdFeSi_2$ taken at T = 4.2 K show only a splitting due to transferred hyperfine fields (Malaman, 1990a).

The magnetic properties of $RCoSi_2$ compounds were studied by magnetic susceptibility measurements between 2 and 250 K (Pellizone et al., 1982). The Ce and Y compounds show an essentially temperature-independent Pauli paramagnetism. The compounds with R = Nd, Sm, Gd, Tb, Dy, Ho, Er, or Tm are antiferromagnetically ordered below 20 K. The effective rare-earth moments in the paramagnetic state agree well with the free-ion values. For the heavy rare earths, the Néel temperatures do not obey the De Gennes rule. In $HoCoSi_2$ and $DyCoSi_2$, neutron diffraction data indicate the presence of an antiferromagnetic structure of the G type (Bertaut, 1968; Szytuła et al., 1989). In $TbCoSi_2$, a complex spiral structure is observed (see Figure 4.3e) (Szytuła et al., 1989). The magnitudes of the magnetic moments at 4.2 K localized on the R^{3+} ions in $RCoSi_2$ (R = Tb, Dy, or Ho) are much smaller than the free-ion values. This may be due to the strong crystalline electric field anisotropy, which probably is also responsible for the complex magnetic ordering scheme found in $TbCoSi_2$.

The magnetic properties of a large number of $RNiX_2$ compounds (X = Si, Ge) were investigated. $CeNiSi_2$ is a mixed-valence system and does not

TABLE 4.3
Magnetic Properties of RTX$_2$ Compounds

Compound	Crystal structure	Type of magnetic ordering	$T_{C,N}(K)$	$\Theta_p(K)$	$\mu_{eff}(\mu_B)$	$\mu_R(\mu_B)$	$\mu_{Mn}(\mu_B)$	Ref.
LaMnSi$_2$	TbFeSi$_2$	F	386	395	2.6	—	2.07	1, 2
CeMnSi$_2$	TbFeSi$_2$	F	398	420	3.52	0.23	2.24	1, 2
PrMnSi$_2$	TbFeSi$_2$	F,AF	434, 35	450	4.29	2.04	2.35	1, 2
NdMnSi$_2$	TbFeSi$_2$	F	441	460	4.41	1.8	2.29	1, 2
SmMnSi$_2$	TbFeSi$_2$	F	464	482	3.02			1, 2
NdMnGe$_2$	CeNiSi$_2$	AF	34	−23.5	5.08	1.55		3
SmMnGe$_2$	CeNiSi$_2$			−57.2	3.70			3
GdMnGe$_2$	CeNiSi$_2$	AF	35	−30.5	8.75			3
TbMnGe$_2$	CeNiSi$_2$	AF	28	−37.1	9.72	7.0		3
DyMnGe$_2$	CeNiSi$_2$	AF	30	−17.8	11.1			3
HoMnGe$_2$	CeNiSi$_2$	AF	7	−12.8	11.2	7.3		3
ErMnGe$_2$	CeNiSi$_2$			−10.7	9.95			3
NdCoSi$_2$	CeNiSi$_2$	AF	2.5	−28	3.0			4
SmCoSi$_2$	CeNiSi$_2$	AF	4					4
GdCoSi$_2$	CeNiSi$_2$	AF	7.5	−8	7.5			4
TbCoSi$_2$	CeNiSi$_2$	AF	18.5	−18	9.8	8.62		4, 5
DyCoSi$_2$	CeNiSi$_2$	AF	10.5	−8	10.7	5.74		4, 5
HoCoSi$_2$	CeNiSi$_2$	AF	6.3	−6	10.5	6.72		4, 5
ErCoSi$_2$	CeNiSi$_2$	AF	4.5	−17	9.6			4
TmCoSi$_2$	CeNiSi$_2$	AF	2.5	−76	7.8			4
PrNiSi$_2$	CeNiSi$_2$	F	20	20	3.54	3.09		6, 7
NdNiSi$_2$	CeNiSi$_2$	F	9.5	8	3.57	2.38		6, 7
SmNiSi$_2$	CeNiSi$_2$	AF	9.8	−4	0.88			7
GdNiSi$_2$	CeNiSi$_2$	AF	21	−33	9.4			7
TbNiSi$_2$	CeNiSi$_2$	AF	37.6	−17	9.84	8.7		7, 8
DyNiSi$_2$	CeNiSi$_2$	AF	25	−12	11.1			7
HoNiSi$_2$	CeNiSi$_2$	AF	10	3	10.6			7
ErNiSi$_2$	CeNiSi$_2$	AF	3	4	9.56			7
CeNiGe$_2$	CeNiSi$_2$	AF	3.9	−20.8	2.5			9
PrNiGe$_2$	CeNiSi$_2$	F	15	5	3.2			7
NdNiGe$_2$	CeNiSi$_2$	F	15	1	3.74			7
SmNiGe$_2$	CeNiSi$_2$			−22	1.17			7
GdNiGe$_2$	CeNiSi$_2$	AF	24.5	−33	8.6			7
TbNiGe$_2$	CeNiSi$_2$	AF	42	−31.5	10.57	8.8		7, 10
DyNiGe$_2$	CeNiSi$_2$	AF	22	−12	10.96			7
HoNiGe$_2$	CeNiSi$_2$	AF	7.6	−5	10.71	6.4		7, 10
ErNiGe$_2$	CeNiSi$_2$	AF	2.5	−2	9.69			7
CeCuSi$_2$	CeNiSi$_2$	F	11					11
TbCuGe$_2$	CeNiSi$_2$	AF	39		10.4	8.82		12
CePdSi$_2$	CeNiSi$_2$	AF	7					11
CePtSi$_2$	CeNiSi$_2$			−17	2.56			13
CeMnSn$_2$	CeNiSi$_2$	F	320	117	4.03			14
TbRu$_2$B$_2$	LuRuB$_2$	F	49					15
TmRu$_2$B$_2$	LuRuB$_2$	F	4					16
NdNiC$_2$	CeNiC$_2$	AF	7					17, 18
GdNiC$_2$	CeNiC$_2$	AF	14					17
TbNiC$_2$	CeNiC$_2$	AF	25			6.8		17, 19

TABLE 4.3 (continued)
Magnetic Properties of RTX$_2$ Compounds

Compound	Crystal structure	Type of magnetic ordering	$T_{C,N}(K)$	$\Theta_p(K)$	$\mu_{eff}(\mu_B)$	$\mu_R(\mu_B)$	$\mu_{Mn}(\mu_B)$	Ref.
DyNiC$_2$	CeNiC$_2$	AF	10					17
ErNiC$_2$	CeNiC$_2$	AF	8			8.7		17, 20
TmNiC$_2$	CeNiC$_2$	AF	8					17, 18
YNiC$_2$	CeNiC$_2$	Pauli paramagnetic						17

References: 1. Venturini et al., 1986; 2. Malaman et al., 1990; 3. Gil et al., 1993a; 4. Pelizzone et al., 1982; 5. Szytuła et al., 1989; 6. Schobinger-Papamantellos and Buschow, 1992a; 7. Gil et al., 1993b; 8. Schobinger-Papamantellos and Buschow, 1991; 9. Percharsky et al., 1991; 10. Bażela et al., 1992; 11. Adroja and Rainford, 1992; 12. Schobinger-Papamantellos and Buschow, 1992a; 13. Lee et al., 1990; 14. Weitzer et al., 1992; 15. Weidner et al., 1985; 16. Ku and Shelton, 1981; 17. Kotsanidis et al., 1989; 18. Yakinthos et al., 1990; 19. Yakinthos et al., 1989; 20. Yakinthos et al., 1991.

order magnetically down to 1.4 K, while CeNiGe$_2$ has two-step antiferromagnetic phase transitions at T_N^I = 3.9 and T_N^{II} = 3.2 K (Pecharsky et al., 1991). Two antiferromagnetic phase transitions are also observed on specific-heat curves (Geibel et al., 1992).

The compounds with R = Pr and Nd are ferromagnets (Gil et al., 1993a; Schobinger-Papamantellos and Buschow, 1992), while those containing R = Gd, Tb, Dy, Ho, and Er are antiferromagnets with Néel temperatures at T_N = 2.5 K in ErNiGe$_2$ and T_N = 42 K in TbNiGe$_2$ (see Table 4.3) (Gil et al., 1993a). For both the RNiSi$_2$ and RNiGe$_2$ groups of compounds, the Néel temperatures do not follow De Gennes scaling. This behavior indicates the influence of the crystalline electric field effects. The values of the effective magnetic moments are nearly the free R^{3+} ion values.

Neutron diffraction studies show that PrNiSi$_2$ and NdNiSi$_2$ order ferromagnetically, while TbNiSi$_2$ is an antiferromagnet (Schobinger-Papamantellos and Buschow, 1991, 1992). The magnetic moments prefer direction along the c axis in all cases. The magnetic ordering in TbNiSi$_2$ corresponds to the G mode. A similar antiferromagnetic structure was found in TbNiGe$_2$ and HoNi$_{0.64}$Ge$_2$ (Bażela et al., 1992).

CeCu$_{1.54}$Si$_{1.46}$ crystallizes in the CeNiSi$_2$-type structure. Magnetic, thermal, and transport data indicate that it is an antiferromagnetic heavy fermion system with a Néel temperature of T_N = 7 K (Takabatake et al., 1988). Magnetic and neutron diffraction measurements carried out for the TbCu$_{0.4}$Ge$_2$ compound indicate that the rare-earth moments order antiferromagnetically below T_N = 39 K, with the magnetic structure described by the G mode and the magnetic moments parallel to the c axis (Schobinger-Papamantellos and Buschow, 1992).

CePtSi$_2$ behaves like a 100-K paramagnet down to 70 mK. Above 100 K, the inverse molar magnetic susceptibility follows the Curie-Weiss law, leading to an effective moment of 2.56 μ_B per cerium ion and a paramagnetic Curie temperature of -17 K (Lee et al., 1990).

Magnetic data collected for $CeTSi_2$ (T = Co, Ni, Cu, Rh, Pd, Ir, and Pt) and $CeNiGe_2$ compounds show the following properties (Adroja et al., 1992):

1. Magnetic susceptibility of $CeRhSi_2$, $CeIrSi_2$, and $CeCoSi_2$ shows a broad maximum, indicating the valence fluctuating nature of the Ce ion,
2. $CePtSi_2$ and $CeNiGe_2$ show heavy fermion-like behavior.
3. Susceptibility of $CeCuSi_2$ shows a rapid rise below 11 K, indicating ferromagnetic ordering of the Ce moment,
4. The susceptibility of $CePdSi_2$ exhibits two peaks at 7 and 2.5 K, indicating two antiferromagnetic phase transitions.

$RMn_{1-x}Sn_{2-y}$ (R = La, Ce, Pr, Nd, and Sm) compounds crystallize in the defect $CeNiSi_2$-type structure. $LaMn_{0.24}Sn_{1.79}$ has been found to be a temperature-dependent paramagnet due to the localized manganese moment. $CeMn_{1-x}Sn_{2-y}$ alloys are ferromagnets, whereas (Pr, Nd, Sm)$Mn_{1-x}Sn_{2-y}$ exhibit antiferromagnetic ordering (Weitzer et al., 1992).

The RTB_2 compounds in which R = Y or Lu, and T is Ru or Os, crystallize in an orthorhombic structure with space group Pnma. They are superconductors with critical temperatures of 7.7 K for $YRuB_2$, 1.9 K in $YOsB_2$, 9.9 K in $LuRuB_2$, and 2.4 K in $LuOsB_2$ (Chevalier et al., 1983). $TbRuB_2$ is a ferromagnet with T_c = 49(2) K (Weidner et al., 1985). In the $Tm_{1-x}Lu_xRuB_2$ system, the magnetic susceptibility data allow construction of the magnetic phase diagram. The initial depression of T_s near the composition $LuRuB_2$ is caused by the magnetic Tm^{3+} ions (Ku and Shelton, 1981).

The magnetic data collected for the RTC_2 compounds indicate that the phases with R = Nd, Gd, Tb, Dy, Er, or Tm are all antiferromagnets with ordering temperatures between 7 K for R = Nd and 25 K for R = Tb (Kotsanidis et al., 1989). Neutron diffraction studies of $RNiC_2$ indicate two different types of colinear antiferromagnetic structures with magnetic propagation vectors k = (0, 0, 1) (see Figure 4.3f) in $TmNiC_2$ (Yakinthos et al., 1990) and $ErNiC_2$ (Yakinthos et al., 1991), and (1/2, 0, 1/2) (see Figure 4.3g) in $NdNiC_2$ (Yakinthos et al., 1990) and $TbNiC_2$ (Yakinthos et al., 1989). The magnetic moments are oriented parallel to the a axis in Nd, Er, and Tm, and parallel to the c axis in Tb. The determined values of the rare-earth magnetic moments are smaller than those calculated for free R^{3+} ions. This result indicates the strong influence of the crystalline electric field acting on the rare-earth ions.

The magnetic susceptibility of the $RNiAl_2$ compounds, where R = Y, La, Lu, or Yb, is temperature independent. For the compounds with R = Gd, Tb, Dy, Ho, Er, or Tm the $\chi^{-1}(T)$ function obeys the Curie-Weiss law in the temperature range 78 to 300 K (Romaka et al., 1982b).

The magnetic susceptibility was also measured for the $RNiGa_2$ compounds in the temperature range 78 to 300 K. It is independent of the temperature for R = La, Yb, Lu, or Y. For the other $RNiGa_2$ compounds, the susceptibility obeys the Curie-Weiss law (Romaka et al., 1982a).

The values of the magnetic susceptibility, paramagnetic Curie temperature and effective magnetic moments are shown in Table 4.3.

4.3 RT_2X PHASES

The family of compounds with 1:2:1 stoichiometry crystallizes in two groups of different structure types: in the cubic $L2_1$ structure, also found in Heusler alloys, with the general formula X_2YZ, and in the orthorhombic Fe_3C-type structure.

Lanthanide based Heusler alloys exhibit a number of interesting and diverse phenomena, such as superconductivity, mixed valence, and magnetic phase transitions. The Heusler phases occur in the following lanthanide systems: RAg_2In (Galera et al., 1984), RAu_2In (Besnus et al., 1985b), and RCu_2In (Felner, 1985), as well as in a number of RPd_2Sn (Malik et al., 1985b) and RNi_2Sn (Skolozdra and Komarovskaja, 1983) compounds.

The RCu_2In (R = Sm, Gd, Tb, or Dy) compounds are ordered antiferromagnetically at low temperatures (see Table 4.4), $LaCu_2In$ is a Pauli paramagnet and $LuCu_2In$ is diamagnetic (Felner, 1985).

New data indicate that $CeCu_2In$ is a heavy fermion compound with a specific heat coefficient $\gamma = 1.4$ $Jmol^{-1}$ K^{-2} (Takanayagi et al., 1988). A ^{63}Cu nuclear magnetic resonance study showed the presence of two antiferromagnetic transitions at $T_{N1} = 1.6$ K and $T_{N2} = 1.1$ K (Nakamura et al., 1988a).

The specific-heat data available for $NdCu_2In$ and $SmCu_2In$ compounds reveal antiferromagnetic phase transitions at 2.0 and 3.7 K, respectively, while $PrCu_2In$ remains nonmagnetic at 1.5 K. Only a Schottky-type anomaly has been observed (Sato et al., 1992).

The RAg_2In compounds (R = Ce, Nd, Sm, Gd, Tb or Dy) order antiferromagnetically at low temperatures, while no ordering above 2 K was observed in compounds with R = Pr, Ho, Er, or Tm (Galera et al., 1984). Neutron diffraction measurements revealed in $CeAg_2In$ antiferromagnetic ordering of Ce moments of the first type (see Figure 4.4a). The moment at T = 4.2 K is found to be $\mu = 0.97(7)$ μ_B, which is the result of a small CEF splitting ($\Delta = 18$ K). The neutron diffraction pattern of $NdAg_2In$ at T = 1.8 K exhibits four additional lines that cannot be indexed in a simple manner (Galera et al., 1982b). The compound $PrAg_2In$ remains a Van Vleck type paramagnet down to 1.5 K. In this case, the ground state is the nonmagnetic orbital doublet Γ_3 (Galera et al., 1982a).

Neutron diffraction data indicate that $TbAg_2In$ orders antiferromagnetically at $T_N = 8.3(2)$ K. Below this temperature, magnetic ordering is a

TABLE 4.4
Magnetic Data for RT$_2$X Compounds

Compound	Type of magnetic ordering	T$_N$(K)	Θ_p(K)	μ_{eff}(μ_B)	μ_s(μ_B)	Ref.
LaCu$_2$In	Pauli paramagnet					1
CeCu$_2$In			-30	2.52		1
PrCu$_2$In			-35	3.53		1
NdCu$_2$In			-70	2.60		1
SmCu$_2$In	No Curie-Weiss behavior					1
GdCu$_2$In	AF	12	-31	7.60		1
TbCu$_2$In	AF	6	-33	9.70		1
DyCu$_2$In	AF	3	-21	10.63		1
ErCu$_2$In			-10	9.63		1
LuCu$_2$In	Pauli paramagnet					1
CeAg$_2$In	AF	2.7	-9	2.54		2
	AF	2.5	-9		0.97(7)	3
PrAg$_2$In			-16	3.63		2
NdAg$_2$In	AF	2.5	-24	3.68		2
SmAg$_2$In	AF	4.5	-23			2
GdAg$_2$In	AF	10.0	-56	8.1		2
TbAg$_2$In	AF	8.8	-29	9.62		2
	AF	8.3			5.6(2)	4
DyAg$_2$In	AF	3.5	-20	10.33		2
HoAg$_2$In			-14	10.23		2
ErAg$_2$In			-9.5	9.47		2
TmAg$_2$In			-5	7.33		2
CeAu$_2$In			-15	2.57		5, 6
PrAu$_2$In			-5	3.55		5, 6
NdAu$_2$In			-8	3.70		5, 6
SmAu$_2$In	No Curie-Weiss behavior					5, 6
GdAu$_2$In	AF	11.5	-14	8.0		5, 6
TbAu$_2$In	AF	8.2	-7	9.7		5, 6
DyAu$_2$In	AF	6.0	-5	10.8		5, 6
HoAu$_2$In	AF	2.6	-5	10.8		5, 6
ErAu$_2$In			-4	9.63		5, 6
TmAu$_2$In			-35	7.06		5, 6
YbAu$_2$In	No Curie-Weiss behavior					5, 6
YAu$_2$In	Pauli paramagnet					6
TbPd$_2$Sn	AF	9.0	-8.6	9.95		7, 8
DyPd$_2$Sn	AF	15	-9.3	10.83		7, 8
	AF	7		10.8	6.7(5)	9
HoPd$_2$Sn	AF	4.2	-6.2	10.67	4.4(1)	7, 9
	AF	5			5.81(8)	10
ErPd$_2$Sn	AF	1	-5.8	9.57		11
	AF	1			5.7(5)	12
TmPd$_2$Sn			0	7.4		7
YbPd$_2$Sn			-4.3	4.34		7
	AF	0.23			1.6	13, 14
YbNi$_2$Sn			-38	4.33		15
LuNi$_2$Sn	Pauli paramagnet					15

TABLE 4.4 (continued)
Magnetic Data for RT₂X Compounds

Compound	Type of magnetic ordering	$T_N(K)$	$\Theta_p(K)$	$\mu_{eff}(\mu_B)$	$\mu_s(\mu_B)$	Ref.
CePd₂Si	F	2.3	−11		1.1	16
GdPd₂Si	AF	13.5	14	8.06	6.98	17
TbPd₂Si	AF	21	16	9.9	5.64	17
DyPd₂Si	F	9	6	10.61	7.38	17
HoPd₂Si	F	3.5	2	10.53	7.68	17
ErPd₂Si	F	2.8	1	9.48	6.62	17
NdPd₂Ge	AF	1.7				18
EuPd₂Ge	AF	6.5				18
GdPd₂Ge	AF	7.2				18
TbPd₂Ge	F(?)	9.7				18
DyPd₂Ge	AF	8.1				18
HoPd₂Ge	F(?)	3.6				18
ErPd₂Ge	F(?)	3.0				18
TmPd₂Ge	F(?)	1.2				18

References: 1. Felner, 1985; 2. Galera et al., 1984; 3. Galera et al., 1982b; 4. André et al., 1992; 5. Besnus et al., 1986; 6. Besnus et al., 1985b; 7. Malik et al., 1985b; 8. Umarji et al., 1985; 9. Donaberger and Stager, 1987; 10. Li et al., 1989; 11. Shelton et al., 1986; 12. Stanley et al., 1987; 13. Kierstead et al., 1985; 14. Hodges and Jéhanno, 1988; 15. Skolozdra and Komarovskaya, 1983; 16. Barandiaran et al., 1986a; 17. Gignoux et al., 1984; 18. Jorda et al., 1983.

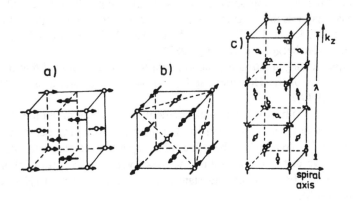

FIGURE 4.4. Magnetic structures of (a) CeAg₂In, (b) HoPd₂Sn, and (c) TbAg₂In.

cycloidal spiral. The magnetic moment of 5.6(2) μ_B at T = 1.6 K is smaller than the free ion Tb³⁺ value (9 μ_B), implying the strong influence of the crystal electric field. With a rise of temperature, a change of the magnetic structure to the sine wave modulation is observed near T = 6 K (see Figure 4.4c). No magnetic ordering was detected in DyAg₂In at T = 1.66 K (André et al., 1992b).

$CeAg_2In$ is a Kondo system in which Ce is trivalent, it orders antiferromagnetically at $T_N = 2.7$ K. The compound $CeCu_2In$ has two Néel temperatures: $T_{N1} = 1.6$ and $T_{N2} = 1.1$ K (Nakamura et al., 1988a). In the $CeAg_{2-x}Cu_xIn$ system, the Néel temperature increases first from 2.7 K for x = 0 to 5.5 K for x = 1.5. Then it drops to below 1.5 K as x increases. $CeCu_2In$ shows incipient valence fluctuations and a large Kondo-like resistivity.

The RAu_2In compounds with R = Gd, Tb, Dy, or Ho order antiferromagnetically at low temperatures (see Table 4.4). Except for $GdAu_2In$, the magnetic moments measured at T = 1.5 K and in fields up to 15 T are reduced compared to the free-ion g_JJ values. This reduction is manifested mostly at both ends of the series (Besnus et al., 1985a, 1986).

The temperature dependence of the magnetic susceptibility of $CeAu_2In$ shows an anomaly at T = 1.2 K which may suggest the Néel temperature (Pleger et al., 1987).

In the RPd_2Sn series, the compounds with Tb, Dy, Ho, Er, or Yb are all antiferromagnets. The compounds with R = Er, Tm, or Yb have also been found to be superconductors (Malik et al., 1985a,b).

Heat capacity, magnetic susceptibility, and resistivity experiments made on $ErPd_2Sn$ indicate that long-range magnetic order coexists with superconductivity ($T_s = 1.17$ K, $T_N = 1.0$ K) (Shelton et al., 1986).

$DyPd_2Sn$ and $HoPd_2Sn$ are antiferromagnets with the magnetic MnO-type structure T_N, which is 15 and 5 K, respectively. The corresponding magnetic structure (type II) is shown in Figure 4.4b. The magnetic unit cell doubles along all three crystallographic directions (wave vector k = [1/2, 1/2, 1/2]). In both compounds the magnetic moments are perpendicular to the (111) axis. Their magnitudes at 1.2 K are 6.7(5) μ_B for $DyPd_2Sn$ and 4.3(5) μ_B for $HoPd_2Sn$ (Donaberger and Stager, 1987).

Supplementary neutron diffraction data for $HoPd_2Sn$ reveal that the magnetic moments of the Ho atoms at T = 0.34 K are as small as 5.81(8) μ_B (Li et al., 1989).

[166]Er and [170]Yb Mössbauer measurements were made down to 0.05 K for $ErPd_2Sn$ and $YbPd_2Sn$. They confirm that magnetic ordering occurs at $T_N = 0.7$ K (Er) and 0.26 K (Yb). The values of the spontaneous rare-earth magnetic moments have distributions around the mean values of 5.6 μ_B (Er) and 1.8 μ_B (Yb). For Yb^{3+}, the ground state is the Kramers doublet (Hodges and Jéhanno, 1988).

The temperature dependence of the magnetic susceptibility in the temperature range from 78 to 300 K indicates that $LuNi_2Sn$ is a Pauli paramagnet, while $\chi_N^{-1}(T)$ for $YbNi_2Sn$ obeys the Curie-Weiss law, with $\Theta_p = -38$ K and $\mu_{eff} = 4.33$ μ_B (Skolozdra and Komarovskaja, 1983).

The RPd_2Si compounds (R = Gd, Tb, Dy, Ho, or Er) and the RPd_2Ge compounds (R = all lanthanides except Pm and Yb) crystallize in the orthorhombic Fe_3C-type structure (Moreau et al., 1982) This structure belongs to the Pnma space group. It is built up by trigonal prisms with the corners occupied by two rare earth atoms (4c site) and four Pd atoms (8d site), while the Si atom is situated in the center.

Resistivity and magnetic measurements performed between 1.7 and 300 K on a polycrystalline sample of $CePd_2Si$ indicate that this compound is ferromagnetic below $T_c = 2.3$ K. Neutron diffraction data show a noncolinear structure with an antiferromagnetic component along the c axis and a ferromagnetic component along the a axis (Barandiaran et al., 1986a).

RPd_2Si compounds (R = Gd, Tb) are antiferromagnetic, with Néel temperatures of 13.5 and 21 K, respectively. For both compounds, a metamagnetic transition is observed at low magnetic fields. In $TbPd_2Si$, a transition between two different antiferromagnetic phases has been found at 8.5 K. The Dy-, Ho-, and Er-based compounds are ferromagnetic, with Curie temperatures of 9, 3.5, and 2.8 K, respectively (Gignoux et al., 1984).

Of the RPd_2Ge compounds, only those with R = Nd, Eu, Gd, Tb, Dy, Ho, Er, or Tm order antiferromagnetically (Jorda et al., 1983).

The magnetic data for RPd_2X (X = Si or Ge) compounds are listed in Table 4.4.

4.4. RTX$_3$ PHASES

The RTX_3 compounds crystallize in a tetragonal $BaNiSn_3$-type structure (space group I4mmm). Although this type is similar to the $ThCr_2Si_2$ type, the lanthanide ions have different nearest neighbors.

The magnetic properties of only some of the RTX_3 compounds have been investigated up to now. The obtained data are summarized in Table 4.5. The ternary $RFeSi_3$ (r = Gd–Tm, Lu, or Y) (Gladyshevskii et al., 1978), $RCoSi_3$ (R = Ce, Tb, or Dy) (Yarovetz, 1978), and $RNiSi_3$ (R = Gd–Lu or Y) (Yarovetz, 1978; Gladyshevskii et al., 1977) compounds are all paramagnets.

The $RTSi_3$ compounds (R = La or Ce; T = Co, Ru, Rh, Ir, or Os) also crystallize in the $BaNiSn_3$-type structure. $LaRhSi_3$ and $LaIrSi_3$ are superconductors with $T_c = 2.5$ K. The temperature dependence of the magnetic susceptibility for $CeTSi_3$ compounds in the temperature range between 1.2 and 293 K indicates that above T = 80 K, the Curie-Weiss law is obeyed, and that there exists an anomalous $\chi^{-1}(T)$ dependence at low temperatures. The $CeCoSi_3$ compound has superconducting properties below $T_s = 1.4$ K (Haen et al., 1985).

$GdIrSi_3$ and $DyIrSi_3$ are ordered antiferromagnetically at 15.5 and 7.5 K, respectively. The [155]Gd Mössbauer data show that the Gd moments are aligned in the basal plane. The [161]Dy Mössbauer data imply the occurrence of an incommensurate modulated spin structure in $DyIrSi_3$. The reduced value of the average Dy moment of 8.7 μ_B is attributed to the crystal field effect. The values of T_N in RTX_3 are smaller than those observed in the RT_2X_2 compounds. This fact indicates that the magnetic interactions depend essentially on the surrounding nearest neighbors (Sanchez et al., 1991).

TABLE 4.5
Magnetic Properties of RTX₃ Compounds

Compound	Type of magnetic ordering	$T_N(K)$	$\Theta_p(K)$	$\mu_{eff}(\mu_B)$	Ref.
GdFeSi₃			35	7.79	1
TbFeSi₃			47	9.36	1
DyFeSi₃			82	9.72	1
HoFeSi₃			49	9.05	1
ErFeSi₃			65	8.09	1
TmFeSi₃			74	7.25	1
LuFeSi₃	Pauli paramagnet				1
YFeSi₃	Pauli paramagnet				1
CeCoSi₃	Pauli paramagnet				2
TbCoSi₃			−22	9.3	3
DyCoSi₃			12	10.11	3
TbNiSi₃			−5	7.92	3, 4
DyNiSi₃			−27	9.54	3, 4
HoNiSi₃			−52	9.03	3, 4
ErNiSi₃			−58	7.98	3, 4
TmNiSi₃			5	6.7	3, 4
YbNiSi₃			−10	3.9	3, 4
YNiSi₃			−117	0.4	3, 4
CeRhSi₃			78	2.62	5
CeOsSi₃			−106	1.03	5
CeIrSi₃			−113	2.59	5
GdIrSi₃	AF	15.5	−30	8.12	6
DyIrSi₃	AF	7.5	15.7	10.41	6
GdRhSi₃			−13	7.95	7
TbRhSi₃			−18	9.95	7

References: 1. Gladyshevskii et al., 1978; 2. Bodak et al., 1977; 3. Yarovetz, 1978; 4. Gladyshevskii et al., 1977; 5. Haen et al., 1985; 6. Sanchez et al., 1991; 7. Szytuła, 1990.

4.5. R₂T₃Si₅ PHASES

4.5.1. R₂Fe₃Si₅ COMPOUNDS

Some of the lanthanide iron silicides crystallize in the tetragonal Sc₂Fe₃Si₅-type structure (Bodak et al., 1977b). The compounds with R = Gd–Er are antiferromagnetic, with Néel temperatures below 11 K (see Table 4.6) (Braun et al., 1981). Above 60 K, the susceptibilities follow the Curie-Weiss law, with effective magnetic moment per rare-earth ion close to the free ion value.

The Néel temperatures do not agree well with the De Gennes factor $(g - 1)^2 J (J + 1)$ (see Figure 4.5) indicating that, apart from the RKKY interactions, other factors also influence the magnetic interactions. Two factors should be taken into consideration:

1.   Dipole-dipole interaction
2.   Crystalline electric field (CEF) effect

TABLE 4.6
Magnetic Data for $R_2T_3Si_5$ Compounds

Compound	Type of magnetic ordering	$T_{C,N}(K)$	$\Theta_p(K)$	$\mu_{eff}(\mu_R)$	$\mu_R(\mu_B)$	Ref.
$R_2Fe_3Si_5$						
R = Gd	AF	8.6	−15.4	7.8	8.7	1, 2
Tb	AF	10.3	1.8	9.6	9.7	1, 2
Dy	AF	4.3	−11.4	10.4		1
Ho	AF	2.85	0.3	10.4		1
Er	AF	2.85	−8.0	9.7	8.2	1, 3
Tm	AF	1.13			6.5	4, 6
$R_2Co_3Si_5$						
R = Y	Pauli paramagnetic					5
Gd			−17.4	7.89		5
Tb			8.8	9.73		5
Dy			8.7	11.01		5
Ho			−5.8	10.7		5
Er			3.1	9.64		5
Tm			−4.1	7.66		5
Lu	Pauli paramagnetic					5

References: 1. Braun et al., 1981; 2. Moodenbaugh et al., 1982; 3. Moodenbaugh et al., 1984; 4. Segre and Braun, 1981; 5. Gorelenko et al., 1985; 6. Moodenbaugh et al., 1985.

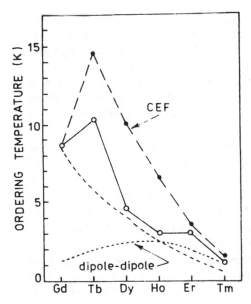

FIGURE 4.5. Magnetic transition temperatures normalized at Gd of heavy $R_2Fe_3Si_5$ compounds as observed (open circles) (Braun et al., 1981; Serge and Braun, 1981) and as predicted by CEF modeling, the De Gennes rule (dotted line), and for nearest-neighbor dipole interactions.

The isotropic portion of the dipole-dipole interaction energy can be estimated by making use of an ordering temperature, T_M, due to the dipole-dipole interaction only, according to

$$kT_M = (\mu_0/4\pi) \sum_n \mu_n^2/r_n^3$$

where the sum is taken over the four nearest rare-earth neighbors with μ_n the effective free-ion moment and r_n the nearest-neighbor distance. The calculated T_Ms are shown in Figure 4.5 by a dashed line. Despite this simplification, the calculated T_Ms are of the same order of magnitude as the measured ordering temperatures. This suggests that dipolar interactions may be important in these compounds, even though they cannot explain the aforementioned trend in the Néel temperatures. Better agreement is obtained by incorporating the CEF terms in the magnetic Hamiltonian interaction (Noakes and Shenoy, 1982). The obtained results are also presented in Figure 4.5.

Neutron diffraction studies performed on $Tb_2Fe_3Si_5$ and $Er_2Fe_3Si_5$ show that in the Tb compound a commensurate magnetic structure with wave vector $k = (1/2, 1/2, 1/2)$ and magnetic moment of the Tb ion directed along the (001) set of directions occurs below $T = 7$ K. Above, an incommensurate structure with the wave vector $k = (1/2, 1/2, 1/2-\tau)$, $\tau = 0.135$, is observed (Moodenbaugh et al., 1982). A specific-heat peak found near 7 K corresponds to the transformation between the two magnetic phases (Vining and Shelton, 1983; Moodenbaugh et al., 1982).

In $Er_2Fe_5Si_3$, magnetic diffraction peaks at positions incommensurate with the chemical cell are observed below 2.85 K. A second set of magnetic peaks, at positions commensurate with the chemical cell but with the tetragonal a axis doubled, have been observed below 2.55 K. Between 2.55 and 2.45 K, the two sets of magnetic peaks coexist, and a hysteresis of about 0.01 K is observed. In this temperature region, peaks due to the incommensurate phase decrease steadily in intensity as the temperature is reduced until, below 2.45 K, only commensurate peaks are present. They increase in intensity as the temperature is reduced to the lowest temperatures. The refinement of data taken at 1.83 K shows that the magnetic intensities could be well-accounted for by a noncolinear magnetic structure whose principal features are as follows: (1) ordering occurs only at Er atom sites, the moment being 8.2 μ_B at 1.83 K, and (2) the moments are ordered in the (001) planes with a noncolinear arrangement along the (110) set of directions, the c-glide planes of the chemical structure being retained in the magnetic cell (space group C_pccm'). For the incommensurate model, all elements of the commensurate arrangement are retained, but a sinusoidal modulation along (001) with a wavelength of about 5.7c is introduced. This wavelength is independent of temperature to within $+3\%$. Specific-heat measurements show three peaks: one at 2.75 K corresponding to the onset of magnetic order, and two others at 2.45 and 2.43 K in the coexistence region of the incommensurate and commensurate

TABLE 4.7

Parameters of the CEF Hamiltonian for R = Tb, Dy, Ho, Er, and Tm in R$_2$Fe$_3$Si$_5$ Compounds, and Their Effect on the R Electronic State

R	B$_2^0$(K)	B$_2^2$(K)	Ground state (J$_z$)	First excited level (K)	Overall splitting (K)
Tb	−2.41	−2.05	±6 doublet	75	312
Dy	−1.45	−1.23	±15/2 doublet	58	297
Ho	−0.49	−0.41	±8 doublet	21	114
Er	0.53	0.45	±1/2(±3/2) doublet	19	109
Tm	2.04	1.73	0 (±2) nearly degenerate ±1 pair	55	264

From Noakes, D. R., Umarji, A. M., and Shenoy, G. H., *J. Magn. Magn. Mater.*, 39, 309, 1983. With permission.

phases (Moodenbaugh et al., 1984). ^{57}Fe, ^{155}Gd, and ^{166}Er Mössbauer effect measurements made on Gd$_2$Fe$_3$Si$_5$ and Er$_2$Fe$_3$Si$_5$ compounds have shown that:

1. The iron atoms carry no moments.
2. The gadolinium and erbium ions exhibit magnetic moments which order at low temperatures.
3. The CEF parameters have been determined, as listed in Table 4.7.

The lanthanide site symmetry in R$_2$Fe$_3$Si$_5$ makes the expression for the complete crystal-field Hamiltonian large and unhandy. Consequently, analysis of the Mössbauer data has been performed with the following Hamiltonian:

$$H_{CEF} = B_2^0 O_2^0 + B_2^2 O_2^2$$

The parameters of CEF obtained using Mössbauer spectroscopy data indicate that the B$_2^0$ and B$_2^2$ coefficients show similar magnitudes. For R = Tb, Dy, and Ho, both coefficients are negative, whereas for R = Er and Tm, they are positive (Noakes et al., 1983b).

The results of a neutron diffraction study of Tm$_2$Fe$_3$Si$_5$ indicate an antiferromagnetic ordering of the magnetic moments, as well as no evidence for the existence of the magnetic moment on the iron ions (Moodenbaugh et al., 1985).

Lu$_2$Fe$_3$Si$_5$ shows a superconducting transition temperature at 6 K, whereas in Tm$_2$Fe$_3$Si$_5$, reentrant superconducting properties have been observed (Segre and Braun, 1981).

4.5.2. R$_2$Co$_3$Si$_5$ COMPOUNDS

Magnetic susceptibility measurements in the temperature range 78 to 293 K indicate that Y$_2$Co$_3$Si$_5$ and Lu$_2$Co$_3$Si$_5$ are Pauli paramagnets, while in

$R_2Co_3Si_5$ compounds (R = Gd–Tm), magnetic susceptibility obeys the Curie-Weiss law. The effective magnetic moments per R atom agree well with the values calculated for free R^{3+} ions (see Table 4.6) (Gorelenko et al., 1985).

4.5.3. $R_2Rh_3Si_5$ COMPOUNDS

Magnetic data for $R_2Rh_3Si_5$ compounds (R = Ce, Nd, Sm, and Gd–Er) indicate antiferromagnetic ordering in compounds with R = Gd and Tb, whereas the reciprocal magnetic susceptibilities of the remaining compounds obey the Curie-Weiss law up to 4.2 K (Chevalier et al., 1983).

4.6. RT_2X_2 PHASES

The family of intermetallic phases with the stoichiometric composition 1:2:2 was found to be particularly numerous. In addition, their physical properties turned out to be very interesting. A vast amount of experimental data indicates that the T component usually does not carry a magnetic moment in most of these phases, with the exception of manganese-containing compounds. Therefore, the latter will be reviewed in a separate chapter.

Taking the magnetic properties into account, the RT_2X_2 compounds can be divided into three groups:

1. Those with R = Y, La, and Lu show temperature-independent magnetic susceptibility, suggesting Pauli paramagnetism. Some of them were found to be superconducting at low temperatures (Braun, 1984; Chevalier et al., 1983).
2. The phases with R = Ce, Eu, and Yb exhibit magnetic properties which are clearly influenced by the electronic instability of the respective R elements (mixed-valence effect).
3. The compounds in which the R element appears at the +3 oxidation state and carries a permanent magnetic moment show various schemes of long-range magnetic ordering at low temperatures.

The latter group, being the most numerous, will be discussed first in the next section.

4.6.1. RT_2Si_2 AND RT_2Ge_2 PHASES WITH R = Pr AND Nd

The magnetization measurements in the varying field at 4.2 K, show a spin flop or a metamagnetic transition at a field of about 1 and 10 kOe.

The magnetic properties of RT_2X_2 compounds with R = Pr and Nd are summarized in Tables 4.8 and 4.9. It can be seen that their magnetic properties are very similar.

$PrFe_2Si_2$ and $PrFe_2Ge_2$ order antiferromagnetically below T_N = 7.7 and 14.2 K, respectively. Neutron diffraction and Mössbauer experiments have shown that both have the AFII-type antiferromagnetic structure with moments

TABLE 4.8
Magnetic Data for PrT_2Si_2 and PrT_2Ge_2 Compounds

T	Type of magnetic ordering	$T_{C,N}(K)$	$\Theta_p(K)$	$\mu_{eff}(\mu_B)$	$\mu_R(\mu_B)$	Ref.
			PrT_2Si_2			
Fe	AF	7.7			1.5	1
Co	AF	30	−40	3.7	3.19	4, 5
Ni	AF	18	−6	3.67	2.6	7
Cu	AF	20.8	−2	3.41	2.51	8, 9
Ru	F	17.8	46	3.49	3.18	11, 12
Rh						
Os	F	12.6	53	3.53		13
Pt			7	3.46		14
Au			−9.2	3.58		15
			PrT_2Ge_2			
Fe	AF	14.2	10	3.5	3.2	2, 3
Co	AF	28	−9	2.9	3.98	6
Cu	AF	16			2.53	9
Ru	F	18	27	3.85	1.82	10
Rh	AF	47	−14	3.90		10

References: 1. Malaman et al., 1992; 2. Leciejewicz et al., 1982; 3. Szytuła et al., 1990; 4. Kolenda et al., 1982; 5. Yakinthos et al., 1984; 6. Szytuła et al., 1980; 7. Barandiaran et al., 1987; 8. Schlabitz et al., 1982; 9. Szytuła et al., 1983; 10. Felner and Nowik, 1985; 11. Hiebl et al., 1983a; 12. Ślaski et al., 1984; 13. Hiebl et al., 1983b; 14. Hiebl and Rogl, 1985; 15. Felner, 1975.

of 1.5 μ_B (Si sample) and 2.75 μ_B (Ge sample) at 4.2 K, aligned along the tetragonal axis. $PrFe_2Ge_2$ shows, however, a magnetic phase transition at 9 K from the colinear-type AFII structure, with wave vector k = (0, 0, 0.5), to an incommensurate structure characterized by the wave vector k = (0, 0, 0.476) (Szytuła et al., 1990) (see Figure 4.6). At 4.2 K, the observed small values of ordered moments on Pr^{3+} were also explained by the CEF effect (Malaman et al., 1992).

At 4.2 K both $NdFe_2Si_2$ and $NdFe_2Ge_2$ exhibit a colinear AFII-type antiferromagnetic ordering (Figure 4.6) that is stable up to the respective Néel temperatures (Pinto and Shaked, 1973; Szytuła et al., 1983). The other magnetic parameters are collected in Table 4.9.

The temperature dependence of the magnetic susceptibility and electrical resistivity in $PrCo_2Si_2$ indicates three transition temperatures: T_1 = 9 K, T_2 = 17 K, and T_3 = 30 K (Kolenda et al., 1982). They are reflected on the specific-heat curve (Takeda et al., 1992). Neutron-diffraction data show that at 4.2 K, $PrCo_2Si_2$ exhibits the colinear AFI-type antiferromagnetic ordering

TABLE 4.9
Magnetic Data for NdT_2Si_2 and NdT_2Ge_2 Compounds

T	Type of magnetic ordering	$T_{C,N}(K)$	$\Theta_p(K)$	$\mu_{eff}(\mu_B)$	$\mu_R(\mu_B)$	Ref.
		NdT_2Si_2				
Fe	AF	15.6			3.1	1
Co	AF	32			3.27	3, 4
Ni	AF	6.2	−4	3.73	2.6	7
Cu			−7	3.7		9
Ru	AF, F	26, 10	−6	3.56	3.23	10, 11
Rh	AF	55	7	3.59	3.25	14
Os	F	2.9	19	3.56		16
Pt			−4.5	3.55		17
Au			−6.1	3.62		18
		NdT_2Ge_2				
Fe	AF	13			3.38	2
Co	AF	26	−5.5	4.09	3.20	5, 6
Ni	AF	20	−10	3.69	3.22	8
Ru	AF, F	21, 7	44	3.44	3.64	12, 13
Rh	AF	46	0	3.82	3.02	12, 15

References: 1. Pinto and Shaked, 1973; 2. Szytuła et al., 1983; 3. Leciejewicz et al., 1983; 4. Shigeoka et al., 1988; 5. Kolenda et al., 1982; 6. André et al., 1990; 7. Barandiaran et al., 1987; 8. Szytuła et al., 1988; 9. Kotsanidis and Yakinthos, 1981; 10. Felner and Nowik, 1984; 11. Chevalier et al., 1985; 12. Felner and Nowik, 1985; 13. Szytuła et al., 1987; 14. Szytuła et al., 1984; 15. Venturini et al., 1988; 16. Hiebl et al., 1983a; 17. Hiebl and Rogl, 1985; 18. Felner, 1975.

scheme (Leciejewicz et al., 1983); however, when the temperature rises, the integral intensity of the magnetic superlattice neutron peaks changes, resulting in a series of magnetic structures with wave vectors $(0, 0, k_z)$ modulated along the c axis (Shigeoka et al., 1989).

High-field magnetization measurements were performed up to 300 kOe on a single $PrCo_2Si_2$ crystal at 4.2 K (Shigeoka et al., 1989). The magnetization along the c axis increased in four steps — at the critical fields of H_{c1} = 12, H_{c2} = 38, H_{c3} = 67, and H_{c4} = 122 kOe — while the a axis magnetization was very small and depended on the applied field H. Above H_{c4} = 122 kOe, the c axis magnetization reached the saturation value of 3.20 μ_B/Pr atom (Shigeoka et al., 1989a).

On the basis of the magnetic and neutron diffraction data, the (H, T) magnetic phase diagram for $PrCo_2Si_2$ was constructed. It is shown in Figure 4.7.

Neutron diffraction measurements carried out in the presence of pulsed magnetic fields indicate three incommensurate magnetic structures with the

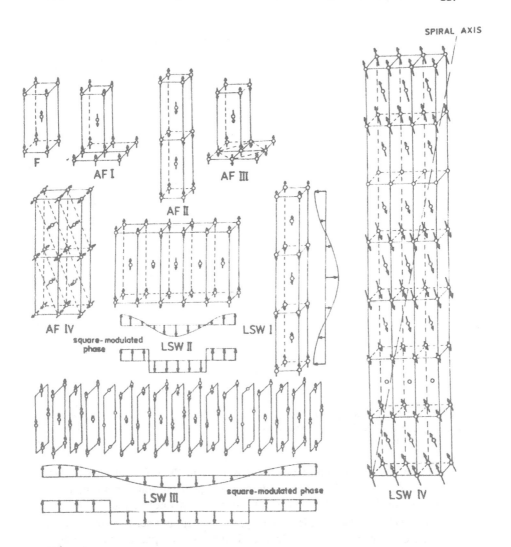

SPIRAL AXIS

FIGURE 4.6. Magnetic structure of the rare earth sublattice in various RT_2X_2 compounds.

wave vector $k = (0, 0, k_z)$. The values of k_z change from $k_z = 13/14$ (H_{c1} $< H < H_{c2}$) and $11/14$ ($H_{c2} < H < H_{c3}$) to $7/9$ ($H_{c3} < H < H_{c4}$). Below H_{c4}, ferromagnetic ordering was detected. The μSR experiment performed with a pulsed magnetic field detected only two incommensurate phases (Nojiri et al., 1992).

Similar properties were also observed in $NdCo_2Si_2$. Three antiferromagnetic phases were found below 32 K. A simple colinear antiferromagnetic structure AFI is present in the temperature range 0 to 15 K (Kolenda et al., 1982). Square-wave structures with the propagation vectors $k = (0, 0, 0.928)$ and $(0, 0, 0.785)$ appear in the ranges 15 K $< T <$ 24 K and 24 K $< T \times$ 32 K, respectively (Shigeoka et al., 1988).

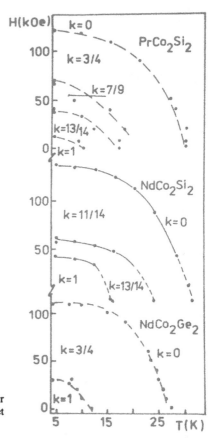

FIGURE 4.7. Magnetic phase diagrams for PrCo$_2$Si$_2$, NdCo$_2$Si$_2$, and NdCo$_2$Ge$_2$ (Vinokurova et al., 1991; Ivanov et al., 1992a).

Different magnetic properties are observed in PrCo$_2$Ge$_2$ and NdCo$_2$Ge$_2$ compounds. In PrCo$_2$Ge$_2$ a sine-modulated magnetic structure with the propagation vector k = (0, 0, 0.73) sets up in the temperature range 4.2 to 28.5 K (Szytuła et al., 1980), while in NdCo$_2$Ge$_2$ the magnetometric measurements show a Néel temperature T_N = 28 K and an additional phase transition at T_1 = 10 K (Kolenda et al., 1982). The specific-heat and resistivity data for NdCo$_2$Ge$_2$ gave only one transition at 28 K (Ślaski et al., 1988). The neutron diffraction data indicate that NdCo$_2$Ge$_2$ has a colinear antiferromagnetic AFI-type structure below T_t = 12 K and above T_t, a modulated structure with the propagation vector k = (0, 0, 0.739) (André et al., 1990).

Magnetic measurements at high magnetic fields and T = 4.2 K demonstrate the presence of a three step transition in NdCo$_2$Si$_2$ (Ivanov et al., 1992a) and a two-step transition in NdCo$_2$Ge$_2$ (Vinokurova et al., 1991). The rise of temperature causes a change in the value of the critical field. The magnetic phase diagrams determined with these data are presented in Figure 4.7.

Below $T_N = 18$ K, $PrNi_2Si_2$ exhibits a modulated antiferromagnetic structure with $k = (0, 0, 0.87)$ (Barandiaran et al., 1986b). Pr moments are along the c axis, with a maximum value of 2.6(1) μ_B at 5.5 K. Magnetic measurements performed on a single crystal sample reveal a large uniaxial magnetocrystalline anisotropy (Blanco et al., 1992).

Below $T_N = 6.2$ K, $NdNi_2Si_2$ shows a modulated magnetic structure similar to that observed in $PrNi_2Si_2$, with the Nd moment perpendicular to the c axis (Barandiaran et al., 1987).

Below $T_N = 20$ K, $NdNi_2Ge_2$ shows a modulated antiferromagnetic structure with the propagation vector $k = (0, 0, 0.805)$ and magnetic moments parallel to the c axis (Szytuła et al., 1988c). At $T = 4.2$ K, the magnetization is a linear function of the magnetic field up to $H_c = 6.0$ T. In this field, it changes to the ferromagnetic state. The magnetic moment at $T = 4.2$ K and $H = 140$ kOe amounts to 2.3 μ_B (Ivanov et al., 1992b).

At $T = 4.2$ K, $PrCu_2Si_2$ and $PrCu_2Ge_2$ exhibit a simple antiferromagnetic AFI-type ordering (Szytuła et al., 1983). Magnetization curves obtained at 4.2 K indicate a one-step metamagnetic process with a critical field of 35 kOe in $PrCu_2Si_2$ and 15 kOe in $PrCu_2Ge_2$. An increase in temperature leads to an increase in the value of the critical field.

Specific heat, electrical resistivity, and magnetic susceptibility data recently obtained for $PrCu_2Ge_2$ indicate a discontinuous phase transition at 4.2 K (Sampathkumaran et al., 1992a).

The results of magnetic susceptibility, electrical resistivity, and heat capacity measurements performed for the pseudoternary solid solutions $Pr_{1-x}R_xCu_2Si_2$ (R = Y, La, Gd), $PrCu_{2-x}Ni_xSi_2$, and $PrCu_2Si_{2-x}Ge_x$ (Das et al., 1992a, Sampathkumaran et al., 1991, 1992b) revealed changes of the Néel temperature. In the $Pr_{1-x}Gd_xCu_2Si_2$ system, the heavy fermion-like behavior has been detected (Sampathkumaran et al., 1992b).

$NdCu_2Si_2$ was found to remain paramagnetic below 4.2 K (Kotsanidis and Yakinthos, 1981).

$PrRu_2Si_2$ (Hiebl et al., 1983a) and $PrRu_2Ge_2$ (Felner and Nowik, 1985) are both ferromagnets with Curie temperatures of 17.8 and 18 K, respectively. A neutron diffraction study indicated that the magnetic moments of Pr atoms align along the c axis (Ślaski et al., 1984). The magnetization curve determined for a single crystal of $PrRu_2Si_2$ indicated a giant magnetic anisotropy with easy magnetization along the c axis, anisotropy constant $K = 5.9 \times 10^8$ erg/cm³, and anisotropy field $H_a = 4000$ kOe (Shigeoka et al., 1992).

$NdRu_2Si_2$ and $NdRu_2Ge_2$ exhibit two magnetic transitions: at 10 and 26 K in $NdRu_2Si_2$ (Felner and Nowik et al., 1984) and at 7 and 21 K in $NdRu_2Ge_2$ (Felner and Nowik et al., 1985).

Neutron diffraction data indicate that $NdRu_2Si_2$ exhibits a complicated magnetic structure: below $T_N = 26$ K, it develops a sine-wave modulation $k = (0.13, 0.13, 0)$ of the magnetic moments, with an amplitude of 3.23 μ_B/Nd atom. A squaring of the magnetic structure occurs at about 15 K. At

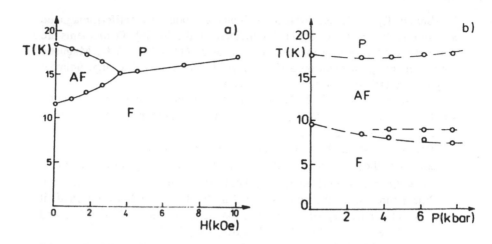

FIGURE 4.8. Magnetic phase diagrams (a) (H, T) and (b) (p, T) for NdRu$_2$Ge$_2$ (Ivanov et al., 1992b,c).

temperatures T $<$ 10 K, a ferromagnetic order sets up, with the moments always along the c axis (Chevalier et al., 1985a). Neutron diffraction studies performed on NdRu$_2$Ge$_2$ showed that, in the temperature range from 17 to 10 K, this compound exhibits two types of sine-wave-modulated magnetic structures, having the wave vectors k = (0.19, 0.05, 0.125) and k = (0.12, 0.12, 0) with the amplitude of the moment along the c axis. At 10 K, a first-order transition to a ferromagnetic state occurs with a moment of 3.64(13) μ_B aligned along the tetragonal axis (Szytuła et al., 1987).

Heat capacity and electrical resistivity studies give evidence of two magnetic phase transitions at 10 and 23 K in NdRu$_2$Si$_2$ and at 10 and 17 K in NdRu$_2$Ge$_2$ (Ślaski et al., 1988).

Magnetization curves of NdRu$_2$Si$_2$ measured along the c axis at various temperatures also indicate the complicated character of its magnetic properties. At 4.2 K, the magnetization increases rapidly at a very low field and almost reaches saturation at 0.3 kOe. The saturation moment is 2.8 μ_B per formula unit. In the temperature range 10 K $<$ T $<$ 15 K, the magnetization process exhibits a one-metamagnetic transition at a very low critical field (H$_c$ ~ 3 kOe at 12 K). Above T = 15 K, the two-step magnetization process is observed.

The critical fields are H$_{c1}$ = 5 kOe and H$_{c2}$ = 7 kOe at 16 K (Shigeoka, 1990). The (H, T) and (p, T) phase diagrams of NdRu$_2$Ge$_2$ are displayed in Figure 4.8. The low-field magnetic data indicate a ferromagnetic ordering below T$_t$ = 10 K and an antiferromagnetic order in the region T$_t$ $<$ T $<$ T$_N$ = 17 K. The values of T$_t$ increase as the magnetic field becomes larger, but T$_N$ falls. At a magnetic field of 3.45 kOe, a threefold critical point appears (Ivanov et al., 1992b). T$_N$ increases slowly, while T$_t$ decreases, when external

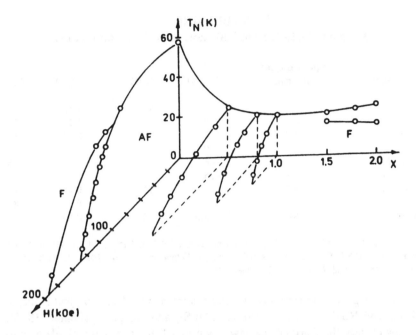

FIGURE 4.9. Magnetic phase diagrams (H, T, x) for $Nd(Rh_{1-x}Ru_x)_2Si_2$ (Ivanov et al., 1993).

pressure is applied. At p > 4.3 kbar, a new magnetic phase was observed (Ivanov et al., 1992c).

$NdRh_2Si_2$ and $NdRh_2Ge_2$ are antiferromagnets with Néel temperatures of 55 K (Szytuła et al., 1984) and 46 K (Felner and Nowik, 1985), respectively. Both compounds exhibit a colinear antiferromagnetic AFI-type structure (Szytuła et al., 1984; Chevalier et al., 1985a; Venturini et al., 1988).

A neutron diffraction study of the $Nd(Rh_{1-x}Ru_x)_2Si_2$ system at 4.2 K revealed the following magnetic structures:

- At x = 0.25, the colinear antiferromagnetic AFI structure and a modulated structure with the propagation vector k = (0, 0, 0.838) coexist.
- At x = 0.4, 0.5, and 0.75, modulated structures appear.
- At x = 0.9, the ferromagnetic and modulated structures coexist.

For all samples, the observed magnetic moment is the same as the free-ion moment $\mu = 3.27 \mu_B$ (Jaworska et al., 1988).

The magnetic phase diagram of $Nd(Rh_{1-x}Ru_x)_2Si_2$ in magnetic fields up to 140 kOe is displayed in Figure 4.9. A two-step metamagnetic process is observed in $NdRh_2Si_2$. As the Ru concentration x (for x > 0.5) increases the magnetization curves change from two step to one step (Ivanov et al., 1993a).

ROs_2Si_2 compounds with R = Pr Nd are ferromagnets with Curie temperatures of 12.6 and 29 K, respectively (Hiebl et al., 1983b).

TABLE 4.10
Magnetic Data for SmT_2Si_2 and SmT_2Ge_2 Compounds

Compound	Type of magnetic ordering	$T_{C,N}(K)$	$\Theta_p(K)$	$\mu_{eff}(\mu_B)$	Ref.
$SmCo_2Ge_2$			-43.8	1.12	1
$SmCu_2Si_2$	AF	9		1.12	2
$SmRu_2Si_2$	AF	7	14.4	0.54	3
$SmRu_2Ge_2$	F	10			4
$SmRh_2Si_2$	AF	46			5, 6
$SmRh_2Ge_2$	AF	35			4
$SmOs_2Si_2$	AF	5	12	0.47	7
$SmPt_2Si_2$			0	0.7	8
$SmAu_2Si_2$	AF	15.9	2.5	0.63	9

References: 1. McCall et al., 1973; 2. Schlabitz et al., 1982; 3. Hiebl et al., 1983b; 4. Felner and Nowik, 1985; 5. Felner and Nowik, 1983; 6. Felner and Nowik, 1984; 7. Hiebl et al., 1983b; 8. Hiebl and Rogl, 1985; 9. Felner, 1975.

The magnetic susceptibility vs. temperature curves of $PrPt_2Si_2$ and $NdPt_2Si_2$ (Hiebl and Rogl, 1985), as well as $PrAu_2Si_2$ and $NdAu_2Si_2$ (Felner, 1975), follow closely the Curie-Weiss law down to 1.8 K. The magnetic data summarized in Tables 4.8 and 4.9 indicate that, in the paramagnetic state, the effective magnetic moments are near the free Pr^{3+} or Nd^{3+} values; however, in the magnetic ordering state, they are smaller. These results demonstrate the influence of the crystal electric field on the magnetic properties.

4.6.2. SmT_2X_2 PHASES

Only a few SmT_2X_2 compounds have been studied to date. The reciprocal magnetic susceptibility of $SmCo_2Ge_2$ obeys the Curie-Weiss law down to 4.2 K (McCall et al., 1973), while $SmCu_2Si_2$ orders antiferromagnetically below $T_N = 9$ K (Schlabitz et al., 1982).

Antiferromagnetism has been observed in $SmRu_2Si_2$ below $T_N = 7$ K (Hiebl et al., 1983), whereas $SmRu_2Ge_2$ is ferromagnetic below $T_C = 10$ K (Felner and Nowik, 1985).

$SmRh_2Si_2$ and $SmRh_2Ge_2$ order antiferromagnetically below 46 K (Felner and Nowik, 1983, 1984) and 7 K (Felner and Nowik, 1985), respectively. Also, $SmOs_2Si_2$ (Hiebl et al., 1983b) and $SmAu_2Si_2$ (Felner, 1975) are antiferromagnets with $T_N = 5$ and 15.9 K, respectively. Magnetic data for $SmPt_2Si_2$ ($\mu_{eff} = 0.7 \mu_B$) show that the Sm ion behaves as an ideal Van Vleck Sm^{3+} ion with the ground state $J = 7/2$ (Hiebl and Rogl, 1985).

The available magnetic data for SmT_2X_2 systems are listed in Table 4.10

4.6.3. GdT_2X_2 PHASES

The studies of compounds with Gd are of great interest since, in this element, the 4f shell is half-filled and its ground state, $^8S_{7/2}$, is insensitive to

TABLE 4.11
Magnetic Data for GdT_2Si_2 and GdT_2Ge_2 Compounds

T	Type of magnetic ordering	$T_{C,N}(K)$	$\Theta_p(K)$	$\mu_{eff}(\mu_B)$	Ref.
		GdT_2Si_2			
Fe	AF	8.4	3	8.18	1, 2
Co	AF	43	−40	8.3	5
Ni	AF	15	−1	7.97	7
Cu	AF	12.5	−21	8.03	1
Ru	AF	44	39	7.96	1, 9
Rh	AF	106	−2.5	8.17	11, 12
Pd	AF	19	−45.5	8.05	11, 12
Ag	AF	10	−29.4	7.98	11, 12
Os	AF	26	27	8.04	11, 12
Ir	AF	81	−4	7.96	11, 12
Pt	AF	9.7	−5	8.14	11, 12
Au	AF	12.5	−35	7.96	11, 12
		GdT_2Ge_2			
Fe	AF	11			3, 4
Co	AF	40	−39.5	6.96	6
Ni	AF	22	−19	8.6	4
Cu	AF	14	−24	8.0	4, 8
Ru	AF	32	39	8.2	10
Rh	AF	90	20	7.6	10

References: 1. Łątka et al., 1979; 2. Noakes et al., 1983; 3. Malik et al., 1976; 4. Felner and Nowik, 1978; 5. Kolenda et al., 1982; 6. McCall et al., 1973; 7. Yakinthos and Ikonomou, 1980; 8. De Vries et al., 1985; 9. Ślaski et al., 1984; 10. Felner and Nowik, 1985; 11. Łątka, 1989; 12. Czjzek et al., 1989.

CEF effects. Thus, the magnetic properties of compounds containing Gd are determined exclusively by the nature of exchange interactions. The magnetic data of the GdT_2X_2 systems are listed in Table 4.11. It follows from this table that all GdT_2X_2 are antiferromagnets. The GdT_2Si_2 (Łątka, 1989; Czjzek et al., 1989) and GdT_2Ge_2 (Felner and Nowik, 1978) series show an oscillatory character of the Néel temperature, which is caused by the increase in the number of d electrons in T elements. The maximum value is observed for the compounds with Co, Rh, or Ir. These results suggest that, while the magnetic interactions may be discussed in terms of the RKKY model, the number of free electrons donated to the conduction band depends on the number of d electrons.

In some of these compounds, on additional phase transition below T_N has been observed, e.g., in $GdRh_2X_2$ (Felner and Nowik, 1983, 1984, 1985), $GdPt_2Si_2$ (Łątka, 1989). The complicated magnetic properties are displayed

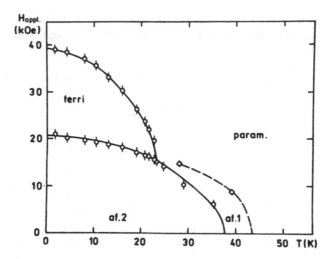

FIGURE 4.10. Magnetic phase diagram of $GdRu_2Si_2$. Circles and solid lines mark phase boundaries derived from magnetic data (Łątka, 1989; Czjzek et al., 1989). Diamonds and the broken line mark results reported by Buschow and De Mooij (1986).

by $GdRu_2Si_2$. The magnetic phase diagram of $GdRu_2Si_2$ is shown in Figure 4.10. Two antiferromagnetic phases and one ferromagnetic phase have been detected (Łątka 1989; Czjzek et al., 1989).

The results of ^{155}Gd Mössbauer measurements provide the values of the hyperfine parameters and angle which define the direction of the magnetic moment. The data presented in Table 4.11 indicate that in the majority of GdT_2X_2 compounds, the easy axis of magnetization is in the basal plane. Only in $GdRu_2Si_2$ and $GdOs_2Si_2$ is a different orientation observed (Łątka 1989; Czjzek et al., 1989).

These results have been verified by neutron diffraction measurements. The experiments were performed only for $GdNi_2Si_2$, $GdCu_2Si_2$ (Barandiaran et al., 1988), and $GdPt_2Si_2$ (Gignoux et al., 1991). All exhibit antiferromagnetic structures described by the following wave vectors: (0.207, 0, 0.903) in $GdNi_2Si_2$, (1/2, 0, 1/2) in $GdCu_2Si_2$, and (1/3, 1/3, 1/2) in $GdPt_2Si_2$. The Gd moment are aligned in the basal planes, in agreement with the results of the Mössbauer experiments.

The ^{57}Fe Mössbauer spectra of $GdFe_2Si_2$ taken at 4.2 K display Zeeman hyperfine interactions. The hyperfine magnetic field $H_{hf} = 10$ kOe is transferred from Gd spins (Łątka, 1989). Similar effects are observed in the case of other RFe_2X_2 compounds (Umarji et al., 1983; Noakes et al., 1983; Görlich et al., 1982).

The ^{61}Ni Mössbauer spectra in $GdNi_2Si_2$ show no hyperfine splitting down to 1.2 K, which confirms that the 3d moment is zero (Łątka et al., 1979).

TABLE 4.12
Magnetic Data for TbT_2Si_2 and TbT_2Ge_2 Compounds

T	Type of magnetic ordering	$T_{C,N}(K)$	$\Theta_p(K)$	$\mu_{eff}(\mu_B)$	$\mu_R(\mu_B)$	Ref.
			TbT_2Si_2			
Fe	AF	8.4	3	8.18	7.9	1, 2
Co	AF	45	−73	10.2	8.9	5, 6
Ni	AF	15	16	9.93	8.8	7
Cu	AF	12	−12	9.35	8.5	10, 11
Ru	AF	58	10	9.65	8.77	14, 15
Rh	AF	92	−26	8.4	8.92	18, 19
Pd	AF	20	−1	9.8	9.3	20, 21
Os	AF	41	20	10.5	8.85	23
Ir P	AF	13	13	9.78		24
I	AF	80	42	9.97	8.1	18, 24
Pt	AF	6.5	58.5	9.55		25
Au	AF	14.5	−14	9.78	8.7	26
			TbT_2Ge_2			
Fe	AF	7.5				3, 4
Co	AF	32	−22	10.3	8.3	5
Ni	AF	17				8, 9
Cu	AF	15	−21.5	9.7	8.48	12, 13
Ru	AF	30	−3	10.9	9.1	16, 17
Rh	AF	77	−12	10.3	9.4	16, 17
Pd	AF	4.2			8.4	22

References: 1. Łątka et al., 1979; 2. Bażela et al., 1988; 3. Malik et al., 1976; 4. Felner and Nowik, 1978; 5. Leciejewicz et al., 1982; 6. Yakinthos et al., 1984; 7. Barandiaran et al., 1987; 8. Szytuła, 1992; 9. Pinto et al., 1985; 10. Oesterreicher, 1976; 11. Leciejewicz et al., 1986; 12. Kotsanidis and Yakinthos, 1981; 13. Schobinger-Papamantellos et al., 1984; 14. Hiebl et al., 1983; 15. Ślaski et al., 1984; 16. Felner and Nowik, 1985; 17. Szytuła et al., 1987; 18. Ślaski et al., 1983; 19. Felner and Nowik, 1984; 20. Yakinthos and Gamari-Seale, 1982; 21. Szytuła et al., 1986; 22. Szytuła et al., 1988a; 23. Kolenda et al., 1985; 24. Hirjak et al., 1984; 25. Hiebl and Rogl, 1985; 26. Ohashi et al., 1992.

4.6.4. RT_2X_2 COMPOUNDS WITH HEAVY LANTHANIDES (R = Tb–Tm)

$GdCr_2Si_2$, $TbCr_2Si_2$, and $DyCr_2Si_2$ are antiferromagnets, while those with R = Ho, Er, and Tm remain paramagnetic up to 2 K (Dommann et al., 1988).

Some RFe_2X_2 compounds show antiferromagnetic ordering at low temperatures (see Tables 4.12 to 4.16). Neutron diffraction studies indicate that $TbFe_2Si_2$ (Bażela et al., 1988), $DyFe_2Si_2$ (Bourée-Vigneron et al., 1990), and $HoFe_2Si_2$ (Leciejewicz and Szytuła, 1985a) exhibit rather complex magnetic

TABLE 4.13
Magnetic Data for DyT_2Si_2 and DyT_2Ge_2 Compounds

T	Type of magnetic ordering	$T_{C,N}(K)$	$\Theta_p(K)$	$\mu_{eff}(\mu_B)$	$\mu_R(\mu_B)$	Ref.
		DyT_2Si_2				
Fe	AF	3.8			7.52	1
Co	AF	19.5	−27	10.5	9.4	3–5
Ni	AF	7	−4	10.7		8
Cu	AF	11	−4	10.5	8.3	5, 9
Ru	AF	29	12	11.4	9.33	12
Rh	AF	55	7	10.6	9.9	14–16
Pd	AF	6	−4	9.93	9.6	18
Os	AF	23	101	10.35		19
Ir I	AF	40				20
Pt	AF	7	37	10.59		2, 21
Au	AF	7.5	−3.6	10.6		2, 22
		DyT_2Ge_2				
Fe	AF	6				2
Co	AF	14	−4.7	10.17	9.5	6, 7
Ni	AF	11				2
Cu	AF	8	−15	10.6	8.0	10, 11
Ru	AF	22	−8	10.5		13
Rh	AF	45	32	10.61	9.27	13, 17

References: 1. Bourée-Vigneron et al., 1990; 2. Nowik et al., 1983; 3. Kolenda et al., 1982; 4. Leciejewicz and Szytuła, 1983; 5. Pinto et al., 1985; 6. McCall et al., 1973; 7. Pinto et al., 1985; 8. Yakinthos and Ikonomou, 1980; 9. Kotsanidis and Yakinthos, 1981; 10. Kotsanidis and Yakinthos, 1981a; 11. Kotsanidis et al., 1984; 12. Ślaski et al., 1984; 13. Felner and Nowik, 1985; 14. Felner and Nowik, 1983; 15. Felner and Nowik, 1984; 16. Malamud et al., 1984; 17. Venturini et al., 1988; 18. Bażela et al., 1991; 19. Hiebl et al., 1983a; 20. Tomala et al., 1992; 21. Hiebl and Rogl, 1985; 22. Felner, 1975.

structures which could be described as helicoidal alignments of magnetic moments with a propagation vector $k = (0, k_y, k_z)$. The magnetic ordering is represented by a static linear transverse spin wave displayed in Figure 4.6. (LSW 4). The magnetic moment localized on rare-earth atoms is parallel to the c axis in $TbFe_2Si_2$ and $DyFe_2Si_2$ and from angle 15(5)° with the c axis in $HoFe_2Si_2$.

In $ErFe_2Si_2$, a colinear antiferromagnetic structure of the $+ - - +$ type was found. The magnetic moment of Er atoms is perpendicular to the c axis (Leciejewicz et al., 1984).

$TbFe_2Ge_2$ is antiferromagnetic below $T_N = 7.5\,K$ with an incommensurate magnetic structure (Pinto et al., 1985).

All RCo_2X_2 compounds order antiferromagnetically (see Tables 4.12 to 4.16). The temperature dependence of their magnetic susceptibility shows

127

TABLE 4.14
Magnetic Data for HoT_2Si_2 and HoT_2Ge_2 Compounds

T	Type of magnetic ordering	$T_{C,N}(K)$	$\Theta_p(K)$	$\mu_{eff}(\mu_B)$	$\mu_R(\mu_B)$	Ref.
			HoT_2Si_2			
Fe	AF	2.3			7.4	1
Co	AF	11.2	−10	10.5	10.0	2, 3
Ni	AF	4.3	0	10.58	9.6	5
Cu	AF	8	−3	10.3	8.2	7, 8
Ru	AF	19	11.6	11.1	9.25	11
Rh	AF	27	3	9.3	8.8	14–16
Pd	AF	21	−5	10.5	7.7	17, 18
Os	AF	19	6.5	10.9	9.9	19
Pt			11	10.56		20
Au	F	14.8	−2.1	10.6		21
			HoT_2Ge_2			
Co	AF	8	−10	10.0	8.64	2, 4
Ni	AF	6			8.0	6
Cu	AF	6.4	−6	10.6	6.5	9, 10
Ru	AF	3	−64	11.1	6.6	12, 13
Rh	AF	40	−27	10.8		12

References: 1. Leciejewicz and Szytuła, 1985a; 2. Kolenda et al., 1982; 3. Leciejewicz et al., 1983; 4. Szytuła et al., 1980; 5. Baranadiaran et al., 1987; 6. Pinto et al., 1985; 7. Kotsanidis and Yakinthos, 1981; 8. Leciejewicz et al., 1986; 9. Kotsanidis and Yakinthos, 1981a; 10. Schobinger-Papamantellos et al., 1984; 11. Ślaski et al., 1984; 12. Felner and Nowik, 1985; 13. Yakinthos and Roudaut, 1987; 14. Felner and Nowik, 1983; 15. Felner and Nowik, 1984; 16. Ślaski et al., 1983; 17. Yakinthos and Gamari-Seale, 1982; 18. Leciejewicz and Szytuła, 1985; 19. Kolenda et al., 1985; 20. Hiebl and Rogl, 1985; 21. Felner, 1975.

characteristic maxima connected with the Néel temperatures. Above them, the reciprocal magnetic susceptibility obeys the Curie-Weiss law.

Neutron diffraction investigations were performed for the majority of RCo_2Si_2 and RCo_2Ge_2 compounds (Szytuła et al., 1980; Leciejewicz and Szytuła, 1983; Yakinthos et al., 1983, 1984; Nguyen et al., 1983; Schobinger-Papamantellos et al., 1983; Pinto et al. 1979, 1983, 1985). At low temperatures, indexable reflections, assuming a magnetic unit cell of the same dimensions as the crystallographic one, were observed (see Figure 4.11). A simple antiferromagnetic structure was deduced. Magnetic moments localized on the lanthanide atoms form ferromagnetic basal planes. The magnetic coupling between adjacent planes is antiferromagnetic. The magnetic structure could be displayed as a piling up of the ferromagnetic sheets along the c axis with the sequence + − + −, etc. This type of magnetic structure is usually denoted as an AFI type (see Figure 4.6).

TABLE 4.15
Magnetic Data for ErT_2Si_2 and ErT_2Ge_2 Compounds

T	Type of magnetic ordering	$T_{C,N}(K)$	$\Theta_p(K)$	$\mu_{eff}(\mu_B)$	$\mu_R(\mu_B)$	Ref.
			ErT_2Si_2			
Fe	AF	2.9			7.4	1
Co	AF	6	−9	10.4	8.7	2, 3
Ni	AF	3	4	9.58	6.4	5
Cu	AF	1.6		9.49		6
Ru	AF	5	2	9.48	6.6	9
Rh	AF	12.8	−2	9.45	7.74	10
Pd	AF	4	0	9.66	8.8	11
Os	AF	4	−16	9.58	8.24	12
Ir			−27	6.8		13
Pt			26	9.57		14
Au	F	40	13	9.0		15
			ErT_2Ge_2			
Co	AF	4.2	−8	8.7	7.32	4
Cu	AF	1.9	−4.5	10.7	8.0	7, 8

References: 1. Leciejewicz et al., 1984; 2. Yakinthos et al., 1980; 3. Leciejewicz et al., 1983; 4. Leciejewicz et al., 1983a; 5. Barandiaran et al., 1987; 6. Schlabitz et al., 1982; 7. Kotsanidis and Yakinthos, 1981a; 8. Yakinthos, 1985; 9. Ślaski et al., 1984; 10. Szytuła et al., 1986; 11. Bażela et al., 1991; 12. Kolenda et al., 1985; 13. Ślaski and Szytuła, 1982; 14. Hiebel and Rogl, 1985; 15. Felner, 1975.

In the case of RCo_2X_2 compounds where R is Tb, Dy, and Ho, the absence of (001) reflections indicates that the magnetic moments are parallel to the c axis. In the case of R = Er and Tm compounds, the presence of magnetic reflections (001) indicates that the magnetic moments on Er or Tm atoms lie in the basal plane, i.e., perpendicular to the tetragonal axis (Leciejewicz et al., 1983; Leciejewicz and Szytuła, 1983). Slightly different parameters were obtained from an independent neutron diffraction study by Yakinthos et al. (1983). The resulting magnetic moment of Er is 6.75 μ_B at an angle of 56.2° with the c axis.

The magnetization curves for $TbCo_2Si_2$, $DyCo_2Si_2$, $TbCo_2Ge_2$, and $DyCo_2Ge_2$ show a two-step metamagnetic character. The values of critical fields were determined from the field dependence of the differential magnetization dM/dH. The temperature dependencies of the transition fields are shown in Figure 4.12. The magnetic phase diagrams were determined using these data. Below H_{c1}, the colinear antiferromagnetic structure AFI is observed. In the intermediate region ($H_{c1} < H < H_{c2}$) either a ferromagnetic order with the sequence + + + − (Iwata et al., 1990) or a modulated structure

TABLE 4.16

Magnetic Data for TmT$_2$Si$_2$ and TmT$_2$Ge$_2$ Compounds

T	Type of magnetic ordering	$T_{C,N}(K)$	$\Theta_p(K)$	$\mu_{eff}(\mu_B)$	$\mu_R(\mu_B)$	Ref.
		TmT$_2$Si$_2$				
Co	AF	3	−8	6.9	6.2	1, 2
Ni	AF	1.7	6.5	7.64	5.6	4
Cu	AF	−2.8	−2	7.4	3.2	5, 6
Ru	F	1	65.7	7.38		8
Rh	AF	4.2			4.2	10
Os	F	1	64	7.34		11
Pt			65	7.32		12
		TmT$_2$Ge$_2$				
Fe			−0.8	5.89		3
Cu			−2	7.5		7
Ru			−6	7.1		9

References: 1. Kolenda et al., 1982; 2. Leciejewicz and Szytuła, 1983; 3. McCall et al., 1973; 4. Barandiaran et al., 1987; 5. Kotsanidis and Yakinthos, 1981; 6. Kozłowski et al., 1987; 7. Kotsanidis and Yakinthos, 1981a; 8. Hiebl et al., 1983; 9. Felner and Nowik, 1985; 10. Yakinthos, 1986a; 11. Hiebl et al., 1983a; 12. Hiebl and Rogl, 1985.

(Blanco et al., 1991), have been observed. For $H > H_{c2}$, ferromagnetic ordering becomes stable. The variation of the magnetization as a function of the external magnetic fields for other RCo$_2$X$_2$ compounds does not exhibit any anomaly (Vinokurova et al., 1991).

Magnetic measurements carried out on RNi$_2$Si$_2$ (R = Tb–Tm) compounds indicate that all of them are antiferromagnets at low temperatures (see Tables 4.12 to 4.16) (Yakinthos and Ikonomou, 1980; Barandiaran et al., 1987). All show incommensurate antiferromagnetic ordering. In terbium-containing phases, the magnetic moments are parallel to the c axis, while the magnetic moments of R = Ho, Er, and Tm are aligned in the basal plane (Barandiaran et al., 1987).

In the case of TbNi$_2$Si$_2$, the temperature dependence of the magnetic susceptibility and resistivity indicates that the Néel transition occurs at $T_N = 15$ K and an additional phase transition at $T_t = 9$ K. Neutron diffraction showed that below 9 K, the colinear antiferromagnetic structure is of the AFIII type. Above this temperature, an incommensurate structure is present with wave vector k = (1/2 + τ, 1/2 − τ, 0), $\tau = 0.074$ (Barandiaran et al., 1987).

The specific-heat measurements indicate only a transition at T_N. The magnetization curve obtained at 4.2 K for a single crystal sample shows a three-step metamagnetic process along the c axis (Blanco et al., 1992).

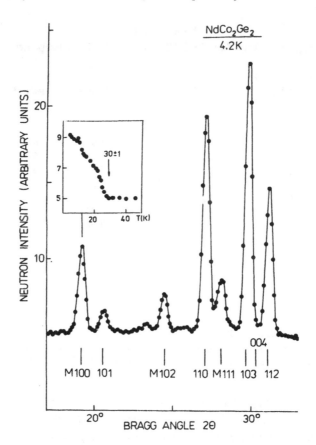

FIGURE 4.11. Neutron diffraction pattern of $NdCo_2Ge_2$ taken at 4.2 K. The temperature dependence of the M(100) reflection is shown in the inset (Leciejewicz et al., 1983).

Neutron diffraction experiments performed in the presence of a magnetic field lead to the magnetic phase diagram of $TbNi_2Si_2$ presented in Figure 4.13. The following magnetic phases have been found: I — modulated, II — colinear AFIII, III — modulated with the wave vector $k_3 = (1/2 - 1, 1/2 + 1, 0)$, and IV — modulated with the k_3 wave vector and $k_4 = (0.78, 0, 0)$ (Blanco et al., 1991).

The low-temperature (at 4.2 and 2.1 K) neutron diffraction patterns obtained for $TbNi_2Ge_2$ and $HoNi_2Ge_2$ show incommensurate structures. The magnetic reflections yield two transition temperatures in each compound: at 4.2 and 3.1 K in $HoNi_2Si_2$ and at 16 and 9 K in $TbNi_2Ge_2$. In $HoNi_2Ge_2$, the incommensurate magnetic structure at 21 K is consistent with a spiral described by the propagation vector $k = (0, 0, 0.76)$ and a turning axis along this propagation vector. The magnetic moment is perpendicular to the c axis (Pinto et al., 1985). New neutron diffraction data indicate that $TbNi_2Ge_2$ exhibits two magnetic phase transition temperatures: at $T_t = 10.25$ K and

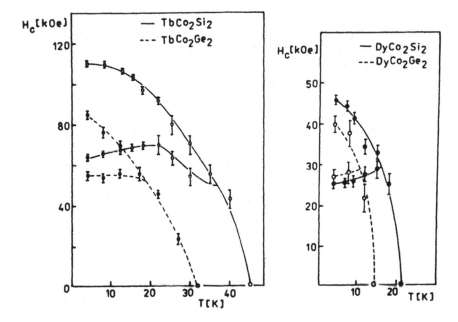

FIGURE 4.12. Temperature dependencies of the transition fields for TbCo$_2$Si$_2$, TbCo$_2$Ge$_2$, DyCo$_2$Si$_2$, and DyCo$_2$Ge$_2$ (Vinokurova et al., 1992).

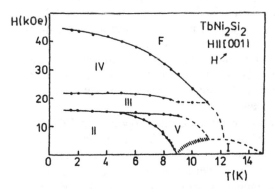

FIGURE 4.13. Magnetic phase diagram of TbNi$_2$Si$_2$ along the c axis. Phase F is induced ferromagnetic, phase I is amplitude modulated, phase II is simple antiferromagnetic, phase III is long-period commensurate, phase IV is amplitude modulated, and zone V is a complex magnetic phase (Blanco et al., 1991).

$T_N = 17$ K. The magnetic reflections in the temperature range $T_t < T < T_N$ could be indexed with the propagation vector k = (00, 1 − k$_z$) where k$_z$ = 1/4. Below T_t, a square-modulated structure is observed (Bourée-Vigneron, 1991).

The temperature dependence of the magnetic susceptibility, the resistivity, and the magnetostriction performed for the RCu$_2$Si$_2$ compounds (R = Tb–Tm) indicate that all order antiferromagnetically at low temperatures (Schlabitz et al., 1982).

The saturated magnetostriction and Δ(c/a) anomalies (Schlabitz et al., 1982) imply that the axis of the magnetic moment is perpendicular to the c axis in Pr, Nd, Tb, Dy, and Ho, but parallel to it in Sm, Er, and Tm.

The electrical resistivity measurements carried out between 1.3 and 300 K for RCu_2Si_2 (R = Tb–Tm) (Cattano and Wohlleben, 1981) indicate that the disorder resistivities and the Néel temperatures deviate strongly from the De Gennes factor.

The temperature dependence of the specific heat of $TmCu_2Si_2$ gives a magnetic ordering temperature at about 2.8 K and shows that the Schottky anomaly corresponds to the two lowest singlet levels. The separation of these two levels amounts to about 7 K, while the distance to the higher excited levels is approximately 10 times larger (Kozłowski, 1986).

The magnetic susceptibility and specific-heat measurements performed in the temperature range 1 to 100 K for $Tm_{1-x}Lu_xCu_2Si_2$ (x = 0, 0.025, 0.05, 0.1, 0.25, 0.5, and 1.0) samples indicate that the Néel temperature decreases from T_N = 2.80(5) K when x = 0 to T_N = 2.30(5) K for x = 0.1. Other samples are paramagnetic down to 1 K. The CEF level scheme was also determined (Kozłowski et al., 1987).

$DyCu_2Si_2$ (Pinto et al., 1983), $TbCu_2Si_2$, and $HoCu_2Si_2$ (Pinto et al., 1985; Leciejewicz et al., 1986) were also studied by neutron diffraction, and their magnetic structures were determined at low temperatures. The magnetic ordering may be described as ferromagnetic (101) planes of R^{3+} ions coupled antiferromagnetically with the sequence + − + −. Such a structure requires the doubling of unit cell dimensions along the a and c axes, which results in an orthorhombic unit cell. The magnetic moments are perpendicular to the c axis, but form different angles with the a axis: 23.4(1.3)° in $TbCu_2Si_2$, 11.1(1.1)° in $DyCu_2Si_2$, and 7.5(3.1)° in $HoCu_2Si_2$. This type of magnetic ordering is denoted as the AFIII type. It is displayed in Figure 4.6.

$TmCu_2Si_2$ was studied by Mössbauer spectroscopy (Stewart and Zukrowski, 1982). The CEF-level scheme of the Tm^{3+} ion was determined. The Mössbauer data suggest that in the ordered state below 2.8 K, it is a magnetically ordered 2-singlet system.

Large amounts of data concerning the magnetic properties of RCu_2Ge_2 compounds have been collected (Oesterreicher, 1977a; Kotsanidis and Yakinthos, 1981). The temperature dependence of the reciprocal magnetic susceptibility of RCu_2Ge_2 follows the Curie-Weiss law, with negative paramagnetic Curie temperatures. The compounds with R = Tb, Dy, Ho, Er, and Tm have been found to be antiferromagnetic. Newly published neutron diffraction data (Kotsanidis et al., 1984; Schobinger-Papamantellos et al., 1984; Yakinthos 1985) provided information about the magnetic ordering in $TbCu_2Ge_2$, $DyCu_2Ge_2$, $HoCu_2Ge_2$, and $ErCu_2Ge_2$. They are antiferromagnetics at 4.2 K, except $ErCu_2Ge_2$, which shows antiferromagnetic ordering at 1.9 K. The magnetic unit cell is four times larger than the chemical one. The structure is described by the wave vector k = (1/2, 0, 1/2). The magnetic moments of R^{3+} ions belonging to the same (101) plane of the magnetic unit cell are coupled ferromagnetically, while those belonging to the adjacent (101) planes are coupled antiferromagnetically. The moment directions and their values at low temperatures are as follows:

- In $TbCu_2Ge_2$, $\mu = 8.46(6)$ μ_B. It is aligned along the (111) direction of the tetragonal crystallographic unit cell.
- In $DyCu_2Ge_2$, $\mu = 8$ μ_B. It makes an angle of 30° with the a axis and 70° with the c axis of the tetragonal unit cell.
- In $HoCu_2Ge_2$, $\mu = 6.5(1)$ μ_B. It makes an angle of 80° with the a axis and 81.4° with the c axis.
- In $ErCu_2Ge_2$, $\mu = 8.0(4)$ μ_B. It is aligned along the tetragonal axis.

The magnetic data determined for the RCu_2Si_2 and RCu_2Ge_2 systems are gathered in Tables 4.12 to 4.16.

(Tb, Dy, Ho, Er)Ru_2Si_2 (Hiebl et al., 1983a; Ślaski and Szytuła, 1982; Felner and Nowik, 1984) and (Tb, Dy, Ho)$RuGe_2$ (Felner and Nowik, 1985) systems were reported to be antiferromagnetic at low temperatures. These results were verified by neutron diffraction experiments.

Neutron diffraction data for (Tb, Dy, Ho, Er)Ru_2Si_2 collected at 4.2 K are interpreted in terms of a helicoidal magnetic ordering of a linear transverse wave mode. The static moment propagates along the b axis with a wave vector $k = (0, k_y, 0)$. It is polarized in the direction of the tetragonal c axis in the case of (Tb, Dy, Ho)$RuSi_2$ and is perpendicular to the axis in the case of $ErRu_2Si_2$ (Ślaski et al., 1983, 1984). In $TbRu_2Si_2$ at low temperatures, a squaring of the modulation develops, giving rise to a square-wave modulation at $T = 3.1$ K (Chevalier et al., 1985a) and $T = 4.2$ K (Szytuła et al., 1987).

Magnetic and neutron diffraction measurements performed on a single crystal of $TbRu_2Si_2$ indicate two-step metamagnetism at low temperatures (4.2 K). A magnetic phase with $k = (0, 0.23, 0)$ persists up to $H_{c1} = 21$ kOe. In the range $2.1 < H < 29$ kOe, a ferromagnetic phase with antiferromagnetic components of $k = (0, 0.23, 0)$ and $(0.23, 0.23, 0)$ exists. Above $H_{c2} = 29$ kOe, a ferromagnetic phase with the full Tb moment is stable. The temperature variation of peak intensities implies that squaring is observed up to 40 K (Shigeoka et al., 1992). The temperature dependence of the electrical resistivity and thermopower of $TbRu_2Si_2$ shows the Néel temperature at 59 K. The increase of $d\rho/dT$ below 14 K seems to be associated with squaring-up effects (Pinto et al., 1992).

Neutron diffraction measurements performed on $TbRu_2Ge_2$ indicate that the magnetic structure of the compound is sine modulated below 32 K (the Néel point). It becomes square modulated at 4.2 K, with the terbium magnetic moment amounting to 9.06 μ_B and the propagation vector $k = (0.2331, 0, 0)$. The magnetic moment is directed along the c axis (Yakinthos, 1986; Szytuła et al., 1987).

At the Néel temperature ($T_N = 20$ K), holmium moments order antiferromagnetically in $HoRu_2Ge_2$. This compound exhibits a square-modulated structure with the propagation vector $k = (0.2216, 0.0111, 0)$ and a holmium magnetic moment of 6.6 μ_B, parallel to the c axis (Yakinthos and Roudaut,

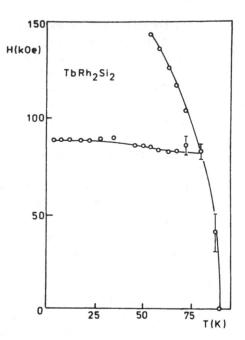

FIGURE 4.14. Magnetic phase diagram of $TbRh_2Si_2$ (Ivanov et al., 1993b).

1987). The [161]Dy Mössbauer spectra in $DyRu_2Ge_2$ were measured in the temperature range from 4.2 to 63 K. $DyRu_2Ge_2$ is an antiferromagnet with a Néel temperature of 22 K. Above T_N, relaxation spectra are observed. The H_{eff} value at 4.2 K gives a magnetic moment of about 10 μ_B. The ground state of the Dy^{3+} ion in this compound must be close to $\pm(15/2)$. The first excited state, $\pm(13/2)$, occurs at a distance of 209,9 cm^{-1} from the ground state, $\pm(15/2)$ (Yakinthos et al., 1987).

The temperature dependence of the magnetic susceptibility of RRh_2Si_2 and RRh_2Ge_2 (R = Tb–Tm) compounds indicates that an additional phase transition below the Néel temperature takes place (Felner and Nowik, 1983, 1984, 1985). In the temperature range from 150 to 300 K, the magnetic susceptibility of a single crystal sample of $TbRh_2Si_2$ obeys the Curie-Weiss law, with the following parameters:

$$\mu_{eff}^c = 10.3 \ \mu_B/Tb \text{ atom } \Theta_p^c = +50(5)K$$

and

$$\mu_{eff}^a = 10.5 \ \mu_B/Tb \text{ atom } \Theta_p^a = -70(5)K$$

The change of the type of magnetic ordering as a function of the external magnetic fields has been detected in all compounds. The metamagnetic process exhibits a two-step character (Szytuła et al., 1992). The values of critical field H_{c1} and H_{c2} decrease with an increasing number of f electrons. The magnetic phase diagram of $TbRh_2Si_2$ is shown in Figure 4.14 (Ivanov et al.,

1992b). Application of external pressure does not influence the value of the Néel point of TbRh$_2$Si$_2$ (Szytuła et al., 1986a). An anomaly of the temperature dependence of the lattice constants at the Néel temperature was observed in TbRh$_2$Si$_2$ (Szytuła et al., 1986a).

Specific heat has been measured for polycrystalline LaRh$_2$Si$_2$ and Ho-Rh$_2$Si$_2$ (Sekizawa et al., 1987). Above 1.3 K, LaRh$_2$Si$_2$ does not show any magnetic order; however, HoRh$_2$Si$_2$ exhibits two peaks in the specific heat vs. temperature curve: a broad peak at 11 K and a very sharp one at 27 K. The magnetic part of the specific heat, obtained by subtraction of the specific heat of LaRh$_2$Si$_2$ from that of HoRh$_2$Si$_2$, provides a total entropy of S = 24 J/mol K, which is close to Rℓn(2J + 1) for J = 8. This fact suggests that the excess entropy can be attributed to the ground state multiplet of the Ho^{3+} ion (^{5}I$_8$). The temperature dependence of the magnetic heat capacity of HoRh$_2$Si$_2$ can be reproduced remarkably well using three crystal field parameters and one isotropic exchange constant with appropriate values of the parameters (Takano et al., 1987). The lower transition temperature (11 K) is the temperature at which the perpendicular component of the magnetization disappears. The higher transition point is that one at which the parallel component of the magnetization disappears (Néel point). A good fit to the experimental data has been obtained for the following set of parameters: B$_2^0$ = -0.14 K, B$_4^0$ = $+0.0004$ K, B$_4^4$ = -0.002 K.

Neutron diffraction studies (Ślaski et al., 1983; Szytuła et al., 1984; Quezel et al., 1984; Melamud et al., 1984; Chevalier et al., 1985a) carried out for Nd, Tb, Dy, Ho, and Er compounds show that at 4.2 K, all of them evidence antiferromagnetic ordering of the AFI type [wave vector k = (0, 0, 1)], similarly as in the RCo$_2$Si$_2$ system (see Figure 4.6). In TbRh$_2$Si$_2$, the magnetic moment localized on the R atom is aligned along the fourfold symmetry axis; in ErRh$_2$Si$_2$, it is perpendicular to the axis, while in HoRh$_2$Si$_2$ it makes an angle of 28 + 3°. Below T$_N$ = 52 K, the magnetic structure of DyRh$_2$Si$_2$ is of the AFI type, with the magnetic moment on Dy^{3+} amounting to 9.9(1) μ_B. It is aligned along the fourfold symmetry axis; however, neutron diffraction data indicate that below 18 K, it makes an angle of approximately 19° with the tetragonal axis (Melamud et al., 1984).

Similar magnetic ordering schemes have been observed in isostructural RRh$_2$Ge$_2$ compounds (Szytuła et al., 1987; Venturini et al., 1988).

The ^{161}Dy Mössbauer data (temperature dependence of the hyperfine field and quadrupole interaction) of DyRh$_2$Si$_2$ have been analyzed with the Hamiltonian

$$H = H_{CEF} + N(g_J\mu_B)^2(J_x \langle J_x \rangle + J_z\langle J_z \rangle)$$

where J$_x$ and J$_z$ are the angular momentum operators and $\langle J_x \rangle$ and $\langle J_z \rangle$ their thermal averages. From fits of the experimental data to this Hamiltonian, the parameters of the crystalline electric field B$_n^m$ and the value of the angle with the c axis equal to 24.9° were determined (Tomala et al., 1992).

In $TbRh_{2-x}T_xSi_2$ (T = Ru or Ir) systems, the magnetic susceptibilities satisfy the Curie-Weiss law at temperatures higher than 130 K. Different behaviors were observed in the concentration dependence of the Néel temperature. In the $TbRh_{2-x}Ir_xSi_2$ system, the Néel temperature decreases from 92 K for x = 0 to 75 K for x = 0.5. At higher concentrations, it remains constant. In the $TbRh_{2-x}Ru_xSi_2$ system, T_N decreases to 15 K for x = 0.8 and thereafter increases to 55 K for x = 2 (Jaworska and Szytuła, 1987).

The results of the neutron diffraction studies of the $TbRh_{2-x}Ru_xSi_2$ system indicate that the samples with x < 0.5 exhibit a simple colinear antiferromagnetic ordering of the AFI type. Neutron diffraction patterns of a sample with x = 0.8 at 4.2 K are characteristic of spin glass or micromagnetic behavior. A sample with x = 1 has a sine-modulated magnetic structure below 27 K, while samples with x > 1.2 show square-modulated structures. The propagation vector k decreases with Ru concentration (Jaworska et al., 1989).

These results were interpreted in terms of the classical model, in which interactions between the first and second neighbors are considered. $TbRh_2Si_2$ is colinear antiferromagnetic, with an AFI-type structure implying the following exchange constants between nearest-neighbor moments: J^a/k_B = 8(2) K and J^c/k_B = −9(2) K (Chevalier et al., 1985b), where J^a is an exchange constant operating within the (001) planes and J^c is an interplane constant. $TbRu_2Si_2$ has a colinear square-modulated structure with the propagation vector k = $(k_x, 0, 0)$. The interaction integrals J^a and $J^{aa'}$ between the first and second neighbor atoms, belonging to the same (001) plane, must be positive.

Substitution of Ru for Rh leads to a change in magnetic interactions, and the observed change in the magnetic structure is a result of the frustration of the exchange integrals. The exchange integral frustration causes the appearance of spin-glass or micromagnetic behavior in compounds with x = 0.8.

The magnetic phase diagram in high magnetic fields is shown in Figure 4.15. Two different regions are observed. In the first (x < 1), the magnetization increases slowly as the magnetic field rises, and attains saturation in the fields above 120 kOe. In the second (x > 1), saturation is reached in the fields above 20 kOe (Ivanov et al., 1993b).

In the region x < 0.6, the specific heat and magnetization measurements of the $DyRh_{2-x}Co_xSi_2$ system show two transition temperatures. The values of T_N and T_t decrease linearly with an increase of the concentration z (Takano et al., 1992).

Magnetometric measurements revealed antiferromagnetic properties of $TbPd_2Si_2$ below 20 K, while (Dy, Ho, Er)Pd_2Si_2 were reported to remain paramagnetic at 4.2 K (Yakinthos and Gemari-Seale, 1982).

Neutron diffraction studies indicate a helicoidal character of the magnetic ordering in $TbPd_2Si_2$ (Szytuła et al., 1986); $TbPd_2Ge_2$ (Szytuła et al., 1988); $HoPd_2Si_2$ (Leciejewicz and Szytuła, 1985); $DyPd_2Si_2$, and $ErPd_2Si_2$ (Bażela

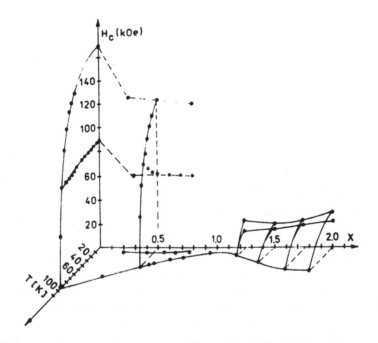

FIGURE 4.15. Magnetic phase diagram (H, T, x) in high magnetic fields for $Tb(Rh_{1-x}Ru_x)_2Si_2$ (Ivanov et al., 1993b).

et al., 1991). In the helicoidal magnetic order, the configuration of magnetic moments is represented by a static linear transverse spin wave with the wave vector $k = (k_x, 0, k_z)$. The magnetic moments localized on the rare-earth atoms make angles ϕ with the tetragonal axis: $TbPd_2Si_2$ ($\mu = 7.9\ \mu_B$ at 4.2 K, $\phi = 90°$), $TbPd_2Ge_2$ ($\mu = 8.4(2)\ \mu_B$ at 1.8 K, $\phi = 62°$), $DyPd_2Si_2$ ($\mu = 9.6(4)\ \mu_B$ at 2.0 K, $\phi = 73.5°$), $HoPd_2Si_2$ ($\mu = 10.2(2)\ \mu_B$ at 4.2 K, $\phi = 63(5)°$), $ErPd_2Si_2$ ($\mu = 8.8(4)\ \mu_B$ at 2.0 K, $\phi = 13.8°$).

An inspection of the Néel points in the series $(Gd-Er)Pd_2Si_2$ reveals a random dependence on the number of f electrons (see Figure 4.16). This fact suggests that the influence of a crystalline electric field (CEF) should also be considered in explaining the magnetism of RPd_2Si_2 compounds (Bażela et al., 1991). However, the observed directions of the magnetic moments do not show any correlations with the signs of the B_2^0 coefficients. The higher-order CEF terms may show a fairly strong effect in the RPd_2Si_2 series.

ROs_2Si_2 compounds with R = Ho, Er, and Tm are ferromagnets, while the compounds with R = Tb and Dy show an antiferromagnetic behavior at low temperatures (Hiebl et al., 1983). Magnetometric measurements performed at 4.2 K for $TbOs_2Si_2$ and $HoOs_2Si_2$ indicate that they become ferromagnetic at critical fields of 10 and 2 kOe, respectively (Kolenda et al., 1985).

Neutron diffraction experiments reveal that at low temperatures $TbOs_2Si_2$, $HoOs_2Si_2$, and $ErOs_2Si_2$ exhibit a magnetic structure which can be described

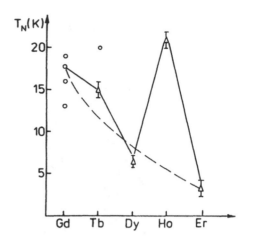

FIGURE 4.16. Transition temperatures into magnetic ordered phases of RPd_2Si_2 (R = Gd–Er) compounds ($\triangle$, neutron diffraction data, $\circ$, magnetic susceptibility data). The dotted line represents temperatures predicted by De Gennes rule (Bażela et al., 1991).

as an incommensurate longitudinal static magnetization wave. The magnetic moments in $TbOs_2Si_2$ and $HoOs_2Si_2$ are aligned along the c axis, but the magnetic wave propagates along the b axis. In the case of $ErOs_2Si_2$, the magnetic moments are aligned along the b axis. The propagation wave vector is k = (0, 1/3, 0) in all three cases (Kolenda et al., 1985).

Ternary RIr_2Si_2 silicides have two allotropic forms. At low temperatures, they crystallize in the $ThCr_2Si_2$-type structure. The stable form at high temperatures is isostructural with $CaBe_2Ge_2$ (space group P4/nmm). Both structural types are ordered ternary derivatives of the binary $BaAl_4$-type structure (Parthé et al., 1983).

Both forms of $TbIr_2Si_2$ are antiferromagnetic: an I form with T_N = 80 K and Θ_p = +42 K, and a P form with T_N = 13 K and Θ_p = +13 K (Hirjak et al., 1984). Since the Tb^{3+}–Tb^{3+} intra- and interplane distances are similar in both phases, a change in the density of the states at the Fermi level (RKKY-type interaction) could explain the small T_N values observed in the P form.

Only the I-$TbIr_2Si_2$ phase has been studied by neutron diffraction (Ślaski et al., 1983). A colinear, antiferromagnetic ordering of the AFI type with the moments aligned along the tetragonal axis [wave vector k = (0, 0, 1)] was discovered.

The magnetic susceptibility of $DyIr_2Si_2$, $HoIr_2Si_2$, and $ErIr_2Si_2$ follows the Curie-Weiss law down to 4.2 K, indicating the absence of a long-range ordering above 4.2 K (Zygmunt and Szytuła, 1984). Magnetization and Mössbauer spectroscopy measurements with ^{161}Dy and ^{193}Ir nuclei indicate that I-$DyIr_2Si_2$ compounds order antiferromagnetically at 40 K and exhibit a metamagnetic transition at the critical field H_c of about 25 kOe. Analysis of the ^{161}Dy hyperfine interaction data indicated that the Dy moments of 10 μ_B are parallel to the tetragonal c axis. Given the absence of any transferred field at the ^{193}Ir nuclei, it was anticipated that the magnetic structure would be of the AFI type (Tomala et al., 1992).

The field dependence of magnetization is linear up to 50 kOe in $TbIr_2Si_2$, $DyIr_2Si_2$, $HoIr_2Si_2$, and $ErIr_2Si_2$, which exhibit typical behavior for paramagnetic compounds (Zygmunt and Szytuła, 1984).

The crystal structure of all ternary RPt_2Si_2 silicides is of the $CaBe_2Ge_2$ type. The magnetic properties were determined in the temperature range from 1.5 to 1100 K in the presence of external magnetic fields up to 13 kOe, revealing a typical paramagnetism of free R^{3+} ions at temperatures larger than 200 K (Hiebl and Rogl, 1985).

Antiferromagnetic ordering is observed below 6.5 K in the case of $TbPt_2Si_2$, whereas (Dy, Ho, Er, Tm)Pt_2Si_2 are ferromagnets below 3 K (Hiebl and Rogl, 1985). The magnetic structure of $TbPt_2Si_2$ is described by the wave vector k = (1/3, 1/3, 1/2). The magnetic moment of the Tb ion is parallel to the a axis (Zerguine, 1988).

$DyPt_2Si_2$ is antiferromagnetic below 7 K (Nowik et al., 1983). The long-range magnetic order at 1.8 K was found by neutron diffraction in $HoPt_2Si_2$ (Leciejewicz et al., 1984a).

Magnetic parameters were determined for all RAu_2Si_2 compounds with R = Tb–Er (Felner, 1975). Antiferromagnetic ordering was found in compounds with R = Tb and Dy. The Néel points range from 5.7 to 15.9 K. $EuAu_2Si_2$ and $ErAu_2Si_2$ order ferromagnetically at 4.2 K.

The Mössbauer data obtained for $DyAu_2Si_2$ confirm the presence of a long-range magnetic order below 7 K (Nowik et al., 1983).

$TbAu_2Si_2$ is an antiferromagnet with the Néel temperature T_N = 14 K and a metamagnetic transition around 40 to 60 kOe at 4.2 K (Sakurada et al., 1990). Neutron diffraction study of a powdered sample of $TbAu_2Si_2$ indicates an antiferromagnetic structure characterized by the wave vector k = (1/2, 0, 1/2). This structure is described as ferromagnetic (101) planes of Tb^{3+} moments coupled antiferromagnetically with the sequence $+ - + -$. The magnetic moment of the Tb^{3+} ion is 8.7 μ_B. It aligns along the (110) direction of the tetragonal unit cell (Ohashi et al., 1992).

The following general conclusions can be deduced from the data presented above:

1. Compounds with nd elements with the same number of d electrons exhibit the same type of magnetic structure. For example, Co, Rh, and Ir have a colinear antiferromagnetic structure of the AFI type; Ni and Pd have a modulated structure, while Cu and Au have an AFIV type.

2. The magnetic moment is parallel to the c axis in compounds with R = Tb–Ho, and perpendicular to the c axis in compounds with R = Er and Tm. The opposite orientation of the magnetic moments is observed in compounds with T = Cu, Au, and Pt. This effect follows the prediction indicated by the sign of the B_2^0 coefficients (see Section 4.8.3).

3. Analysis of the magnitudes of the magnetic moments localized on rare-earth ions shows that in the case of R = Tb–Ho, the experimentally

determined values are almost the same as the free R^{3+} ion values ($g_J\mu_B J$). In the case of Er and Tm, a reduction of the free-ion values is observed due to the CEF effects.

4.6.5. CeT$_2$X$_2$ PHASES

In the RT$_2$X$_2$ series, the Ce-based compounds exhibit unique properties mainly associated with the instability of the Ce shell. The magnetic behavior of Ce compounds may thus be dominated by valence fluctuations between Ce^{3+} and Ce^{4+}, with corresponding moment fluctuations between 2.54 μ_B(4f^1) and 0 μ_B per Ce atom (4f^0).

In the CeT$_2$X$_2$ compounds, where T = 3d, the magnetic order has been observed in CeMn$_2$X$_2$ phases in which the long-range ordering is shown only by the Mn moments (Siek et al., 1978; Narasimhan et al., 1975, 1976).

XAFS data show a correlation between the valence and the strength of the local Ce-Si overlap when T represents a 3d or 4d transition metal (Godart et al., 1987). The mixing of the 4f state with conduction-band states can produce various phenomena such as superconductivity, which was found in CeCu$_2$Si$_2$ (Steglich et al., 1979). The temperature dependence of the specific heat indicates that CeCu$_2$Si$_2$ is a superconductor below 0.5 K. The electronic specific-heat constant, $\gamma = 1.1$ Jmol^{-1} K^{-2}, suggests that CeCu$_2$Si$_2$ is a heavy fermion system with an effective mass of approximately 100 m$_0$ (Stewart, 1984).

The band structure of CeCu$_2$Si$_2$ and the isostructural LaCu$_2$Si$_2$ was calculated using the self-consistent, semirelativistic, linear muffin-tin orbital method (Anderson, 1975; Jarlborg and Arbman, 1977). The Ce-4f levels are situated mainly above the Fermi energy E$_F$. The density of states at E$_F$ is large and heavily concentrated around the Ce-4f band (Jarlborg et al., 1983).

No indication of any long-range magnetic order was found by neutron diffraction in CeCu$_2$Si$_2$, even at the mK range. Only transitions between the Kramers ground state and two excited crystal field doublets of Ce^{4+} ions were determined by inelastic scattering of high-energy neutrons. Consequently, the crystal field-level scheme was determined (Horn et al., 1981).

The susceptibility of CeCu$_2$Si$_2$ follows the Curie-Weiss law above 50 K, with $\Theta_p = -164$ K and $\mu_{eff} = 2.62$ μ_B (Sales and Viswanathan, 1976).

NMR studies of the CeCu$_2$Si$_2$ superconductor gave a phase diagram in which a superconducting region for H < 2.5T and an antiferromagnetic region for 2.5 T < H < 7 T were observed (Nakamura et al., 1988a).

Muon spin relaxation measurements on CeCu$_2$Si$_2$ have shown a static magnetic ordering below T$_N \sim 0.8$ K, which coexists with a superconducting state below T$_s \sim 0.7$ K. The observed muon spin depolarization suggests the existence of a spin-glass or an incommensurate spin-density wave state, with a small averaged static moment on the order of 0.1 μ_B per formula unit at T = 0 K (Uemura et al., 1989).

A change from the mixed-valence state to trivalent or "nearly" trivalent cerium occurs for the same number of d electrons when T represents a 3d or 5d element. By contrast, this behavior is not found in the 4d series, where only trivalent cerium compounds have been identified. The CeT_2Si_2 compounds, with T = Fe, Co, or Ni, are Pauli paramagnets (Ammarguellat et al., 1987).

Investigations of solid solutions between various ternary CeT_2Si_2 systems also provide interesting information on the magnetic state of Ce atoms.

In the $CeCu_2Si_{2-x}Ge_x$ system ($0 < x < 2$), the evolution of a heavy fermion into an antiferromagnetic state with an increase of Ge content was observed (Rambabu and Malik, 1987). $CeCu_2Ge_2$ is a Kondo lattice system which orders antiferromagnetically below T_N = 4.15 K (De Boer et al., 1987). It displays an incommensurate magnetic ordering with the wave vector k = (0.28, 0.28, 0.54) and a small value of the magnetic moment μ = 0.74 μ_B (Knopp et al., 1989).

The evolution of heavy-fermion behavior in $CeCu_2Si_2$ from a strong mixed valence in $CeNi_2Si_2$ was demonstrated in the $CeCu_{2-x}Ni_xSi_2$ system. Heavy-fermion behavior, as probed by low-temperature susceptibility, $\chi(0)$, was destroyed at x = 0.65 (Sampathkumaran and Vijayaraghavan, 1986).

This problem was also studied by magnetic susceptibility (4.2 to 300 K) and resistance (1.2 to 300 K) measurements made for the $CeCu_{2-x}Co_xSi_2$ series (Dhar et al., 1987). There is a significant drop in low-temperature susceptibility $\chi(0)$ for x = 0.5. For higher concentrations of Ce, χ is nearly temperature independent, which is typical for strongly mixed-valence systems.

In $Ce(Cu_{1-x}Ni_x)_2Ge_2$, a monotonic increase of the Kondo temperature from 7 K (x = 0) to 30 K (x = 1) is accompanied by a drastic change of ground state properties:

- For x < 0.2, a modulated magnetic structure with the wave vector k_1 = (0.28, 0.28, 0.54) (T_N = 4.1) K) has been observed. It becomes strongly depressed as x increases.
- At x = 0.2, a different modulation develops below $T_{N2}(x)$ which attains a maximum of 4 K for x = 0.5
- A nonmagnetic heavy fermion ground state exists for x > 0.75 (Steglich et al., 1990).

Magnetic measurements performed on $Ce(Mn_{1-x}T_x)_2Si_2$ (T = Fe, Co, Cu) solid solutions (Siek and Szytuła, 1979; Szytuła and Siek, 1982a) indicate that the compositions with x < 0.5 are ordered antiferromagnetically. The Néel temperatures decrease with an increasing concentration of x. The magnetization dependence on the applied external field obtained at 4.2 K for $CeMnFeSi_2$ and $CeMnCuSi_2$ compounds shows transitions of the metamagnetic type at the critical fields of 12.5 and 60 kOe, respectively.

In the case of $Ce(Mn_{1-x}Cr_x)_2Si_2$, there is strong Mn 3d antiferromagnetism and a decrease of T_N with an increase of the Cr concentration (Liang et al., 1988).

In the case of $CeRu_2Si_2$, the temperature dependence of the magnetic susceptibility above 20 K can be described by the Van Vleck paramagnetism of widely spaced multiplets. The determined effective paramagnetic moment, $\mu_{eff} = 2.27 \ \mu_B$, shows that about 85% of the cerium ions are in the Ce^{3+} state. At low temperatures, $CeRu_2Si_2$ is a nonmagnetic heavy fermion ($\gamma = 355$ mJ/mol K^2) (Thompson et al., 1985). This system does not exhibit superconductivity down to 40 mK (Gupta et al., 1983). The magnetization measured at low temperature and at high magnetic fields shows a saturation effect (Besnus et al., 1985). $CeRu_2Ge_2$ is a ferromagnet with a Curie temperature of 11 K (Umarji et al., 1986). A neutron diffraction study indicates a spontaneous magnetization of 1.98(2) μ_B along the c axis (Besnus et al., 1991).

Recent magnetic susceptibility studies of the pseudoternary $Ce Ru_2Si_{2-x}Ge_x$ series show that as x increases, the system progresses from a heavy electron paramagnet (x = 0) to an antiferromagnet and eventually, at x = 2, it becomes ferromagnetic. The unit cell volume of these compounds also increases with the increase in x. In $CeRu_2Si_{2-x}Ge_x$ compounds, magnetic ordering is observed for samples with x = 0.5, 1.0, and 1.5. Their Néel or Curie temperatures are $T_N = 8$ K, 9.2 K and $T_c = 10$ K, respectively (Umarji et al., 1986; Godart et al., 1987).

Neutron diffraction data indicate that in $CeRu_2Si_{2-x}Ge_x$ for $0.1 < x < 2.0$, a sinusoidally modulated magnetic structure, with the moments along the c axis and the modulation wave vector k = (0.33, 0, 0), is observed. The magnetic moments of the Ce atoms vary smoothly from 1.76 μ_B at x = 2.0 to 0.35 μ_B at x = 0.1 (Dakun et al., 1992).

$Ce_{1-x}R_xRu_2Si_2$ compounds (where R is La or Y) were also investigated (Besnus et al., 1987). When Ce ions are replaced by Y or La ions, the internal pressure changes because the atomic radii are different: R(Ce) = 1.8247 Å and R(La) = 1.8791 Å. The substitution of Ce atoms by La atoms causes a negative lattice pressure, while yttrium atoms substituted in the place of Ce atoms produce a positive lattice pressure. The $Ce_xLa_{1-x}Ru_2Si_2$ compounds, where $0.1 < x < 0.85$, are antiferromagnetics, with the Néel temperatures changing from 0.9 K for x = 0.1 to 6.3 K for x = 0.85. The $Ce_xY_{1-x}Ru_2Si_2$ system does not show any magnetic ordering.

Neutron diffraction study of the $Ce_{1-x}La_xRu_2Si_2$ solid solutions for $0.8 < x < 0.92$ revealed a modulated magnetic structure with the wave vector k = (0.309, 0, 0). The magnetic moment and the transition temperature ($\mu = 1.2 \ \mu_B$ and $T_N = 5.8$ K for x = 0.8) increase continuously with x (Quezel et al., 1988).

The magnetic susceptibility of a number of compositions in the $Ce-Os_xRu_{2-x}Si_2$ system was also studied (Hiebl et al., 1984). For all concentrations, the temperature dependence of the cerium valence state could be satisfactorily represented using the ICF model (Sales and Wohlleben, 1975).

$CeRh_2Si_2$ exhibits a magnetic structure described by the wave vector k = (1/2, 0, 1/2) (see the AFIII type in Figure 4.6) (Quezel et al., 1984). $CeRh_2Ge_2$ exhibits an incommensurate longitudinal static magnetization wave [k = (0, 0, 0.842)] with the moment and propagation direction along the c axis [(μ = 2.2(2) μ_B)] (Venturini et al., 1988).

The influence of various chemical substituents at the Rh site on T_N in the $CeRh_{1.8}T_{0.2}Si_2$ series (T = Co, Ni, Cu, Ru, Pd, Os, and Au) was also investigated by measuring magnetic susceptibility (Sampathkumaran et al., 1987). The results suggested that all these substituents have a general tendency to suppress T_N, but the degree of suppression varies from one substituent to another.

The magnetic properties of $Ce(Rh_{1-x}Ru_x)_2Si_2$ solid solutions have been investigated by magnetometric, specific heat, and neutron diffraction methods (Lloret et al., 1987). Magnetic ordering of the AFII type is observed up to x = 0.4. Magnetic moments are always found to be parallel to the c axis and their values are strongly depressed when x increases (μ = 1.5 μ_B per Ce atom for x = 0 and μ = 0.65 μ_B per Ce atom for x = 0.4). The variation of T_N with the ruthenium concentration is anomalous (see Figure 4.17). For small concentrations (0 < x < 0.15), a sharp decrease in T_N is observed (from 36 to 12 K), whereas for 0.15 < x < 0.4 T_N maintains a nearly constant value of about 11 K.

The results of specific-heat measurements of the $Ce(Rh_{1-x}Ru_x)_2Si_2$ system at low temperatures revealed that in the x = 0.45 sample, long-range antiferromagnetic ordering disappears. At this critical concentration, the low temperature c/T exhibits a maximum (600 mJ/K^{2}mol) (Calemczuk et al., 1990). The magnetic phase diagram of $Ce(Rh_{1-x}Ru_x)_2Si_2$ established on the basis of specific heat and susceptibility data is displayed in Figure 4.17. The region denoted by I corresponds to the magnetic structure determined by Lloret et al. (1987). Regions II and IV correspond to the nonmagnetic state. Region III (0.95 < x < 0.75) is characterized by a new antiferromagnetic phase. Its magnetic structure is still not known (Miyako et al., 1992).

$CePd_2Si_2$ was reported to order antiferromagnetically below T_N = 10 K (Murgai et al., 1982), but Hiebl et al. (1986) did not find any magnetic transition in $CePd_2Si_2$. On the other hand, neutron diffraction data for $CePd_2Si_2$ indicate a colinear antiferromagnetic ordering of the AFIII type (see Figure 4.6). The magnetic moment configuration is described as ferromagnetic planes with the moment perpendicular to the (110) plane. The magnitude of the Ce moment was determined to be 0.66(3) μ_B at 4.2 K (Grier et al., 1984).

Results of specific-heat susceptibility and magnetization experiments show that $CePd_2Ge_2$ orders antiferromagnetically with the Néel temperature T_N = 5.1 K (Besnus et al., 1992). The effective magnetic moment on the Ce ion is 1.79 μ_B. The magnetic susceptibility and specific heat data indicate that

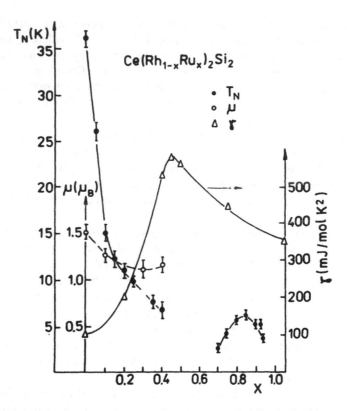

FIGURE 4.17. Magnetic phase diagram of $Ce(Rh_{1-x}Ru_x)_2Si_2$. Concentration dependence of the Néel temperature, value of the magnetic moment and electronic specific heat (Miyako et al., 1992; Lloret et al., 1987).

the ground state is represented by the wavefunction $0.88/ \pm 5/2 > + 0.48/ \mp 3/2 >$, which correlates well with the small value of the magnetic moment.

In the case of the $Ce_{1-x}Y_xPd_2Si_2$ $(0 < x < 1)$ system, the results of heat capacity, resistivity, and susceptibility measurements (Besnus et al., 1987a) show that a partial substitution of Ce by Y leads to a rapid decrease in the Néel temperature. Spontaneous magnetism and the Kondo effect coexist up to $x = 0.4$ $(T_N = 3$ K$)$, where T_K is estimated as 3 T_N. With higher Ce dilution, the magnetic moment of the ground state doublet becomes completely quenched and a genuine heavy fermion behavior is evident, e.g., $\gamma = 0.875$ J/K^2 Ce atom and $T_K = 10$ K at $x = 0.5$, and $\gamma = 0.58$ J/K^2 Ce atom and $T_K = 15$ K at $x = 0.75$.

In the pseudoternary system $Ce_xLa_{1-x}Pd_2Si_2$, replacement of Ce by La leads to a decrease of the Néel temperatures (Besnus et al., 1991), whereas in $CePd_2Si_{2-x}Ge_x$, the values of the Néel temperature decrease linearly with x (Das and Sampathkumaran, 1991). In these systems, the Néel temperatures are dependent on the unit-cell volume, as shown in Figure 4.18 (Das and

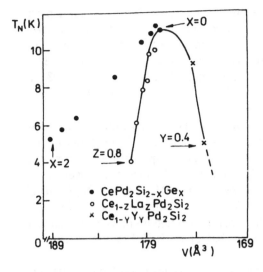

FIGURE 4.18. Néel temperature as a function of unit cell volume for Ce-Pd$_2$Si$_{2-x}$Ge$_x$, Ce$_{1-z}$La$_z$Pd$_2$Si$_2$ and Ce$_{1-y}$Y$_y$Pd$_2$Si$_2$ (Das and Sampathkumaran, 1992).

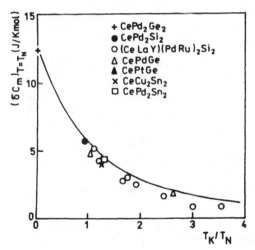

FIGURE 4.19. Variation of the δC_m of the specific heat at the magnetic transition as a function of T_K/T_N. Solid line, calculated specific heat jump at T_N for a doublet ground state system (Besnus et al., 1992).

Sampathkumaran, 1992). The linear dependence for CePd$_2$Si$_{2-x}$Ge$_x$ is evident. A close relationship between the specific-heat jump at the ordering temperature and the ratio of the two characteristic temperatures T_K and T_N is observed (see Figure 4.19.). This result indicates the dominant role of the Kondo effect in reducing the ordered moment values (Besnus et al., 1992).

In CeAg$_2$Si$_2$, the resistivity and susceptibility measurements reveal a transition to the antiferromagnetic state near 10 K and a small ferromagnetic component of gJ = 0.03 μ_B below this point (Murgai et al., 1982). The neutron diffraction data indicate that the magnetic ordering in CeAg$_2$Si$_2$ at 4.2 K may be described as an incommensurate longitudinal wave of magnetization propagating along the a axis (Grier et al., 1984).

The temperature dependence of the resistivity of $CeAg_2Ge_2$ has a peak near 8 K, below which $\rho(T)$ decreases rather steeply, indicating that magnetic ordering occurs in this compound. Measurements of the d.c. susceptibility suggest an antiferromagnetic phase transition (Rauchschwalbe et al., 1985).

The determined paramagnetic moment of Ce in $CeOs_2Si_2$, $\mu = 0.98$ μ_B (Hiebl et al., 1983a) or 1.53 μ_B (Hiebl et al., 1986), is compatible with a rather small concentration of Ce^{3+} ions.

Pressure exerts a considerable influence on the Néel temperature in several CeT_2Si_2 phases. The weak linear pressure dependence of T_N in $CeAg_2Si_2$ and $CeAu_2Si_2$ ($dT_N/dp = +0.1$ and -0.04 K/kbar, respectively) confirms the suggestion that in these materials, T_K is much smaller than T_N. However, the strong nonlinear decrease in T_N with pressure in $CePd_2Si_2$ and $CeRh_2Si_2$ ($dT_N/dp = 1.4$ and -5 K/kbar, respectively) suggests the opposite regime. Changes in $T_N(p)$ agree qualitatively with Doniachs phase diagram in which the energy of a Kondo singlet is compared with that of an RKKY-antiferromagnetic ground state (see Figure 4.20a).

For I-$CeIr_2Si_2$, the inverse susceptibility deviates considerably from the Curie-Weiss law at temperatures below 200 K. The P-$CeIr_2Si_2$ exhibits a broad maximum in the $\chi^{-1}(T)$ curve below 700 K, a fact commonly observed in intermediate-valence systems (Hiebl et al., 1986).

The temperature dependence of the magnetic susceptibility of $CePt_2Si_2$ is similar to the susceptibility of a Kondo lattice and the Fermi liquid system below 2 K (Gignoux et al., 1986). Magnetic measurements indicate that $CeAu_2Si_2$ is an antiferromagnet with $T_N = 10$ K (Palstra et al., 1986). A neutron diffraction study carried out only for $CeAu_2Si_2$ (Grier et al., 1984) indicated that its magnetic structure is of the AFI type. The magnetic moment of $\mu = 1.29(5)$ μ_B is localized on the Ce^{3+} ion and it is parallel to the tetragonal axis.

The magnetic data for $CeT_2X_2(x = Si$ or Ge) compounds are listed in Table 4.17.

CeT_2Sn_2 compounds crystallize in a tetragonal, primitive $CeBe_2Ge_2$ type of structure. The magnetic properties have been determined for only a few of these compounds (see Table 4.18).

The antiferromagnetic Kondo compound $CeNi_2Sn_2$ is characterized by $T_N = 1.8$ K and $T_K = 8$ K. Neutron diffraction measurements show that:

- Below $T_s = 265$ K, a tetragonal-to-monoclinic structural phase transition occurs.
- Below $T' = 6$ K, a two-dimensional ferromagnetic ordering scheme is operating.
- Below 1.8 K, an antiferromagnetic structure with the Ce magnetic moments aligned in the ab plane is present (Ekino et al., 1992).

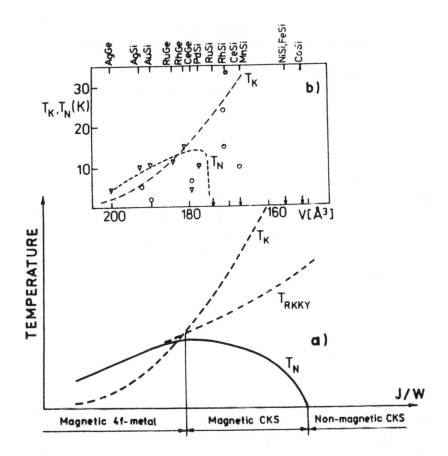

FIGURE 4.20. (a) Classification of concentrated Kondo systems (CKS) by the relationship between two characteristic temperatures: T_K and T_{RKKY}, T_N is the magnetic transition temperature. (b) Kondo temperatures T_K ($\circ$) and magnetic ordering temperatures T_N ($\triangledown$) vs. the unit cell volume for several CeT_2X_2 compounds (Szytuła, 1991).

CeT_2Sn_2 compounds, where T = Cu, Rh, Pd, Ir, and Pt, order antiferromagnetically with the Néel temperature at 1.6, 0.47, 0.50, 4.1, and 0.88 K, respectively (Beyermann et al., 1991b).

The magnetic susceptibility for each member of the series except T = Ir follows Curie-Weiss behavior, with the effective moment $\mu_{eff} \sim 2.5$ μ_B per Ce atom and a small negative paramagnetic Curie temperature.

In $CePt_2Sn_2$, the linear temperature coefficient of the specific heat was found to be approximately 3.5 J/mol K^2, thereby allowing this material to be considered as a heavy-electron system (Beyermann et al., 1991a).

In the above compounds, however, 4f electrons are subject to crystalline field effects and hybridization with conduction electrons. The latter often gives rise to instability of the localized state, in some cases leading to the formation of an itinerant renormalized state near the Fermi level E_F.

TABLE 4.17
Magnetic Data for CeT_2Si_2 and CeT_2Ge_2 Compounds

T	Type of magnetic ordering	$T_{C,N}(K)$	$\Theta_p(K)$	$\mu_{eff}(\mu_B)$	$\mu_R(\mu_B)$	Ref.
			CeT_2Si_2			
Fe	Paramagnetic					
Co	Pauli paramagnetic					
Ni	Pauli paramagnetic					
Cu	Pauli paramagnetic					
Ru			−11	2.27		6
Rh	AF	36	−81	2.91	1,5	7, 9
Pd	AF	10.5	−57	2.55	0.66	11–13
Ag	AF	9.5	−36	2.54	0.93	11, 13
Os			−580	1.53		6
Ir I			−116	2.05		6
P			−360	1.9		6
Pt			−85	2.42		11, 16
Au	AF	10	−18	2.43	1.29	11, 13
			CeT_2Ge_2			
Fe	Pauli paramagnetic					2
Co	Pauli paramagnetic					3
Ni	Pauli paramagnetic					4
Cu	AF	4.1			0.74	5
Ru	F	11	−70	2.3	1,98	8, 15
Rh	AF	15	−25	2.43	2.15	10
Pd	AF	5.1		1.79		14
Ag	AF	6.3			1.5	15
Pt	AF	2.2	−31	2.5		16
Au	AF	16			1.88	15

References: 1. Ammarguellat et al., 1987; 2. Felner et al., 1975; 3. McCall et al., 1973; 4. Knopp et al., 1988; 5. Knopp et al., 1989; 6. Hiebl et al., 1986; 7. Felner and Nowik, 1984; 8. Felner and Nowik, 1985; 9. Quezel et al., 1984; 10. Venturini et al., 1988; 11. Palstra, 1986; 12. Steeman et al., 1988; 13. Grier et al., 1984; 14. Besnus et al., 1992; 15. Loidl et al., 1992; 16. Sampathkumaran et al., 1991.

An important parameter for describing the properties of Ce compounds is the coupling constant J_{sf} between 4f and conduction electrons. It depends on the s-f mixing V_{sf} and the location of the 4f level E_f relative to E_F as $J_{sf} \sim V^{sf}/(E_F - E_f)$. When the 4f levels of Ce^{3+} lie deep below E_F, J_{sf} is small and the 4f electrons stay in the normal localized state with magnetic moments. When E_f approaches E_F, J_{sf} increases and the system begins to exhibit Kondo-like behavior. With the further increase of J_{sf}, the magnetic moment on the Ce ion is reduced and, finally, the system enters the nonmagnetic valence fluctuation region. In the Kondo regime, quasiparticles in the renormalized resonance state developing at E_F behave like heavy fermions with enhanced

TABLE 4.18
Magnetic Data for CeT$_2$Sn$_2$ Compounds

Compound	Type of magnetic ordering	$T_{C,N}(K)$	$\Theta_p(K)$	$\mu_{eff}(\mu_B)$	Ref.
CeNi$_2$Sn$_2$	AF	1.8	−58.4	2.64	1
CeCu$_2$Sn$_2$	AF	1.6	−13.7	2.49	1
CeRh$_2$Sn$_2$	AF	0.47	−13.2	2.53	1
CePd$_2$Sn$_2$	AF	0.5	−6.0	2.50	1
CeAg$_2$Sn$_2$			0.9	1.74	2
CeIr$_2$Sn$_2$	AF	4.1			1
CePt$_2$Sn$_2$	AF	0.88	−25.1	2.59	1
CeAu$_2$Sn$_2$			−0.4	0.89	2

References: 1. Beyermann et al., 1991; 2. Görlich et al., 1992.

effective mass at low temperatures. In addition, the indirect RKKY interactions among Ce ions and the crystalline field substantially modify the electronic structure of Ce in its compounds and produce attractive variations in the low-temperature properties.

Some of them appear to be magnetically ordered, while other are Pauli paramagnets. This variety of properties is considered to be due to two interactions:

1. The indirect magnetic interaction occurs between local moments via conduction electron polarization, the so-called RKKY interaction, whose intensity is proportional to the square of the exchange integral J_{sf}. It can be characterized by a temperature defined as:

$$T_{RKKY} \sim N(E_F)J_{sf}^2$$

2. The other interaction is the Kondo effect, which tends to compensate the local moment on a moment-bearing ion and, consequently, leads to the formation of a nonmagnetic ground state. The Kondo interaction is characterized by a temperature T_K which depends exponentially on J_{sf} in the following way:

$$T_K \sim \exp[-1/N(E_F)J_{sf}]$$

The schematic dependence of the magnetic transition temperature. $T_{C,N}$ on the normalized Kondo coupling constant J_{sf}/W (Figure 4.20a) can be used to classify Ce-based compounds into two groups:

1. Compounds with large J_{sf} values and $T_K \gg T_{RKKY}$ may be called nonmagnetic concentrated Kondo systems (CKS).

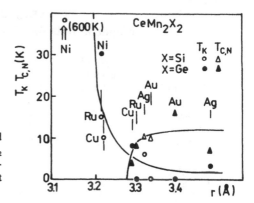

FIGURE 4.21. Kondo temperature T_K and magnetic ordering temperature $T_{C,N}$ in CeT_2X_2 compounds vs. the Ce-transition metal (T) distance (T = Cu, Ag, Au, Ni, Ru) (Steglich et al., 1989).

2.   The intermediate group with $T_K < T_{RKKY}$ corresponds to the magnetic ground state modified essentially by the Kondo compensation of the magnetic moment on the cerium ion.

Such a division correlates well with the available experimental data. Figure 4.20b shows the lattice Kondo temperature T_K and magnetic ordering temperature as functions of the unit-cell volumes of CeT_2X_2 (X = Si, Ge). The results are in good agreement with the Doniach model. They also indicate that ''chemical pressure'' influences the magnitude of the magnetic moment localized on the Ce atom. In compounds with large unit-cell volumes, the RKKY interaction between well-localized f electrons dominates, so that ordinary magnetic properties are observed, whereas the Kondo effect plays a minor role. In the case of small unit-cell volumes and low temperatures, the presence of extended 4f electrons is suggested. For example, $CeNi_2Si_2$ (V = 155.6 Å³) is a typical intermediate-valence system, whereas in a narrow range of unit-cell volumes (165 Å³ < V < 185 Å³), heavy-fermion systems have been found. This fact indicates that the heavy-fermion state results from a delicate balance between the magnitude of the binding energy of the RKKY antiferromagnetic state and the Kondo singlet (Loidl et al., 1989).

In the second step, the experimental data were analyzed in terms of the hybridization model. Figure 4.21 shows the dependence of the Kondo temperature T_K and magnetic ordering temperature $T_{C,N}$ vs. r, the Ce-transition metal (T) distance (Steglich et al., 1990). The Ce-ligand separation, r, is assumed to correlate with the hybridization strength, which can be expressed by the local exchange coupling constant $g = N(E_F)J$, where $N(E_F)$ is the conduction band density of states (DOS) at the Fermi level E_F and the exchange integral J < O. In the extreme case of very weak (/g/<< 1) and very strong (/g/>1) hybridization, the magnetic ordering between stable local f moments and an intermediate valence state is observed. Heavy fermions develop between these two limiting cases (3.2 Å < r < 3.3 Å, see Figure 4.21). The above results indicate that the magnetic state is strongly influenced by changes in the 4f-5d hybridization.

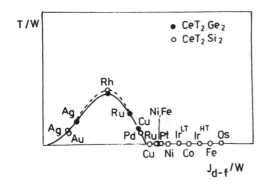

FIGURE 4.22. Schematic phase diagram for the Kondo lattice with the magnetic ordering temperatures of the CeT_2X_2 (X = Si, Ge) compounds (Endstra, 1992).

Schematic phase diagrams for CeT_2Si_2 and CeT_2Ge_2 are presented in Figure 4.22 (Endstra, 1992), in which the normalized values of the critical temperature T/W (W $\sim$ $^1/_{N(o)}$) are plotted against J_{df}/W, where J_{df} is an exchange coupling of f-d hybridization.

The following conclusions can be drawn from this model:

1. The f-ligand hybridization in CeT_2X_2 compounds is governed by f-d hybridization, while the conductions exhibit a strong d character.
2. W is assumed to be constant in the particular $Ce(nd)_2X_2$ series.

The magnitudes of magnetic moments localized on Ce ions in a number of CeT_2X_2 phases can be correlated with the volumes of their unit cells (see Figure 4.23), indicating the influence of the mixed-valence effect.

4.6.6. EuT_2X_2 PHASES

EuT_2X_2 compounds are most suitable to study valence fluctuations because ^{151}Eu is a Mössbauer nucleus and the isomer shift has significantly different values for divalent (δ_{IS} = -10.6mm/s) and trivalent (δ_{IS} = $+0.6$mm/s) ions. Isomer shift values and atomic volumes determined at room temperature are shown in Figure 4.24. In compounds with germanium, Eu ions are divalent, which is associated with the large volume available for Eu atoms in the unit cell. The case of silicides is much more complex: depending on the volume of the unit cell, Eu^{2+} or Eu^{3+} ions or a mixed-valence state are present. Only $EuRh_2Si_2$ does not fulfill the above dependence. Therefore, the mixed-valence effect is observed only in three compounds of the EuT_2Si_2 family, with T = Cu, Pd, and Ir. In these compounds, a temperature dependence of an isomer shift was observed, which is a sign of valence fluctuations (Nowik, 1983; Bauminger et al., 1973). The first nearest neighbors of Eu in EuT_2X_2 are eight Si or Ge atoms. They are most decisive in fixing the Eu valence.

The Mössbauer effect data show that in $EuFe_2Si_2$, the Eu ion is trivalent. Its magnetic properties are still not known (Bauminger et al., 1973). On the other hand, $EuCo_2Ge_2$, $EuNi_2Ge_2$, and $EuCu_2Ge_2$ have been found to order

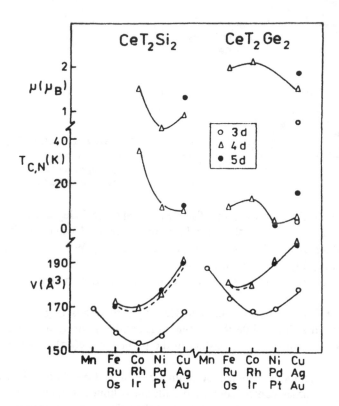

FIGURE 4.23.   Dependence of the magnetic moment μ, the magnetic ordering temperatures $T_{C,N}$, and atomic volumes on CeT_2Si_2 and CeT_2Ge_2 compounds.

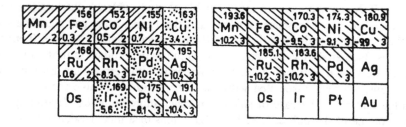

FIGURE 4.24.   The unit-cell volume (in Å³), and isomer shift (mm/s) of the main ¹⁵¹Eu Mössbauer absorption and valence in EuT_2Si_2 and EuT_2Ge_2 compounds at 300 K.

antiferromagnetically, with Néel temperatures of 23, 30, and 13 K, respectively. The Mössbauer effect data imply that the easy axis of magnetization is parallel to the c axis (Felner and Nowik, 1978).

The temperature dependence of the magnetic susceptibility gives no evidence of the magnetic order in $EuCu_2Si_2$. The observed value of the susceptibility in $EuCu_2Si_2$ is intermediate between the values characteristic for Eu^{2+} and Eu^{3+}.

TABLE 4.19
Magnetic Data for EuT_2Si_2 and EuT_2Ge_2 Compounds

T	Type of magnetic ordering	$T_{C,N}(K)$	$\Theta_p(K)$	$\mu_{eff}(\mu_B)$	$\mu_R(\mu_B)$	Ref.
		EuT_2Si_2				
Co						
Ni						
Cu	Van Vleck susceptibility					2
Ru	AF	78	−63	3.95		3
Rh	AF	25	22	7.2		3
Pd			3	7.48		5
Au	F	15.5	−5	6.7		6
		EuT_2Ge_2				
Co	AF	23	−19			1
Ni	AF	30	−8	7.7		1
Cu	AF	13	−20	8.0		1
Ru	AF	62	62	7.53		4
Rh	AF	14	13	7.94		4

References: 1. Felner and Nowik, 1975; 2. Sales and Viswanathan, 1976; 3. Felner and Nowik, 1984; 4. Felner and Nowik, 1985; 5. Palenzona et al., 1987; 6. Felner, 1985.

$EuRu_2Si_2$ and $EuRu_2Ge_2$ are antiferromagnets with Néel temperatures of 78 and 63 K, respectively (Felner and Nowik, 1984, 1985). The values of the hyperfine interaction parameters for $EuRu_2Ge_2$ show that at 4.1 K, the Eu spin is in the basal plane. Due to low magnetic anisotropy, small fields are sufficient to saturate the magnetization (Felner and Nowik, 1985).

The Mössbauer studies performed on ^{151}Eu nuclei embedded in $EuRh_2Si_2$ and $EuRh_2Ge_2$ show that in both compounds, the Eu ions are in the divalent state, and the magnetic moments of the Eu^{2+} ion order antiferromagnetically, with $T_N = 25$ K in $EuRh_2Si_2$ and 14 K in $EuRh_2Ge_2$ (Felner and Nowik, 1984, 1985). The temperature variation of the lattice parameters, magnetic susceptibility, and ^{151}Eu Mössbauer measurements suggest the occurrence of mixed valency in $EuPd_2Si_2$ with a continuous valence transition near 140 K. From these studies, the average valence of the Eu ion in $EuPd_2Si_2$ was calculated to be ~2.3 at 300 K and ~2.7 at 120 K (Sampathkumaran et al., 1981).

Measurements of the ^{151}Eu Mössbauer spectra and magnetic susceptibility show that the 4f configuration of the Eu ion in $EuIr_2Si_2$ changes continuously with temperature from $4f^{6.7}$ at 290 K (Chevalier et al., 1986).

$EuAu_2Si_2$ order ferromagnetically with $T_C = 15.5$ K (Felner, 1975). The magnetic data for EuT_2X_2 compounds are collected in Table 4.19.

X-ray absorption edge measurements indicate two valence states in $EuCu_2Si_2$ and one in $EuCu_2Ge_2$ (Hatwar, 1981). XPS data for $EuCu_2Si_2$ also show that the major component of the spectrum is comparable to the theoretical Eu^{2+} and Eu^{3+} multiplet states. From the energy of separation of the two final multiplet states occurring in the spectrum, an estimate of the Coulomb correlation energy U_{eff} of 4f electrons could be made. The value of U_{eff} in $EuCu_2Si_2$ is 6 to 7 eV (Hatwar, 1981).

Measurements of quadrupole interactions of ^{63}Cu (NMR method) were carried out to determine whether the valence fluctuations in $EuCu_2Si_2$ are between $4f^7$ and $4f^65d$ or between $4f^7$ and $4f^6 + e_c$ (where e_c is a delocalized conduction electron). The obtained results confirmed the latter case (Sampathkumaran et al., 1979).

In $EuPd_2Si_2$, the majority of Eu ions are trivalent at low temperatures. At about $T_v = 150$ K, a valence phase transition takes place and the Eu ions become divalent. The nature of this transition was studied using many techniques (Sampathkumaran et al., 1981). Measurements of ^{151}Eu Mössbauer spectra and magnetic susceptibility show that the 4f configuration of europium in $EuIr_2Si_2$ changes continuously as a function of temperature, from $4f^{6.2}$ at 4.2 K to $4f^{6.7}$ at 290 K (Chevalier et al., 1986).

Pseudoternary $EuT_{2-x}T'_xSi_{2-y}Ge_y$ systems were also investigated. In some compounds with $x = 0$ or $y = 0$, many nonequivalent Eu sites of intermediate valence are formed. In $EuCu_{2-x}Fe_xSi_2$, the average valence of Eu changes nonmonotonically toward Eu^{3+}. In $EuCu_2Si_{2-y}Ge_y$, the Eu valence decreases quickly toward Eu^{2+}. In $EuCo_2Si_{2-y}Ge_y$ ($0 < y < 2$), two narrow Mössbauer absorption lines are observed. They correspond to isomer shifts, -10.6 mm/s and $+0.6$ mm/s, of stable divalent and trivalent europium, respectively. In $EuCo_2Si_{1.5}Ge_{0.5}$, the relative spectral areas of Eu^{2+} and Eu^{3+} are strongly temperature dependent. In $EuNi_2Si_{2-y}Ge_y$ ($y = 1.5$), an intermediate valence was observed [Nowik and Felner, 1986]. In $EuPd_{2-x}Rh_xSi_2$, the valence electron transition temperature T_v falls when x increases. On the other hand, in $EuPd_{2-x}Ru_xSi_2$, unexpectedly, the Eu ions are predominantly divalent and T_v is low. Only for $x > 1.5$ does the Eu trivalent component dominate again (Felner and Nowik, 1985). In the $EuPd_{2-x}Au_xSi_2$ system, the effect of Au alloying on the Eu mixed-valence state in $EuPd_2Si_2$ was investigated. Mössbauer and magnetic susceptibility measurements indicated that for certain values of x, the mixed-valence transition is of the first order (Sauer et al., 1987). The full phase diagram of $EuPd_{2-x}Au_xSi_2$ has been constructed (Segre et al., 1982) and the critical value $x_c = 0.004$ was determined. Above x_c, the first-order valence phase transition takes place. For x higher than $x = 0.175$ the system orders magnetically at $T_N \sim 32$ K.

In the case of EuT_2X_2 compounds, the valence is probably determined by the local chemical environment as well as by the nature of the first and second nearest neighbors, their spatial distribution and distances from Eu ions. In addition, conduction electron density effects may also contribute to setting up the Eu valence (Felner and Nowik, 1985).

4.6.7 YbT₂X₂ PHASES

The valence state of Yb ions in YbT_2X_2 compounds, where T = Fe, Co, Ni, and Cu and X = Si, Ge was studied by the X-ray absorption edge technique. For all the compounds, the mixed-valence state exists. Their valence is almost temperature independent in a wide temperature range (Groshev et al., 1987). ¹⁷⁶Yb Mössbauer measurements show that magnetic ordering occurs within the Yb sublattices of $YbCo_2Si_2$ and $YbFe_2Si_2$ at 1.7 and 0.75 K, respectively. The respective saturation magnetic moments are 1.4 μ_B and 2.0 μ_B, and they point in directions approximately perpendicular to the local tetragonal symmetry axis (Hodges, 1987).

$YbCu_2Si_2$ exhibits unstable 4f shell intermediate-valence effect. At high temperatures, the susceptibility of $YbCu_2Si_2$ follows the Curie-Weiss law, with $\Theta_p = -90$ K and an effective moment $\mu_{eff} = 4.19$ μ_B, which is smaller than the theoretical value of $\mu_{eff} = 4.54$ μ_B resulting from the Hund rule. There is no evidence of magnetic order at low temperatures. The low-temperature magnetization curves are linear. The susceptibility of a single crystal was found to be highly anisotropic. At room temperature, the susceptibility for H\a was $\chi_a = 4.6 \times 10^{-3}$ emu/mol while for H\c, it was $\chi_c = 6.48 \times 10^{-3}$ emu/mol. For both orientations, the susceptibility follows the Curie-Weiss law at high temperatures, with $\mu_{eff} = 4.19$ μ_B. The Curie-Weiss temperatures are -160 K for H\a and -42 K for H\c. Below 75 K, the susceptibility curve flattens in both directions (Sales and Viswanathan, 1976).

The temperature dependence of the magnetic susceptibility of $YbPd_2Si_2$ attains a peak at $T_{max} = 30$ K and tends to a constant value below 10 K. Above 120 K, χ^{-1} is a linear function of temperature. The determined effective magnetic moment $\mu_{eff} = 4.5(1)$ μ_B is close to the moment expected for the free Yb^{3+} ion ($\mu_{eff} = 4.54$ μ_B). From an extrapolation of the high-temperature course of x^{-1}, the paramagnetic Curie temperature $\Theta_p = -70$ K is determined. Such a large negative value is characteristic of homogeneously mixed-valence compounds (Sampathkumaran et al., 1984). Below 16 K, a T^2 dependence of the susceptibility reflects the Fermi liquid nature of the ground state (Dhar et al., 1988). This compound is a superconductor below 1.17 K (Hull et al., 1981). The substitution of Si by Ge induces a negative chemical pressure effect, and the valence of Yb decreases to 2.15 in $YbPd_2Ge_2$ (Umarji et al., 1986).

The magnetic susceptibility vs. temperature curve for $YbPt_2Si_2$ follows closely the Curie-Weiss law down to 1.8 K. The magnetic data for YbT_2X_2 compounds are collected in Table 4.20.

In contrast to compounds with Ce and Eu, in the case of Yb, the mixed-valence state is observed for silicides as well as germanides. The valence of Yb is not temperature dependent. It depends on neither the T-metal nor the volume of the unit cell.

TABLE 4.20
Magnetic Data for YbT_2Si_2 Compounds

Compound	Type of magnetic ordering	$T_{C,N}(K)$	$\Theta_p(K)$	$\mu_{eff}(\mu_B)$	$\mu_R(\mu_B)$	Ref.
$YbFe_2Si_2$	AF	0.75			2.0	1
$YbCo_2Si_2$	AF	1.7	−16.5	4.47	1.7	1, 3
$YbCu_2Si_2$			−90	4.19		2
$YbRu_2Si_2$			−18	3.54		4
$YbPd_2Si_2$			−38	4.28		5
$YbOs_2Si_2$			−21	4.38		6
$YbPt_2Si_2$			−180	3.5		7
$YbCo_2Ge_2$			−91	4.69		3

References: 1. Hodges, 1987; 2. Sales and Viswanathan, 1976; 3. Kolenda and Szytuła, 1989; 4. Hiebl et al., 1983; 5. Palenzona et al., 1987; 6. Hiebl et al., 1983a; 7. Hiebl and Rogl, 1985.

4.6.8. OTHER RT_2X_2 COMPOUNDS

A number of RCo_2B_2 compounds (R = Y, La, Pr, Nd, Sm, Gd, Tb, Dy, Ho, or Er) have been synthesized and studied (see Table 4.21) (Felner, 1984; Jurczyk et al., 1987; Rupp et al., 1987).

The magnetic properties of these alloys were studied in the temperature range 1.5 K < T < 900 K and in the fields up to 10 T. The magnetic susceptibilities of RCo_2B_2 (R = Y or La) are practically temperature independent. The magnetic results for $SmCo_2B_2(\mu_{eff} = 1.76~\mu_B)$, compare well with the ideal Van Vleck behavior of Sm^{3+} ions with a J = 5/2 ground state and a low-lying excited first level J = 7/2. The magnetic susceptibilities of the other compounds satisfy the Curie-Weiss law, with values of the effective magnetic moments close to free R^{3+} ions at temperatures above T = 200 K. At temperatures below 30 K, antiferromagnetic ordering is found for R = Tb (T_N = 13 K), Dy (9.3 K), Ho (8.5 K) and Er (3.3 K), whereas the RCo_2B_2-borides with R = Pr; Nd or Gd order ferromagnetically at T_C = 19.5 K, and 26.5 K, respectively (Rupp et al., 1987).

The hyperfine interaction parameters were determined from ^{155}Gd Mössbauer studies of $GdCo_2B_2$. The direction of the magnetization was found to be in the basal plane (Felner, 1984).

In $TbCo_2B_2$, a neutron diffraction study revealed two magnetic phases below T_N = 19 K. The magnetic cell is commensurate (4a, 4a, c) below T_t = 10 K. At T_t = 10 K, a first-order transition occurs and the magnetic structure becomes incommensurate with the wave vector k = (0.244, 0.244.0) (André et al., 1991).

Also, $CeAl_2Ga_2$ crystallizes in a $ThCr_2Si_2$-type crystal structure. The temperature dependence of the magnetic susceptibility, resistivity, and specific heat indicate that this compound is antiferromagnetic with a Néel temperature

TABLE 4.21
Magnetic Data of RT_2X_2 (X = B, P, As, or Sn) and RAl_2X_2 (X = Si or Ge) Compounds

Compound	Type of magnetic ordering	$T_{C,N}(K)$	$\Theta_p(K)$	$\mu_{eff}(\mu_B)$	$\mu_s(\mu_B)$	Ref.
YCo_2B_2	Pauli paramagnetic					1
$LaCo_2B_2$	Pauli paramagnetic					1
$PrCo_2B_2$	F	19.5	25	3.57	3.0	1, 2
$NdCo_2B_2$	F	32	0	3.27	1.26	3
$SmCo_2B_2$	No Curie-Weiss behavior					1
$GdCo_2B_2$	F	26	80	7.22	6.42	3
$TbCo_2B_2$	AF	19			8.5	4
$TbCo_{1.92}B_2$	AF	13	76	9.76		1
$TbCo_{1.8}B_2$	AF	19	65	9.60	3.5	2
$DyCo_2B_2$	AF	9.3	67	10.87		1
$HoCo_{1.96}B_2$	AF	8.5	28	10.57		1
$ErCo_2B_2$	AF	3.3	22	9.63		1
$LaFe_2P_2$	Pauli paramagnetic					5
$EuFe_2P_2$	F	27	39	7.74		5
$LaCo_2P_2$	F	125	137	2.04		5
$EuCo_2P_2$	AF	67	20	8.10		5
$LaNi_2P_2$	Pauli paramagnetic					6
$CeNi_2P_2$	No Curie-Weiss behavior					6
$PrNi_2P_2$			12	3.4		6
$NdNi_2P_2$			10	3.47		6
$SmNi_2P_2$	Van Vleck paramagnetism					6
$EuNi_2P_2$			-124	7.43		5, 6
$GdNi_2P_2$	AF	10.5	-5	7.86		6
$TbNi_2P_2$			5	9.61		6
$DyNi_2P_2$			5	10.55		6
$HoNi_2P_2$			1	10.49		6
$ErNi_2P_2$			4	9.38		6
$TmNi_2P_2$			5	7.41		6
$YbNi_2P_2$	No Curie-Weiss behavior					6
$LaNi_2As_2$	LT Pauli paramagnetism					7
	HT Pauli paramagnetism					7
$CeNi_2As_2$	LT AF	5.2	-10	1.89		7
	HT		-6	2.46		7
$PrNi_2As_2$	LT AF	11	3	3.4		7
	HT AF	9.5	8	3.37		7
$NdNi_2As_2$			-8	3.52		7
$SmNi_2As_2$	F	9		1.5		7
$EuNi_2As_2$	AF	14	-12	7.41		7
$LaNi_2Sn_2$	Pauli paramagnetism					8
$PrNi_2Sn_2$			0	3.59		8
$NdNi_2Sn_2$			-40	3.73		8
$SmNi_2Sn_2$	Van Vleck paramagnetism					8
$EuAl_2Si_2$	AF	35.5	30	7.2	6.0	9
$EuAl_2Ge_2$	AF	27				9
$GdAl_2Si_2$	AF	22				9

References: 1. Rupp et al., 1987; 2. Jurczyk et al., 1987; 3. Felner, 1984; 4. André et al., 1991; 5. Mörsen et al., 1988; 6. Jeitschko and Reehuis, 1987; 7. Ghadraoui et al., 1988; 8. Skolozdra et al., 1981; 9. Schobinger-Papamantellos and Hulliger, 1989.

of 9 K. Neutron diffraction data indicate that it has a long period commensurate colinear structure, with the wave vector $k = (0, 0, k_z)$ and $k_z = 6/13$, corresponding to a $+ + - - + + + - - + + - -$ sequence of Ce moments which are oriented perpendicular to the fourfold axis. The magnetic moment is 1.18 ± 0.07 μ_B (Gignoux et al., 1988b; Zerguine, 1988). From the anisotropy of the susceptibility at high temperatures, one can deduce the magnitude of the crystalline electric field (CEF) parameters ($B_2^0 = 3.64$ K, $B_4^0 = 0.09$ K, and $B_4^4 = 0.13$ K) and conclude from these parameters that the ground state is the $\pm 1/2$ doublet (Gignoux et al., 1988a).

RNi_2P_2 (R = Ce, Eu, or Yb) compounds also crystallize in the $ThCr_2Si_2$ type. The results of ^{31}P NMR and magnetic susceptibility measurements in the temperature interval 4.2 to 300 K indicate that all these compounds exhibit nonmagnetic ground states (Nambudipad et al., 1986).

$LaNi_2P_2$ is a Pauli paramagnet. At temperatures above 100 K, the compounds RNi_2P_2 with R = Pr, Nd, Gd, Tb, Dy, Ho, Er, or Tm show normal Curie-Weiss behavior, with magnetic moments slightly lower than those of the free R^{3+} ions. The paramagnetic Curie temperatures of these compounds vary between -5 and $+12$ K (see Table 4.21). $GdNi_2P_2$ is antiferromagnetic with a Néel temperature of $T_N = 10.5$ K. $SmNi_2P_2$ shows Van Vleck paramagnetism. $CeNi_2P_2$ and $YbNi_2P_2$ exhibit temperature-dependent paramagnetism, which may be explained with a mixed valence of cerium and ytterbium (Jeitschko and Reehuis, 1987).

The magnetic properties of the RT_2P_2 compounds (R = La or Eu, T = Fe or Co) were studied by ^{57}Fe and ^{151}Eu Mössbauer spectroscopy and by magnetic susceptibility measurements. $LaFe_2P_2$ is a Pauli paramagnet. $LaCo_2P_2$ and $EuFe_2P_2$ are ferromagnets with Curie temperatures of $T_C = 125$ (3) and 27.5(5) K, respectively. $EuCo_2P_2$ orders antiferromagnetically at the Néel temperature, $T_N = 66.5(5)$ K. Mössbauer data demonstrate that the iron sublattice does not participate in the magnetic ordering of the europium-containing compound (Morsen et al., 1988).

The ternary RNi_2As_2 compounds (R = La–Gd) crystallize in two forms: the body-centered $ThCr_2Si_2$ type (L.T. form) and a primitive $CaBe_2Ge_2$ type (H.T. form). Electrical and magnetic measurements indicate that these compounds exhibit metallic conductivity and magnetic ordering at low temperatures when R = Ce, Pr, Sm, or Eu (Ghadraoui et al., 1988).

The magnetic susceptibility in the temperature range between 300 and 1250 K determined for RRh_2P_2 and RRh_2As_2 (R = Ce, Pr, or Nd) follows the Curie-Weiss law, with a magnetic moment close to the free R^{3+} ion values. No long-range magnetic ordering was observed at 4.5 K in all the aforementioned compounds (Madar et al., 1985).

RAl_2Si_2 (R = Y, Pr, or Tb–Lu) and RAl_2Ge_2 (R = Y, La–Nd, or Sm–Tm) compounds crystallize in a hexagonal $CaAl_2Si_2$-type structure (space group $P\bar{3}m1$).

EuAl$_2$Si$_2$, EuAl$_2$Ge$_2$, and GdAl$_2$Si$_2$ order antiferromagnetically below T$_N$ = 33.5 27, and 22 K, respectively. In the ordered state, a spin-flip transition is observed.

Neutron diffraction data showed that EuAl$_2$Si$_2$ orders antiferromagnetically with the wave vector k = (0, 0, 1/2), the europium moments being oriented parallel within the (001) planes (Schobinger-Papamantellos and Hulliger, 1989).

The magnetic data for the systems are summarized in Table 4.22.

4.7. RMn$_2$X$_2$ PHASES

The results of magnetometric measurements for a large number of RT$_2$X$_2$ with R = La, Lu, and Y reveal the existence of Pauli paramagnetism. Only in the manganese-containing systems has the magnetic ordering of the Mn sublattices been observed.

Magnetometric measurements have indicated that RMn$_2$X$_2$ compounds exhibit two critical magnetic ordering temperatures (Szytuła and Szott, 1981):

- At low temperatures, the magnetic moment-localized atoms become ordered.
- At high temperatures, only the magnetic moments on Mn atoms undergo either ferromagnetic or antiferromagnetic ordering.

Ferromagnetic ordering is observed in LaMn$_2$Si$_2$ (Narasimhan et al., 1975) and in the RMn$_2$Ge$_2$ compounds where R is La, Ce, Nd, or Eu (Narasimhan et al., 1976). For LaMn$_2$Ge$_2$ and PrMn$_2$Ge$_2$, measurements of magnetization were made on single-crystal samples. In both cases, the easy axis of magnetization of the Mn sublattice is the (001) axis. For LaMn$_2$Ge$_2$, the values of the magnetic moment and the anisotropy constant K$_1$ were estimated to be 1.55 μ_B per Mn atom and 2.25 $\times$ 10^6 erg/cm^{-3} at 0 K (Shigeoka et al., 1985). For PrMn$_2$Ge$_2$, the magnetization and magnetocrystalline anisotropy constant at 0 K were estimated to be 5.9 μ_B/f.u. (formula unit) and 5.3 $\times$ 10^7 erg/cm^{-3} (Iwata et al., 1986a).

Anomalous temperature dependence of the magnetization is observed in PrMn$_2$Ge$_2$ and NdMn$_2$Ge$_2$ (Iwata et al., 1986; Shigeoka et al., 1988). The spontaneous magnetization increases sharply below T = 100 K. The magnetization curves measured in the (100) and (001) directions indicate changes of the easy axis of magnetization in this temperature region. At low temperatures, the resultant magnetic moment is parallel to the (100) axis. With the temperature increase, the easy axis of magnetization aligns along the (001) axis. The change in the easy axis of magnetization is understandable as the result of competition between Nd and Mn anisotropies. Due to the crystal-field effect (B$_2^0$ > 0), the Nd moments prefer the (100) direction, whereas

TABLE 4.22
Magnetic Data of the Ternary Silicides RMn_2Si_2 and Germanides RMn_2Ge_2

R	Type of magnetic ordering	$T_{C,N}(K)$	$\Theta_p(K)$	$\mu_{eff}(\mu_B)$	$\mu_R(\mu_B)$	$\mu_{Mn}(\mu_B)$	Ref.
				RMn_2Si_2			
Y	AF	460	385	3.5		2.4	1, 2
La	F	310	310	4.5		1.54	1, 3
Ce	AF	379	330	5.1		2.3	1, 5
Pr	AF	348	290	5.1		2.48	1, 2
Nd	AF	365	290	5.1		2.57	1, 2
Sm	AF	398	301	3.3			1
Gd	F, AF	65, 453		4.37			1
Tb	F, AF	65, 550			9.0	1.7	12
Dy	F, AF	37, 473		11.0			1, 15
Ho	AF	453		10.5			1
Er	F, AF	10, 508		10.2	8.9	2.3	1, 17
Tm	AF	498		3.7			1
Yb	AF	513		4.6			1
Lu	AF	464		4.2			1
				RMn_2Ge_2			
Y	AF	396	385	3.8		2.95	1, 2
La	F	306	270	3.5		1.55	4
Ce	F	316		3.1			6
Pr	F	329		5.9	2.77	1.55	7
Nd	F	336			2.6	1.55	8
Sm	F, AF, F	64, 196, 348		4.18			9
	F, AF, F	105, 142, 346					10
Eu	F, AF	9, 330	52				11
Gd	F, AF	97, 365					1
Tb	F, AF	95, 414	100	10.4	8.0	1.7	13
	F, AF	110, 413			7.8	2.3	14
Dy	F, AF	34, 438	100	10.8	10.2	1.95	13, 16
Ho	F, AF	37, 403		10.0			1, 13
Er	F, AF	8.5, 390		9.0	7.7	2.3	1, 17
Tm	AF	458		6.2			1
Yb	AF	338		5.8			1
Lu	AF	453		3.8			1

References: 1. Szytuła and Szott, 1981; 2. Siek et al., 1981; 3. Sampathkumaran et al., 1983; 4. Narasimhan et al., 1976; 5. Siek et al., 1978; 6. Narasimhan et al., 1975; 7. Iwata et al., 1986; 8. Shigeoka et al., 1988; 9. Fujii et al., 1985; 10. Duraj et al., 1988; 11. Felner and Nowik, 1978; 12. Shigeoka et al., 1986; 13. Shigeoka, 1984; 14. Leciejewicz and Szytuła, 1984; 15. Onodera et al., 1992; 16. Venturini et al., 1992; 17. Leciejewicz et al., 1984.

Mn moments prefer the (001) direction ($K_1 > 0$). The anisotropy of Nd is dominant at low temperatures. With an increase of the temperature, the anisotropy of Nd becomes smaller, whereas K_1 decrease slowly so that the anisotropy of Mn becomes dominant at high temperatures. As a result, the easy axis of magnetization changes from the (100) to the (001) direction.

The effect of pressure on the Curie temperature was studied for $LaMn_2Ga_2$ (Kaneko et al., 1992), $PrMn_2Ge_2$, and $NdMn_2Ge_2$ (Kawashima et al., 1990). For all these compounds, the Curie temperatures fall with an increase of pressure. The value of dT_c/dp is -0.2 K/kbar for $LaMn_2Ge_2$ and $PrMn_2Ge_2$ and -0.6 K/kbar for $NdMn_2Ge_2$. Measurement of the pressure effects exerted on the lattice parameters were performed for RMn_2Ge_2 (R = Y, La, Nd) and $LaMn_2Si_2$ compounds. The coefficients of linear compressibilities have been determined (Kanomata et al., 1990).

Measurements of the thermal expansion near T_c have been carried out for $LaMn_2Ge_2$. Both a and c lattice constants have positive exchange strictions (Kaneko et al., 1992). Other RMn_2X_2 compounds are antiferromagnets, as determined by neutron diffraction investigations carried out for a number of RMn_2X_2 systems. In $CeMn_2Si_2$, the Mn moments, forming an antiferromagnetic sublattice, are stable between 80 and 379 K (the Néel point) with a magnetic moment value of 2.3 μ_B at 80 K (Siek et al., 1978). The magnetic moment lie along the fourfold axis with the sequence $+ - + -$. No information about the ordering of Ce moments below 80 K is available. The absence of magnetic ordering in the Pr and Nd sublattices at 1.8 K was established by neutron diffraction studies of $PrMn_2Si_2$ and $NdMn_2Si_2$ (Siek et al., 1981). By contrast, results of magnetic measurements pointed to a ferromagnetic coupling between the Mn and Nd sublattices (Narasimhan et al., 1975). In $PrMn_2Si_2$ and $NdMn_2Si_2$ as well as in YMn_2Si_2 and YMn_2Ge_2, the Mn moments order antiferromagnetically in the same way as in $CeMn_2Si_2$.

Interesting results were obtained from magnetic measurements performed on a single-crystal sample of $SmMn_2Ge_2$. Ferromagnetic behavior has been found between 196 K and the Curie point at 348 K. At temperatures lower than 196 K, a colinear antiferromagnetic ordering becomes stabilized. It disappears at 64 K and reentrant ferromagnetism is observed below this temperature (Fujii et al., 1985). The easy axis of magnetic moment is parallel to the (110) direction below 196 K, while at higher temperatures it is parallel to the (100) direction.

In the temperature range 153 K $<$ T $<$ 341 K, $SmMn_2Ge_2$ is ferromagnetic under normal atmospheric pressure, but antiferromagnetic between 106.5 and 153 K. It becomes a reentrant ferromagnet below 106.5 K (Duraj et al., 1987, 1988; Gyorgy et al., 1987). The magnetic susceptibility measurements carried out at pressures up to 1.5 GPa revealed that in $SmMn_2Ge_2$ the external pressure changes the magnetic phase transition temperature. The (p, T) diagram for $SmMn_2Ge_2$ is shown in Figure 4.25. The Curie temperature decreases linearly with pressure, while the temperature of the ferro-antiferromagnetic transition

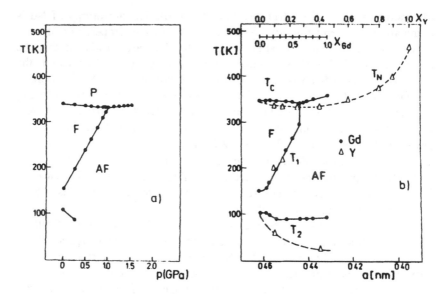

FIGURE 4.25. (a) The (p, T) diagram of $SmMn_2Ge_2$ and (b) (x, T) diagram of $Sm_{1-x}R_xMn_2Ge_2$ systems (R=Gd, Y) (Duraj et al., 1988; Szytuła, 1992).

increases almost linearly for the first phase transition and decreases for the second transition (Duraj et al., 1988).

The temperature dependence of the lattice parameters of $SmMn_2Ge_2$ at atmospheric pressure indicates that at both temperatures of the phase transition, i.e., from the ferro- to the antiferromagnetic phase as well as from the antiferro- to the ferromagnetic, the jump of the value of lattice constant a is observed ($\Delta a/a \simeq 0.2\%$). Lattice parameter c undergoes only a small change in this region ($\Delta c/c \simeq 0.2\%$) (Duraj et al., 1988).

Measurements with a differential scanning calorimeter show that the phase transition near 150 K is endothermic with the transition heat $L = 0.04(1)$ cal $\cdot$ g^{-1} (Gyorgy et al., 1987).

The replacement of Sm atoms by other rare-earth atom (Nd, Gd, or Y) with different atomic radii brings about a change in the lattice constant which may cause a change of the magnetic properties. The replacements mentioned above led to an increase of the lattice constants with x in the $Nd_xSm_{1-x}Mn_2Ge_2$ system and to a decrease for the $(Gd, Y)_xSm_{1-x}Mn_2Ge_2$ systems. The change in the lattice constants influences the magnetic properties of these systems: with an increasing concentration of x, the antiferromagnetic ordering disappears for $Nd_xSm_{1-x}Mn_2Ge_2$ and the ferromagnetic phases disappear for $(Gd, Y)_xSm_{1-x}Mn_2Ge_2$ (Duraj and Szytuła, 1989; Duraj et al., 1989).

The magnetic phase diagrams (x, T) for $Sm_{1-x}R_xMn_2Ge_2$ (R = Y, Gd) as functions of the concentration of R and interatomic distances between Mn atoms are shown in Figure 4.25.

At high temperatures, both systems have similar characteristics. An increase in the concentration of x brings about a decrease in the region of ferromagnetic order. The concentration dependencies of the temperatures T_1 denoting the phase transition from ferromagnetism to antiferromagnetism are similar for both systems; moreover, they are similar to those observed for $SmMn_2Ge_2$ under hydrostatic pressure. This suggests that the transitions from the ferro- to the antiferromagnetic phases are strongly dependent on the interatomic distance.

The magnetization curves determined along the easy and hard directions of magnetization for RMn_2Ge_2 compounds suggest that the magnetic anisotropy is large. The anisotropy energy in a tetragonal lattice is expressed as

$$E_A = K_1 \sin^2\Theta + K_2 \sin^4\Theta + K_3 \sin^4\Theta \sin^2\varphi \cos^2\varphi \qquad (4.1)$$

where K_i is the anisotropy constant of the ith order, and Θ and φ are the angles between the magnetization vector and the c and a axes, respectively. The anisotropy constants K_1 and K_2 can be determined from the following relation

$$\frac{H_{eff}}{M} = \frac{2K_1}{M_s^2} + \frac{4K_2}{M_s^4} M^2 \qquad (4.2)$$

when the easy axis is parallel to the c axis, and

$$\frac{H_{eff}}{M} = -2 \frac{K_1 + 2K_2}{M_s^2} + \frac{4K_2}{M_s^4} M^2 \qquad (4.3)$$

when the easy axis is normal to the c axis. H_{eff} is the effective magnetic field, M is the magnetization, and M_s is the saturation magnetization. The values of K_1 and K_2 for RMn_2Ge_2 and $TbMn_2Si_2$ compounds, are listed in Table 4.23. The large anisotropies observed in these compounds originate mainly from the single-ion anisotropy due to the crystal-field effect. Since the Gd^{3+} ion is in the S state, the anisotropy energy in $GdMn_2Ge_2$ is smaller than in the other compounds.

The temperature dependence of the magnetization for RMn_2Ge_2 where R is Gd, Tb, or Dy, indicates that the magnetic transition at T_t is due to a first-order ferromagnetic-antiferromagnetic transformation (Shigeoka, 1984). An analysis of the temperature dependence of the magnetization at low temperatures carried out using the molecular-field approximation technique showed that the Gd–Mn interaction is antiferromagnetic (Iwata et al., 1986).

The temperature dependence of the electrical resistivity and the thermal expansion of the RMn_2Ge_2 compounds show a sharp jump and a bend at the temperatures corresponding to T_t and T_N, respectively (Shigeoka, 1984).

TABLE 4.23
Values of the Anisotropy Constants K_1 and K_2 for RMn_2X_2 Compounds

Compound	$K_1 \times 10^{-7}$ (erg/cm³)	$K_2 \times 10^{-7}$ (erg/cm³)	Direction of magnetic moment	Ref.
$LaMn_2Ge_2$	0.22	10^{-3}	001	1
$PrMn_2Ge_2$	5.3		001	2
$GdMn_2Ge_2$	0.54	0.36	001	3
$TbMn_2Ge_2$	8.4		001	3
$DyMn_2Ge_2$	13.0		001	3
$HoMn_2Ge_2$	13.0		001	3
$ErMn_2Ge_2$	−42.0		110	3
$TbMn_2Si_2$	22.0		001	4

References: 1. Shigeoka et al., 1985; 2. Iwata et al., 1986b; 3. Shigeoka, 1984; 4. Shigeoka et al., 1986.

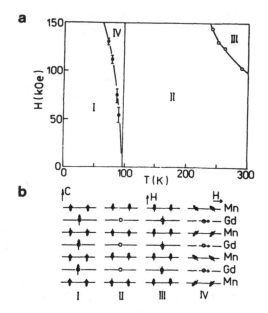

FIGURE 4.26. (a) Magnetic phase diagram of $GdMn_2Ge_2$. (b) Schematic magnetic structure of $GdMn_2Ge_2$ in each phase deduced from magnetization measurements (Kobayashi et al., 1989).

The magnetic properties of the $GdMn_2Ge_2$ compound have been studied by measuring the magnetization of a single crystal in strong fields, up to 150 T. Four types of magnetically ordered phases have been observed (see Figure 4.26). In the presence of small fields, colinear antiferromagnetic properties were observed at temperatures ranging from 96.5 to 365 K. The colinear ferrimagnetic structure becomes stable below 96.5 K, where the magnetic moments of the Gd and Mn atoms are antiparallel to each other. In applied magnetic fields perpendicular to the c axis, the transition from the colinear structure to a canted ferromagnetic structure occurs below 96.5 K. The Gd moments in this canted ferromagnet lie in the c plane, and the Mn moments turn out of the c plane (Kobayashi et al., 1989).

The magnetic structure of $TbMn_2Si_2$ was determined by magnetization and neutron diffraction measurements. The compound was found to exhibit the following magnetic structures:

- A colinear ferromagnetic structure of the Tb sublattice and a canted structure of the Mn sublattice at T = 53 K
- Two colinear magnetic sublattices: a ferromagnetic Tb sublattice and an antiferromagnetic Mn sublattice at 53 K < T < 65 K
- Colinear Mn antiferromagnetic sublattice at 65 K < T < 550 K (Shigeoka et al., 1986).

$TbMn_2Ge_2$ also exhibits interesting magnetic ordering schemes (Leciejewicz and Szytuła, 1984). A neutron diffraction study gave the following results:

- In the temperature range between 4.2 K and T_N = 413 K, the magnetic moments localized on Mn atoms order antiferromagnetically, as in $CeMn_2Si_2$.
- Below T_m = 110 K, a colinear ferromagnetic order of Tb moments along the fourfold axis is observed.
- At T_t = 26 K, the Tb colinear ferromagnetic sublattice transforms into an antiferromagnetic structure, in which the magnetic moments have two components:
  - Ferromagnetic, which is parallel to the c axis
  - Antiferromagnetic, parallel to the (120) direction in a monoclinic unit cell, which is obtained from the tetragonal by the following transformation: $\vec{a}_m = 2\vec{a}_t$, $\vec{b}_m = \vec{a}_t + \vec{c}_t$, $\vec{c}_m = \vec{a}_t$

Mössbauer and neutron diffraction measurements (Venturini et al., 1992) performed for $DyMn_2Ge_2$ show a complicated character of the phase transition between the ferromagnetic and paramagnetic phases. The Mössbauer spectra at 38 and 43 K have two components. The subspectrum with the larger value of the hyperfine field is due to the ferromagnetic phase, while the smaller value of the hyperfine field corresponds to an antiferromagnetic phase in which the dysprosium atoms exhibit magnetic moments. The observed temperature dependence of the hyperfine field in $DyMn_2Ge_2$ indicates a first-order transition interpreted as being due to the crystal electric field effect.

Neutron diffraction data collected below T = 34 K show a colinear ferromagnetic structure with magnetic moments μ_{Dy} = 10 μ_B and μ_{Mn} = 2.0 μ_B. Both are parallel to the c axis. A complicated magnetic structure is observed in the temperature range from 34 to 40 K; it is characterized by the magnetic unit cell with a = a and c' = 3c. The magnetic structure can be

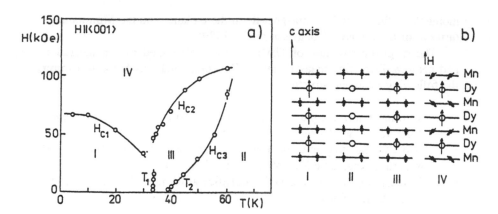

FIGURE 4.27. (a) Magnetic phase diagram of DyMn$_2$Ge$_2$ at low temperatures. (b) Schematic magnetic structure of each phase of DyMn$_2$Ge$_2$ (Kobayashi et al., 1991).

described as ferromagnetic sheets stacked along the c axis with a sequence of $+-++-+$ for Mn atoms and $+\infty+\infty+$ for Dy atoms. Below T = 40 K, DyMn$_2$Ge$_2$ exhibits a simple antiferromagnetic structure of Mn moments, while the Dy moments are disordered. It can be described as a stacking of ferromagnetic Mn planes with the Mn moments aligned along the c axis with the sequence $+-+-$ (Venturini et al., 1992). The magnetic phase diagram constructed on the basis of the magnetic measurements carried out in strong magnetic fields is presented in Figure 4.27. The observed magnetic structure is similar to that discovered in GdMn$_2$Ge$_2$ (see Figure 4.26). (Kobayashi et al., 1992).

The magnetic data for GdMn$_2$Ge$_2$ (Iwata et al., 1986a), DyMn$_2$Ge$_2$ (Venturini et al., 1992), PrMn$_2$Ge$_2$ (Iwata et al., 1986a), and NdMn$_2$Ge$_2$ (Shigeoka et al., 1988) were analyzed in terms of the molecular field approximation with interactions between the R–R, R–Mn and Mn–Mn moments. The Hamiltonian describing the magnetic interactions includes a crystal electric field term. The molecular field acting on the R and Mn moments H$_m$(R) and H$_m$(Mn) can be written as follows:

$$H_m(R) = \frac{(g_J - 1)^2}{g_J\mu_B} J_{R-R}\langle J\rangle + \frac{2(g_J - 1)}{g_J\mu_B} J_{R-Mn}\langle S\rangle$$

$$H_m(Mn) = \frac{(g_J - 1)}{g_J\mu_B} J_{R-Mn}\langle J\rangle + \frac{1}{2g_J\mu_B} J_{Mn-Mn}\langle S\rangle \qquad (4.4)$$

where $\langle J\rangle$ is the thermal average of the total angular momentum of R, $\langle S\rangle$ is the thermal average of the spin angular momentum of Mn, and J_{R-R}, J_{R-Mn}, and J_{Mn-Mn} are the exchange parameters. In the presence of a magnetic field H, the effective Hamiltonians of the respective atoms are given by:

TABLE 4.24
Exchange Constants and Crystalline Field Parameters

	PrMn$_2$Ge$_2$[1]	NdMn$_2$Ge$_2$[2]	GdMn$_2$Ge$_2$[3]	DyMn$_2$Ge$_2$[4]
$J_{R-R}(0)(K)$	33.6	35.0	−6.0	14.0
$J_{R-Mn}(0)(K)$	−74.0	−60.0	−29.6	−30.0
$J_{Mn-Mn}(0)(K)$	480.0	490.0	480.0	504.0
$J_{Mn-Mn}(0)(K)$	—	—	548.0	657.0
$B_2^0(K)$	−0.7	0.45	—	−1.4
$B_4^0(K)$	−0.02	0.0057	—	0.0003
$B_4^4(K)$		−0.02		

References: 1. Iwata et al., 1986b; 2. Shigeoka et al., 1988; 3. Iwata et al., 1986a; 4. Venturini et al., 1992.

$$\mathcal{H}(R) = -g_J\mu_B J[H_m(R) + H] + B_2^0 O_2^0 + B_4^0 O_4^0 + B_4^4 O_4^4$$

$$\mathcal{H}(Mn) = -2\mu_B S[H_m(Mn) + H] \quad\quad (4.5)$$

The free energy per molecule is given as follows:

$$F = -k_B T \ln Z(R) + \frac{1}{2} g_J \mu_B \langle J \rangle H_m(R) - 2k_B T \ln Z(Mn)$$
$$+ 2\mu_B \langle S \rangle H_m(Mn) + 2K_1 \sin\Theta \quad (4.6)$$

with the notation

$$Z(i) = \mathrm{Tr}\, \exp(-\mathcal{H}_i/k_B T) \quad (i = R, Mn) \quad (4.7)$$

The values of $\langle J \rangle$ and $\langle S \rangle$ are given from the stable equilibrium condition that makes a minimum. K_1 is the value of the anisotropy constant of the Mn atom. The values of the parameters such as the temperature dependence of magnetization or hyperfine field and quadrupole splitting, are listed in Table 4.24. This formalism also gives the first-order transition from ferro- to antiferromagnetic states at $T_t = 95$ K in GdMn$_2$Ge$_2$ and $T_t = 40$ K in DyMn$_2$Ge$_2$.

The neutron diffraction study of ErMn$_2$Si$_2$ and ErMn$_2$Ge$_2$ (Leciejewicz et al., 1984b) provided the following results:

- The Mn sublattice orders antiferromagnetically, as in CeMn$_2$Si$_2$.
- The Er sublattice is ferromagnetic with moments perpendicular to the c axis.
- The Curie temperatures are small: 10 K for ErMn$_2$Si$_2$ and 8.5 K for ErMn$_2$Ge$_2$.

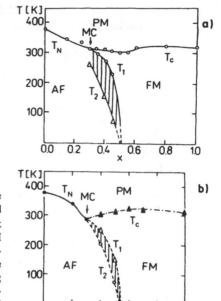

FIGURE 4.28. Temperature vs. concentration phase diagram of the $CeMn_2(Si_{1-x}Ge_x)_2$ system as determined by magnetic measurements after (a) Szytuła and Siek and (b) Liang and Croft (1989). The AF, FM, and PM labels refer to the respective antiferromagnetic, ferromagnetic, and paramagnetic phases for the Mn sublattice order. T_N and T_C represent the Néel and Curie ordering temperatures, respectively, and T_1 and T_2 are temperatures defined in the text. MC represents the multicritical point of the magnetic phase transition.

The specific heat and resistivity measurements of $ErMn_2Ge_2$ showed a phase transition at 5.1 K corresponding to the disordering of magnetic moments in the rare-earth sublattice (Szytuła et al., 1988b).

From the temperature dependence of the magnetization and magnetic susceptibility measurements, the T-x phase transition for $CeMn_2(Si_{1-x}Ge_x)_2$ was determined (Siek and Szytuła, 1979; Liang and Croft, 1989). The T-x magnetic phase diagram (Figure 4.28) has the following regions:

- For $0 < x < 0.3$, the Mn atoms exhibit antiferromagnetic order.
- For $0.3 < x < 0.55$, the samples undergo a transition from the ferromagnetic to the antiferromagnetic state via a transient state. Consequently, two critical temperatures are observed: at T_1, a transition from ferromagnetic to the transient state and at T_2, a transition from the transient to the antiferromagnetic state.
- For $0.55 < x < 1.0$, the samples are ferromagnetic.

The three systems — $La_{1-x}Y_xMn_2Si_2$ (Sampathkumaran et al., 1983), $YMn_2(Si_{1-x}Ge_x)_2$ (Kido et al., 1985a), and $Y_{1-x}La_xMn_2Ge_2$ (Fujii et al., 1986) — show different magnetic structures, depending on the concentration of x.

The magnetic properties of the $Ce_{1-x}La_xMn_2Si_2$ system were investigated by neutron diffraction and magnetic measurements (Szytuła and Siek, 1982b). The samples with a small La concentration (x < 0.5) show antiferromagnetic properties. For x = 0.6, the magnetic ordering changes from antiferromagnetic to ferromagnetic as the temperature increases. An increase in the La content leads to ferromagnetism; a colinear magnetic structure was deduced from neutron diffraction data.

Magnetic measurements performed on the $Ce(Mn_{1-x}T_x)_2Si_2$ solid solution (T = Fe, Co, or Cu) (Siek and Szytuła, 1979; Szytuła and Siek, 1982a) indicate that the compounds with the compositions x < 0.5 are ordered antiferromagnetically. The Néel temperatures decrease with an increasing concentration of x.

In the $Nd(Mn_{1-x}Cr_x)_2Si_2$ system (Obermyer et al., 1979), an increase of the Cr concentration leads to a decrease of the Néel temperature of the ordered Mn sublattice from 380 K for x = 0 to 260 and 200 K for x = 0.2 and 0.3, respectively.

In $Ce(Mn_{1-x}Cr_x)_2Si_2$, there is a strong 3d antiferromagnetism and a decrease of T_N as the Cr concentration increases (Liang et al., 1988).

In the $Ce(Cr_xMn_{1-x})_2Si_2$ series, the T_N values were found to decrease linearly with x at a rate of T_N/x = 6.0 K/at %Cr. A linear rate of depression of T_N, i.e., 6.3 K (at %Cr) similar to that observed in the $Nd(Cr_xMn_{1-x})_2Si_2$ system supports the notion that the rare-earth sublattice plays a minor role in the high-temperature magnetic ordering of these alloys (Obermeyer et al., 1979).

The magnetic transition temperatures of the RMn_2X_2 compounds plotted against the interatomic Mn–Mn distance give a universal curve (Figure 4.29). The critical distance of $R_{Mn-Mn}{}^a$ = 2.85 Å. The coupling between interlayer moments is antiferromagnetic when $R_{Mn-Mn}{}^a$ < 2.85 Å and it becomes ferromagnetic for $R_{Mn-Mn}{}^a$ > 2.85 Å (Szytuła and Siek, 1982; Fujii et al., 1986). Similar critical distances were observed in many other alloys with transition metals (Forrer, 1952). The localization-delocalization effect of the 3d electrons occurs when the critical distance in the Mn compounds reaches 2.85 Å (Goodenough, 1963).

The p_c/p_s ratio has been estimated to be 1.71 for $LaMn_2Ge_2$ (Shigeoka et al., 1985), where p_c and p_s are the magnetic carrier numbers derived from the Curie constant and the spontaneous magnetization moment at OK, respectively. The value lies between the branches of the itinerant and the localized systems (Figure 4.30).

The electronic energy bands of YMn_2Ge_2 and $LaMn_2Ge_2$ calculated using the KKR method show that: (1) the Fermi level for YMn_2Ge_2 is situated at the steep valley of the density of states (DOS) curves of the Mn d bands, and hence YMn_2Ge_2 can be antiferromagnetic like Cr; (2) $LaMn_2Ge_2$ is situated at the peak of the DOS curves, and the valley near the Fermi level is not as steep as in YMn_2Ge_2; and (3) YMn_2Ge_2 has the possibility of being a ferromagnet because the value of the DOS at the Fermi level is 20 states per (atom spin Ryd.), which satisfies the Stoner condition for ferromagnetism (Ishida et al., 1986). Furthermore, it can be expected that the transition from an antiferromagnetic to a ferromagnetic state will be brought about by the variation of temperature and substitution of the constituent elements. The

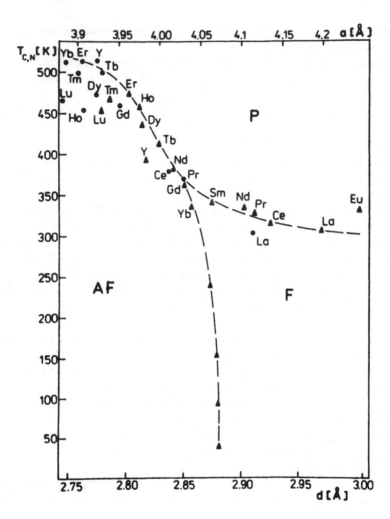

FIGURE 4.29. Dependence of the Néel and Curie temperatures on the intralayer distance for the RMn$_2$X$_2$ compounds (Szytuła, 1992).

interlayer-hopping integrals between the Mn d states are very small and the interaction between the interlayer Mn magnetic moments is small compared to that occurring between the intralayer Mn moments. The band structures of the antiferromagnetic compound YMn$_2$Si$_2$ and ferromagnetic LaMn$_2$Si$_2$ were recently calculated using the scalar-relativistic LMTO method (Kulatov et al., 1990). The calculated band structures are almost the same as in YMn$_2$Ge$_2$ and LaMn$_2$Ge$_2$ (Ishida et al., 1986). In RMn$_2$Ge$_2$ compounds, the antiferromagnetic (AF) state appears for R$^a_{Mn-Mn}$ = 2.85 Å and the ferromagnetic (F) state for R$^a_{Mn-Mn}$ = 2.85 Å. To examine the validity of this criterion as a stability condition of magnetic ordering, the band structure of YMn$_2$Si$_2$ was

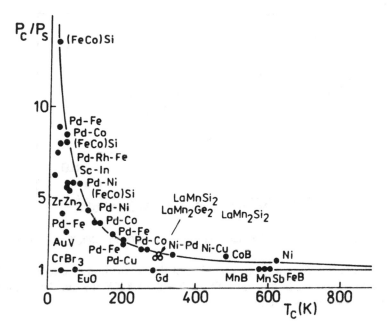

FIGURE 4.30. p_C/p_s vs. T_C, the Rhodes-Wohlfarth plot. p_s is the saturation moment in Bohr magnetons per magnetic atom and p_C is the effective moment deduced from the Curie constant. (Rhodes and Wohlfarth, 1963).

calculated for R^a_{Mn-Mn} = 2.85, 2.775, and 2.91 Å. The results shoved that the Fermi level-state density increases with the increase of R^a_{Mn-Mn}. Consequently, the transition from the AF to the F state is expected to take place at the expanded volume. These theoretical results also suggest that the magnetic properties of YMn_2Si_2 and $LaMn_2Si_2$ are more sensitive to the intralayer than to the interlayer interatomic distance R^a_{Mn-Mn}.

The function representing the variation of the magnitudes of magnetic moments localized on Mn ions with the Mn–Mn distances indicates that the change of magnetic ordering from antiferromagnetic to ferromagnetic can be correlated with a change in moment value (Figure 4.31). It is larger by 0.5 μ_B per Mn ion in the antiferromagnetic than in the ferromagnetic state.

4.8. CONCLUSIONS

4.8.1. MAGNETIC MOMENT

The magnetic moments in the considered ternary compounds are almost exclusively due to localized 4f electrons.

The values of the effective moments were deduced from the temperature dependence of the magnetic susceptibility (see Tables 1 to 16). The disagreement between the experimental moments and the rare earth's free-ion

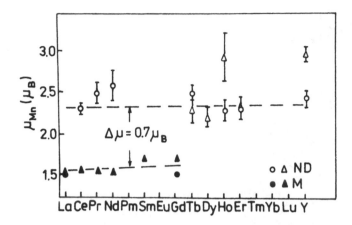

FIGURE 4.31.   Values of the magnetic moment for different RMn_2X_2 compounds.

moments results from a strong coupling of the conduction electrons with the localized moments. In this case, following the RKKY theory, the magnetic effective moment is given by

$$\mu_{eff} = g_J\mu_B\sqrt{J(J + 1)}\left[1 + J_{sf}N(E_F)\frac{g_J - 1}{g_J}\right] \qquad (4.8)$$

where g_J is the Lande factor, $N(E_F)$ is the conduction electron density of states per atom at the Fermi surface for one spin direction, and J_{sf} is the effective s-f exchange interaction due to the direct exchange and s-f mixing. The ratio of experimental effective moments to the free-ion moments as a function of the parameter $(g_J - 1)/g_J$ are plotted in Figure 4.32. The general trend shows that for RRh_2Si_2, J_{sf} is negative, whereas for RRu_2Si_2 it is positive. One obtains from Figure 4.32 that $J_{sf}N(E_F)$ is -0.35 and $+0.05$ for RRh_2Si_2 and RRu_2Si_2, respectively (Felner and Nowik, 1984).

Calculations for RCu_2Si_2 (R = Tb–Tm) give negative values of J_{sf} which do not change with a change of the f element (Budkowski et al., 1987).

Analysis of the experimental results of the effective magnetic moments in the GdT_2Si_2 compounds (T = 3d, 4d, or 5d element) shows that they are higher than the theoretical moments. According to Equation 4.8, this means that J_{sf} is positive (Łątka, 1989). Positive values of J_{sf} were also derived from ESR data for $Gd_{1-x}La_xCu_2Si_2$ (Kwapulińska et al. 1988).

As already mentioned, the excess moment is usually associated with the conduction electron spin polarization. This may be induced by s-f or by d-f exchange interactions (Stewart, 1972). Taking into account the two bands of conduction electrons, one can express the effective moment by

$$\mu_{eff} = g_J\mu_B\sqrt{J(J + 1)}\left[1 + J_{sf}\frac{\chi_s}{\mu_B^2N} + J_{df}\frac{\chi_d}{\mu_B^2N}\right] \qquad (4.9)$$

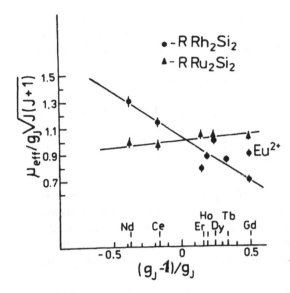

FIGURE 4.32. Ratio of the experimental effective moment to the free-ion moment as a function of the parameter $(g_J - 1)/g_J$ (Felner and Nowik, 1984).

where J_{sf} and J_{df} are s-f and d-f exchange coupling constants, and are the band susceptibilities, respectively, and N denotes the total number of atoms. On the basis of the band-structure calculations (Freeman, 1972) as well as taking into account the results of experiments (Buschow, 1979), it may be concluded that the d band originating from R–5d states contributes to a DOS in the vicinity of the Fermi energy. The 4f-5d exchange parameter is positive and the polarization of the 5d band is enhanced by intraband interaction. Thus, the susceptibility of the 5d electrons seems to play a dominant role in the conduction electron polarization in the vicinity of the R ion. Furthermore, assuming that $/J_{sf}\chi_s/ << /J_{df}\chi_d/$, one obtains the following results from Equation 4.9: $\mu_{eff} = \mu_R^{4f} + \mu_R^{5d}$, where $\mu_R^{4f} = g_J\sqrt{J(J + 1)}\mu_B$ and $\mu_R^{5d} = J_{df}\chi_d/\mu_B^2 N$. Analysis of the magnetic data obtained for the GdT_2Si_2 systems (T = 3d, 4d, or 5d element), in which $GdRh_2Si_2$ appears to have the largest effective moment, gives $\mu_{Gd}^{5d} = 0.28(3)$ μ_B as a maximal value (Łątka, 1989; Czjzek et al., 1989).

In the ternary rare-earth compounds discussed, the magnetic properties are determined by the rare-earth moments. The transition metal atoms T (except Mn) do not carry magnetic moments.

These materials have a metallic character and the interatomic distances between the rare-earth atoms are fairly large. The magnetic interaction between the highly localized 4f electrons is realized by their conduction electrons mediating in an exchange interaction, and by the effect of the crystalline electric field (CEF) acting on the 4f electrons.

4.8.2. EXCHANGE INTERACTIONS

In metallic compounds of rare earths, exchange interactions between rare-earth moments are mediated by the spin polarization of conduction electrons.

This leads, in general, to long-range exchange interactions with an oscillatory dependence of the interaction strength on the distance between the moments. In RKKY theory, the critical temperature of the magnetic ordering T_C, T_N is proportional to the De Gennes factor $(g_J - 1)^2 J(J + 1)$.

For RT_2X_2 compounds, the De Gennes function is not obeyed in phases containing light lanthanide ions. In the case of heavy rare-earth ions, the Néel points, in principle, follow the De Gennes function. However, large discrepancies are observed when T = Cu, Ru, or Os. In all RT_2X_2 series, the oscillatory character of the Néel temperature is caused by an increase in the number of d electrons (Felner and Nowik, 1978; Leciejewicz et al., 1984b; Szytuła and Leciejewicz, 1989).

The variation of T_N with Z_{nd} in GdT_2Si_2 shows similar trends for the three transition-metal rows, with a pronounced maximum for the compounds with Co, Rh, or Ir. For the same column, the lattice parameter a is smaller than for the other compounds. This correlation indicates a dominance of the interactions between Gd moments within the a-b planes (Łątka, 1989; Czjzek et al., 1989).

In the RFe_2Si_2 system, where a local moment on the Fe atom has been found to be absent, the observed transferred [57]Fe hyperfine fields at the Fe sites are in reasonable agreement with the conduction electron polarization due to the rare-earth moments (Noakes et al., 1983).

The difference between the [57]Fe transferred field in $GdFe_2Si_2$ in the antiferromagnetic state at 1.2 K measured in a zero external field (all Gd moments are parallel to the a-b plane) and the field observed in an applied field of 60 kOe at low temperatures indicate anisotropic exchange interactions (Łątka, 1989; Czjzek et al., 1989).

Considerable anisotropy of exchange interactions has also been deduced from single-crystal magnetization data obtained for $TbRh_2Si_2$ (Chevalier et al., 1985b).

Neutron diffraction data were discussed in terms of the RKKY theory. In an isotropic RKKY model based on a spherical Fermi surface, the Fermi vector is strongly dependent on the c/a ratio and on the number of free electrons Z per magnetic ion $k_F = (6\pi^2/a^2c)^{1/3}$. The analysis of the a/c values determined for a large number of RT_2X_2 compounds containing heavy rare-earth ions (R = Tb–Tm) shows that when a/c < 0.415, a simple colinear ordering is observed, while in compounds with a/c > 0.415, an oscillatory magnetic structure occurs (Leciejewicz and Szytuła, 1987).

The stability of the magnetic ordering schemes was discussed in terms of an isotropic RKKY mechanism. Following the RKKY theory, the energy E of a spin system with a screw-like ordering is given by: $E = -N\langle\mu^2\rangle J(k)$, where N is the total number of magnetic ions in the system, $\langle\mu^2\rangle$ is an average value of the magnetic moment, and J(k) represents the Fourier transform of the exchange integral $J(R_1-R_j)$ between i and j magnetic ions, with positions given by the vectors R_i and R_j, respectively. Assuming that the interaction

TABLE 4.25
Values of k_F and Z Obtained for RT_2X_2 Compounds

Compound	k_F	Z	Ref.
$TbRu_2Si_2$	0.233a	3.13	1
$DyRu_2Si_2$	0.22a	3.136	1
$(Ho, Er)Ru_2Si_2$	0.2a	3.16	1
$TbRu_2Ge_2$	0.235a	3.0	2
$TbOs_2Si_2$	0.312a	3.0	3
$(Ho, Er)Os_2Si_2$	0.295a	3.02	3
RCo_2X_2	1.0c	2.75–3.75	4
RRh_2Si_2	1.0c	2.77–3.39	4
$TbRh_2Ge_2$	1.0c	2.84–3.2	5
$NdRu_2Si_2$	0.0c	2.67–3.4	4
$NdFe_2Si_2$	0.5c	1.4	4

References: 1. Ślaski et al., 1984; 2. Yakinthos, 1986a; 3. Kolenda et al., 1985; 4. Leciejewicz and Szytuła, 1987; 5. Szytuła et al., 1987b.

integral J(k) remains constant, the energy E is directly proportional to a function expressed as $-F(k) \sim -[J(k) - J(0)]$. A stable situation for a particular magnetic structure means that function F(k) exhibits a maximum for a nonzero value of the wave vector k of the magnetic structure (Yosida and Watabe, 1962).

The computation of the F(k) function against k for particular values of k_F (or Z) permits selection of these values of k for which F(k) exhibits a maximum. The Z and k values obtained for a number of RT_2X_2 compounds are listed in Table 4.25. In the RT_2X_2 compounds, the observed magnetic ordering schemes require three free electrons per R^{3+} ion. It could indicate that the 6s and 5d electrons of the R^{3+} ion are donated to the conduction band. Valence electrons of the other atoms contribute to the chemical bonding. On the other hand, chemical shift measurements performed by X-ray absorption spectroscopy (XAS) indicate that electrons of all constituents, i.e., of R, T, and X atoms, contribute to the conduction band (Darshan et al., 1984).

Different values of Z were obtained from the analysis of magnetic data (paramagnetic Curie temperature) of GdT_2Si_2 compounds with T = Cu, Ag, or Au. Values of Z equal to 11 were accepted (Buschow and De Mooij, 1986).

Analysis of the values of the paramagnetic Curie temperature θ_p and the effective hyperfine fields obtained for GdT_2Si_2 compounds, including the negative transferred hyperfine field observed experimentally at the Fe position in $GdFe_2Si_2$, was conducted within the framework of the simple RKKY model. It offers an explanation of the experimental results when adopting Z = 13 (Łątka, 1989).

The magnetic properties of $RCu_{1-x}Zn_xSi$ compounds with R = Gd (Kido et al., 1984b), Tb (Bażela and Szytuła, 1989), Dy or Ho (Kido et al., 1985c)

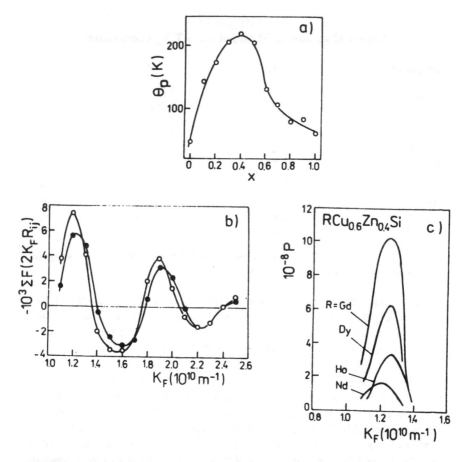

FIGURE 4.33. (a) Paramagnetic Curie temperature Θ_p vs. composition x of $GdCu_{1-x}Zn_xSi$, (b) dependence of lattice sums for GdCuSi (○) and GdZnSi (●) on the Fermi vector, and (c) dependence of the lattice sums for $RCu_{0.6}Zn_{0.4}Si$ on the Fermi wave number (Kido et al., 1984b, 1985c).

were also analyzed in terms of RKKY theory. In these compounds, the paramagnetic Curie temperature plotted as a function of the composition of x exhibit a maximum at x = 0.4 (see Figure 4.33). Assuming an RKKY-type exchange interaction, the paramagnetic Curie temperatures are given by Equation 3.4. The R atoms occupy crystallographically equivalent positions in the $RCu_{1-x}Zn_xSi$ compounds (AlB_2 or Ni_2In-type structure). The summations in Equation 3.4 calculated as a function of k_F for GdCuSi and GdZnSi are shown in Figure 4.33b. The results obtained indicate that the change in Θ_p is primarily due to k_F and secondarily to R_{ij}. Since the Θ_p values in $GdCu_{1-x}Zn_xSi$ are positive, as shown in Figure 4.33a, the sign of the sum functions in Equation 3.4 must be negative. This condition is satisfied when $k_F = 1.25 \times 10^{10}$ and 1.9×10^{10} m^{-1}. The corresponding number of conduction electrons per formula unit N was calculated to be 3.7 and 11.5 for $k_F = 1.25 \times 10^{10}$ and 1.9×10 m^{-1}, respectively. Consequently, one may conclude that in the GdCu$_{1-x}$Zn$_x$Si $(0 < x < 1)$ compounds, N has a value between 3 and 4.

In some RT$_2$X$_2$ compounds, e.g., PrCo$_2$Si$_2$, NdCo$_2$Si$_2$, NdCo$_2$Ge$_2$, PrFe$_2$Ge$_2$, and TbNi$_2$Ge$_2$, a colinear antiferromagnetic structure is observed at low temperatures. An increase in temperature changes a colinear antiferromagnetic structure to an incommensurate one. The former magnetic structure could be displayed as a piling up of the ferromagnetic sheets along the c axis with the sequence $+ - + -$. The distribution of rare-earth ions in the crystal structure of the ThCr$_2$Si$_2$ type and the layered magnetic structure suggests a highly anisotropic character of the magnetic interactions. The anisotropy should be understood in two ways: (1) the magnetic moments are quenched along the c axis, leading to an Ising-like behavior; and (2) the exchange interaction denoted by J$_0$ within the (001) plane is strongly ferromagnetic, whereas the coupling between planes J$_1$, J$_2$, . . . , J$_n$ is weaker. In such cases, the stability of the magnetic structure is described in terms of the ANNNI (anisotropic next-nearest-neighbor-Ising) model. The one-dimensional Ising Hamiltonian is given as:

$$\mathcal{H} = -\sum_{j=1}^{N} \sum_{r=-\infty}^{+\infty} J(r)m_j m_{j+r} - H\sum_{j=1}^{N} m_j + \mathcal{H}_0 \qquad (4.10)$$

where N is the number of layers, $\mathcal{H}$ refers to an isolated plane, J(r) is the effective interlayer interaction, and $/m_j/ = 1$ and H is the magnetic field applied along the (001) direction. In the ANNNI model (Bak and von Boehm, 1980), the interaction between nearest-neighbor layers is positive $(J_1 > 0)$ and the interaction between next-nearest-neighbor layers is negative, i.e., J$_2$ < 0. Mean field calculations produce a number of magnetic structures as a function of a parameter defined as J$_2$/J$_1$. For J$_1 > 0$, the ground state is ferromagnetic if $J_2 > J_1/2$. However, in the case of $J_2 < -J_1/2$, the ground state becomes antiferromagnetic. If $/J_2/J_1/ < 1/4$, the antiferromagnetic structure is of the AFI type $(+ - + -)$. The stability condition for AFII magnetic structure $(+ + - -)$ is $-J_2/J_1 > 1/2$. An ordering of the modulated type with a wave vector k given by $\cos(\pi k) = -J_1/4J_2$ is stable if $/J_2/J_1 / > 1/4$ (Bak and von Boehm, 1980). The observed magnetic structures are in fair agreement with these predictions.

In the case of NdRu$_2$Si$_2$ and NdRu$_2$Ge$_2$, two magnetic phase transitions have been observed at 10 and 26 K in the former and at 10 and 17 K in the latter (Ślaski et al., 1988). At low temperatures, ferromagnetic ordering is observed. It changes to a sine wave-modulated structure with k = (0.13, 0.13, 0) in NdRu$_2$Si$_2$ (Chevalier et al., 1985a) and k = (0.12, 0.12, 0) in NdRu$_2$Ge$_2$ (Szytuła et al., 1987a) as the temperature increases.

The magnetic structures of these compounds are described by wave vectors which maximize J(k). Taking into account the nearest-neighbors (J$_1$) and the

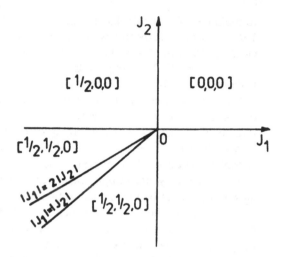

FIGURE 4.34. Stability conditions of the RT_2X_2 ternary compounds (Yakinthos and Roudaut, 1987).

next-nearest-neighbors (J_2) between atoms belonging to the same (110) plane, the Fourier transform of the exchange integral has the following form (Yakinthos and Roudaut, 1987)

$$J(k) = 2J_1(cos2k_x + cos2k_y) + 4J_2(cos2k_x + cos2k_y)$$

The extreme condition of this relation gives:

1.   A ferromagnetic structure with $k_1 = (0, 0, 0)$ and $J_1(k_1) = 4(J_1 + J_2)$ if $J_1 > 0$ and $J_2 > 0$
2.   An antiferromagnetic structure with $k_2 = (1/2, 0, 0)$ and $J_2(k_2) = -4J_2$ if $J_1 < 0$, $J_2 < 0$, and $J_1 < J_2$
3.   An antiferromagnetic structure with $k_3 = (1/2, 1/2, 0)$ and $J_3(k_3) = -4J_1 + 4J_2$ if $J_1 < 0$ and $J_2 > 0$
4.   An incommensurate structure with $k_4 = (k_x, k_y, 0)$ and $J(k_4) = -J_1/J_2$ if $J_1 < 0$, $J_2 < 0$, and $J_1 > J_2$. The corresponding magnetic diagram is shown in Figure 4.34.

Many interesting properties are observed in $TbCo_2B_2$. Neutron diffraction study reveals two magnetic phases: below $T_N = 19$ K, the magnetic cell is commensurate (4a, 4a, c), at $T_t = 10$ K, a first-order transition occurs and the magnetic structure becomes incommensurate, with wave-vector $k = (0.244, 0.244.0)$ (André et al., 1991). The magnetic phase diagram has been determined taking into account first-, second-, and third-neighbor in-plane interactions. The stability of the commensurate (1/4, 1/4) state at low temperatures, as well as the (q, q) incommensurate phase at higher temperatures, is shown to be enhanced by the interplane interaction (Plumer and Caillé, 1992).

4.8.3. CRYSTALLINE ELECTRIC FIELD

A fairly large amount of experimental data have been up to now collected for the RT_2X_2 compounds, whereas the amount of data available for the remaining ternary compounds is rather limited. The experimental results presented in this chapter will be discussed for each group of compounds.

4.8.3.1. RT_2X_2 Compounds

In all these compounds, the R-ion sites are equivalent and show tetragonal point symmetry.

4.8.3.1.1. CeT_2X_2 Compounds

The CEF Hamiltonian acts on the $J = 5/2$ multiplet of the Ce^{3+} ion, which splits into three doublets:

$$D_{5/2} = \Gamma_6 + 2\Gamma_7$$

$$\Gamma_6 = |\pm 1/2\rangle$$

$$\Gamma_7^{(1)} = a|\pm 5/2\rangle + b|\mp 3/2\rangle;\ b|\pm 3/2\rangle + a|\mp 5/2\rangle$$

$$\Gamma_7^{(2)} = a|\pm 3/2\rangle - b|\mp 5/2\rangle;\ -b|\pm 5/2\rangle + a|\mp 3/2\rangle$$

$$a^2 + b^2 = 1$$

These states correspond to the following energy levels and wave functions:

$$\Gamma_6:\ E_6 = -8B_2^0 + 120\ B_4^0 \qquad |\psi_6\rangle = |\pm 1/2\rangle$$

$$\Gamma_7^{(1)}\ E_7^{(1)} = 4B_2^0 - 60B_4^0 - 6\sqrt{(B_2^0 + 20B_4^0)^2 + 20(B_4^4)^2}$$

$$|\psi_7^{(1)}\rangle = \sin\Theta|\pm 5/2\rangle - \cos\Theta|\pm 3/2\rangle$$

$$\Gamma_7^{(2)}\ E_7^{(2)} = 4B_2^0 - 60B_4^0 + 6\sqrt{(B_2^0 + 20B_4^0)^2 + 20(B_4^4)^2}$$

$$|\psi_7^{(2)}\rangle = \cos\Theta|\pm 5/2\rangle + \sin\Theta|\mp 3/2\rangle$$

$$\text{where}\ \ tg^2\Theta = \frac{2\sqrt{5}B_4^4}{B_2^0 + 20B_4^0}$$

For $B_4^4 = 0$, $a = 1$ and $b = 0$, the three doublets are mutually connected by magnetic dipole transitions. The $\Gamma_6 - \Gamma_7^{(1)}$ transition matrix element vanishes for $B_4^4 = 0$. Thus, in the general case, three transitions can be observed in the inelastic neutron scattering experiment. The sign of B_2^0, B_4^0, and B_4^4 parameters influences the shape of the level scheme. For B_2^0 positive, the ground state is Γ_6 while for negative B_2^0 it is Γ_7 (Figure 4.35). Table 4.26 lists the CEF parameters obtained for CeT_2X_2 compounds by inelastic neutron scattering, CEF-level schemes are displayed in Figure 4.35. In all compounds with the crystal structure of the $ThCr_2Si_2$ type, the ground state is Γ_7 which corresponds to the negative value of B_2^0 parameter. In the case of $CePt_2Si_2$,

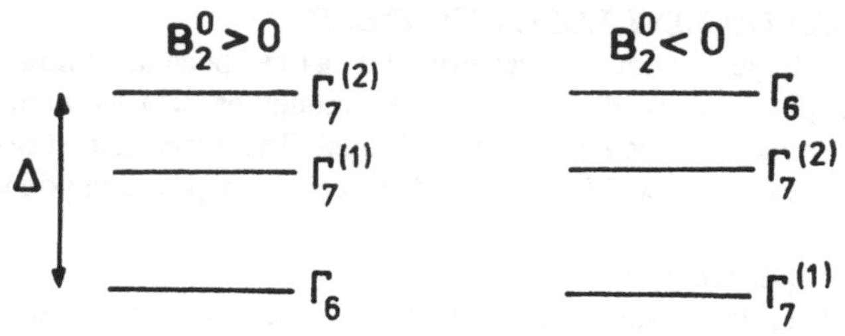

FIGURE 4.35. Energy levels of CeT_2X_2 for positive and negative values of the B_2^0 parameter.

TABLE 4.26
CEF Level Scheme and Parameters for CeT_2X_2 Compounds

Compound	$E_1(K)$	$E_2(K)$	$B_2^0(K)$	$B_4^0(K)$	$B_4^4(K)$	a	b	$\mu_{calc}(\mu_B)$	Ref.
$CeCu_2Si_2$	145.8	378.5	−3.1	0.41	−6.5	0.83	0.56		1
$CePd_2Si_2$	221.7	247.0	−11.4	−0.0012	−3.2	0.899	0.437		1
$CeAg_2Si_2$	106.0	216.9	−0.5	−0.04	−4.0	0.73	0.684		1
$CePt_2Si_2$			30.7	0.93	19.5				2
$CeAu_2Si_2$	198.8	251.8	−1.08	0.34	−4.6	0.604	0.797		1
$CeCo_2Ge_2$			−14.2	0	0.24				3
$CeCu_2Ge_2$	191.0	191.0	−8.78	−0.054	2.79	0.9	0.436	1.53	4
$CeRu_2Ge_2$	500.0	750.0	−30.5	−1.311	5.41	0.98	0.2	2.02	4
$CePd_2Ge_2$			−4.21	−0.28	−3.42				5
$CeAg_2Ge_2$	0	128	6.5	0.097	1.23	0.285	−0.959	1.85	4
$CeAu_2Ge_2$	127	197	−6.4	−0.27	2.6	0.926	0.379	1.75	4

Note: First and second excited levels E_1 and E_2, CEF parameters B_m^n, and coefficients of the eigen states a and b. Ground state ($E_0 = 0$): $a|\pm 5/2\rangle - b|\pm 3/2\rangle$. First excited state ($E_1$): $|\pm 1/2\rangle$. Second excited state (E_2): $b|\pm 5/2\rangle + a|\mp 3/2\rangle$.

References: 1. Severing et al., 1989; 2. Gignoux et al., 1988; 3. Fujii et al., 1988; 4. Loidl et al., 1992; 5. Besnus et al., 1992.

which crystallizes in the $CaBe_2Ge_2$-type structure, the ground state is Γ_6 and B_2^0 is positive. The directions of the magnetic moment in $CePd_2Si_2$, $CeAg_2Si_2$, $CeAu_2Si_2$, and $CeCu_2Si_2$ agree with the sign of B_2^0. Also, the magnitudes of the ordered magnetic moment calculated for the Ce^{3+} ion at 4.2 K in the mean-field approximation using the CEF-level scheme shown in Figure 4.36, are in fair agreement with the experimentally determined values (Table 4.27). In all cases, large reductions of the moment, compared to the free-ion value for Ce^{3+} ($gJ = 2.55 \mu_B$), are observed. This is an indication of the large influence of the CEF on the magnetic properties of CeT_2X_2 systems.

4.8.3.1.2. PrT_2X_2 Compounds
CEF interactions split the ninefold-degenerated ground state $^3H_4$ of multiplets of the Pr^{3+} ion into five singlets and two doublets:

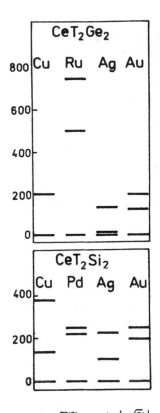

FIGURE 4.36. Crystal field schemes of CeT$_2$Si$_2$ with T = Cu, Pd, Ag, and Au (Severing et al., 1989) and CeT$_2$X$_2$ with T = Cu, Pd, Ag, and Au (Loidl et al., 1992; Besnus et al., 1992).

$\Gamma_{t1}^{(2)}$: $+ (\gamma/|\sqrt{2})| - 4\rangle - \epsilon\sqrt{2}|0\rangle + (\gamma/\sqrt{2})| + 4\rangle$ (singlet)

$\Gamma_{t5}^{(2)}$: $+ \alpha| \pm 3\rangle + \beta| \mp 1\rangle$ (doublet)

$\Gamma_{t5}^{(1)}$: $- \alpha| \pm 1\rangle + \beta| \mp 3\rangle$ (singlet)

Γ_{t4}: $+ 1/\sqrt{2}(| - 2\rangle + | + 2\rangle)$ (singlet)

Γ_{t2}: $+ 1/\sqrt{2}(| - 4\rangle + | + 4\rangle)$ (singlet)

$\Gamma_{t1}^{(1)}$: $+ \epsilon| + 4\rangle + \gamma|0\rangle + \epsilon| - 4\rangle$ (singlet)

Γ_{t3}: $+ 1/\sqrt{2}(|-2\rangle - |+2\rangle)$ (singlet)

These states are written in sequence of the decreasing energy. The ground state $(|-2\rangle - |+z\rangle)$ is a nonmagnetic singlet. In the absence of a magnetic field, the moment associated with this state is zero. The CEF parameters for PrT$_2$X$_2$ compounds have been determined using a variety of experimental techniques. Thus, Mössbauer data obtained for PrFe$_2$Si$_2$ and PrFe$_2$Ge$_2$ give the CEF energy-level scheme shown in Table 4.28. Its characteristic feature is the singlet $\Gamma_{t1}^{(1)}$ ground state with the excited singlet Γ_{t2} and the doublet $\Gamma_{t5}^{(1)}$ close to it (50 mK at most). This scheme indicates the negative sign of

TABLE 4.27
Comparison of the Calculated and Observed Magnetic Moments from Some CeT_2X_2 Compounds

Compound	$\mu_{calc}(\mu_B)$	$\mu_{obs}(\mu_B)$	Direction of the magnetic moments	Ref.
$CeAl_2Ga_2$	1.3	1.18(7)	c	1
$CeCu_2Ge_2$	1.53	0.77(5)	c	2, 3
$CeRu_2Ge_2$	2.02	1.98(3)	c	2
$CeAg_2Ge_2$	1.47	1.50(10)	c	2
$CeAu_2Ge_2$	1.75	1.88(5)	c	2

References: 1. Gignoux et al., 1988a; 2. Loidl et al., 1992; 3. Knopp et al., 1989.

TABLE 4.28
Crystal Field Energy Levels (in K) for PrT_2X_2

Compound	$\Gamma_{t1}^{(1)}$	Γ_{t2}	$\Gamma_{t3}^{(1)}$	Γ_{t4}	Γ_{t3}	$\Gamma_{t3}^{(2)}$	$\Gamma_{t1}^{(2)}$	$\mu_{Pr}(\mu_B)$	Ref.
$PrFe_2Si_2$	0	20.6	46.9	102	324	336	347	1.65	1
$PrFe_2Ge_2$	0	22.0	22.0	42.3	252	261	272	2.75	1
$PrCo_2Si_2$	0			127	302	355	382	3.2	2
$PrNi_2Si_2$	0	37.9	58.4						3

References:1. Melaman et al., 1992; 2. Shigeoka et al., 1989; 3. Blanco et al., 1992.

B_2^0. The overall splitting is on the order of 350 and 270 K for X = Si and Ge, respectively. The excitation energies of the singlet and doublet and their mutual positions depend critically on the higher-order CEF parameters, particularly B_4^4. The proximity of Γ_{t2} and $\Gamma_{t1}^{(2)}$ leads to large values of the ordered moment when one allows for the presence of the molecular field in the ordered state in a self-consistent way. The different behavior of these two compounds can thus be related principally to the different size of the exchange interaction together with the energy difference between the two low-lying CEF levels. The CEF parameters obtained by the simple model consistently give μ_s = 1.65 μ_B and 2.75 μ_B for the Si and Ge compounds, respectively. In the first case, the singlet is the first excited state, whereas it is the doublet in $PrFe_2Ge_2$. An ordered moment of 1.55 μ_B can easily be accounted for by an increase of the B_4 value by a few percentage points. In turn, this results in an increase of the singlet-singlet separation and a decrease of the excitation energy of the doublet, showing that the relative position of the subset of low-lying CEF levels is really critical.

Analysis of the magnetization curves of $PrMn_2Ge_2$ at 4.2 K gives the following CEF parameters: $B_2^0 = -0.7$ K and $B_4^0 = 0.02$ K (Table 4.29). Magnetic susceptibility data obtained for $PrCo_2Si_2$ show CEF splitting giving rise to a doublet ground state although the Pr^{3+} ion is a non-Kramers ion.

TABLE 4.29
Values of the B_n^m Crystal Field Parameters in PrT_2X_2 Compounds

Compound	$B_2^0(K)$	$B_4^0(K)$	$B_4^4(K)$	$B_6^0(K)$	$B_6^4(K)$	Ref.
$PrMn_2Ge_2$	−0.7	−2.0				1
$PrCo_2Si_2$	−8.0	−0.0136	0.05	0.0026	−0.0024	2
$PrCo_2Ge_2$	−6.5	−2.0				3
$PrNi_2Si_2$	−3.99	0.0016	0.156	0.00013	−0.00032	4
$PrRu_2Si_2$	−14.7					3

References: 1. Iwata et al., 1986; 2. Shigeoka et al., 1989; 3. Shigeoka et al., 1992; 4. Barandiaran et al., 1986.

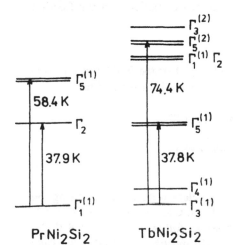

FIGURE 4.37. Low energy levels of $PrNi_2Si_2$ and $TbNi_2Si_2$ (Blanco et al., 1992).

The first excited state is also a doublet which lies 127 K above the ground state, the overall splitting being 382 K. This CEF-level scheme leads to the full moment (3.2 μ_B) of the theoretical Pr^{3+} free ion and the Ising behavior observed at low temperatures. The experimentally obtained CEF parameters and those calculated based on the point-charge model show good agreement. Thus, if one assumes the valencies of Pr, Co, and Si as 3.0, 0.67, and 0, respectively, the calculated second-order parameter B_2^0 will agree with the experimentally determined value. The other higher-order parameters deduced from this model are as follows: $B_4^0 = -1.25 \times 10^{-2}$ K, $B_4^4 = 3.73 \times 10^{-2}$ K, $B_6^0 = 3.00 \times 10^{-3}$ K, and $B_6^6 = -2.82 \times 10^{-3}$ K. The shielding effect has not been taken into account. Thus, in this case, the point-charge model gives a fairly good description of the CEF effect despite its simplicity. The low energy levels of $PrNi_2Si_2$ are shown in Figure 4.37.

Tables 4.28 and 4.29 show the level schemes and CEF parameters reported to date for PrT_2X_2 compounds deduced from magnetization, specific heat, and inelastic neutron scattering data. The observed magnitudes of the magnetic

TABLE 4.30

Values of the B_n^m Crystal Field Parameters in NdT_2X_2 Compounds

Compound	$B_2^0(K)$	$B_4^0(K)$	$B_4^4(K)$	$B_6^0(K)$	$B_6^4(K)$	Ref.
$NdMn_2Ge_2$	0.45	0.0057	−0.020			1
$NdCo_2Ge_2$	−1.6	0	−0.011	−0.0013	−0.0001	2
$NdCu_2Si_2$	−0.37	0.013	−0.016	−0.00038	−0.0008	3
$NdRh_2Si_2$	−0.915	0.0189	−0.0344			4

References: 1. Shigeoka et al., 1988a; 2. Fujii et al., 1988; 3. Goremychkin et al., 1992; 4. Takano et al., 1987.

moment localized on the Pr ions are in all cases smaller than the free-ion values, indicating the effect of the CEF.

4.8.3.1.3. NdT_2X_2 Compounds

CEF splits the $J = 9/2$ ground state multiplet of the Nd^{3+} ion into five Kramers doublets. The CEF parameters have been determined from:

* Magnetization data ($NdMn_2Ge_2$ and $NdRh_2Si_2$)
* Magnetic susceptibility data ($NdCo_2Ge_2$)
* Neutron inelastic scattering data ($NdCu_2Si_2$)

These parameters are collected in Table 4.30.

In the case of $NdCo_2Ge_2$, the CEF parameters were estimated using the point-charge model assuming the valence of Nd, Co, and Ge to be 3.0, 0.67, and 0, respectively. The obtained values are as follows: $B_2^0 = -1.8$ K, $B_4 = 0$ K, $B_4^0 = -0.011$ K, $B_6^0 = -0.0013$ K, and $B_6^6 = -0.0001$ K. They are in fair agreement with the experimental data. This indicates that in $NdCo_2Ge_2$, the anisotropy of magnetic susceptibility can be qualitatively explained as due to the CEF effect. The CEF-level scheme and the f electron wave functions obtained for $NdCu_2Si_2$ are shown in Figure 4.38.

4.8.3.1.4. TbT_2X_2 Compounds

The results obtained for this group of compounds constitute proof for the existence of a CEF singlet ground state associated with the one-dimensional Γ_3 or Γ_4 irreducible representation of the D_{4h} group. The corresponding basis vectors are $\alpha(|+6\rangle + |-6\rangle) + \beta(|+2\rangle + |-2\rangle)$. Magnetization and neutron diffraction measurements show that the Γ_3 and Γ_4 singlets are close to each other, and that is close to unity. Indeed:

* In zero field, the magnetic moment as well as the magnetization at large magnetic fields are close to the maximum value for the Tb^{3+} ion. This indicates that the ground state is close to $|+6\rangle$.
* The sign and value of the second-order CEF term (Table 4.31) would lead, if alone, to the $|6\rangle$ doublet ground state.

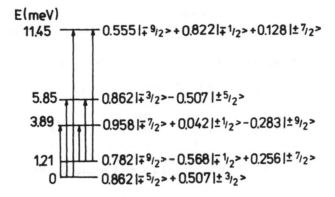

E(meV)

11.45 —— $0.555 | \mp ^9/_2 > + 0.822 | \mp ^1/_2 > + 0.128 | \pm ^7/_2 >$

5.85 —— $0.862 | \mp ^3/_2 > - 0.507 | \pm ^5/_2 >$

3.89 —— $0.958 | \mp ^7/_2 > + 0.042 | \pm ^1/_2 > - 0.283 | \pm ^9/_2 >$

1.21 —— $0.782 | \mp ^9/_2 > - 0.568 | \mp ^1/_2 > + 0.256 | \pm ^7/_2 >$

0 —— $0.862 | \mp ^5/_2 > + 0.507 | \pm ^3/_2 >$

FIGURE 4.38. Crystal field-level scheme and f electron wave functions of the Nd^{3+} ion in $NdCu_2Si_2$. The arrows show the observed transitions at 20 K (Goremychkin et al., 1992).

TABLE 4.31
Values of the B_n^m Crystal Field Parameters in TbT_2X_2 Compounds

Compound	$B_2^0(K)$	$B_4^0(K)$	$B_4^4(K)$	$B_6^0(K)$	$B_6^4(K)$	Ref.
$TbNi_2Si_2$	−0.96	0.000557	−0.0116	−0.000008	0.000937	1
$TbRh_2Si_2$	−1.09	−0.0046	−0.00854			2

References: 1. Blanco et al., 1992; 2. Takano et al., 1987.

The small B_4^4 and B_6^4 terms responsible for the splitting of this doublet can be deduced from the small difference in the M vs. H plot observed along the (100) and (110) directions. The CEF parameters have only been determined for two compounds: $TbNi_2Si_2$ and $TbRh_2Si_2$. The level scheme determined for $TbNi_2Si_2$ is displayed in Figure 4.37.

4.8.3.1.5. DyT_2X_2 Compounds

A fairly large amount of experimental data have been obtained for these compounds, mainly by the Mössbauer spectroscopy method. The results indicate the Kramers doublet as the CEF ground state with almost $| \pm 15/2 \rangle$ character. The CEF parameters for the $DyRh_{2-x}Co_xSi_2$ solid solution were obtained from specific heat and magnetization data. Only small changes have been found in the CEF parameters as x increases. The CEF parameters are shown in Table 4.32.

4.8.3.1.6. HoT_2X_2 Compounds

The CEF parameters are only known for $HoRh_2Si_2$:

• $B_2^0 = -0.12$ K, $B_4^0 = 0.0004$ K, and $B_4^4 = -0.002$ K from the specific heat measurements

TABLE 4.32
Values of the B_n^m Crystal Field Parameters in DyT_2X_2 Compounds

Compound	$B_2^0(K)$	$B_4^0(K)$	$B_4^4(K)$	Ref.
$DyMn_2Ge_2$	-1.4	0.0003	0.033	1
$DyFe_2Si_2$	-1.8	-0.0039		2
$DyCo_2Si_2$	-1.75	0.003		3
$DyRh_2Si_2$	-1.93	0.0043		4
$DyRu_2Si_2$	-4.94	0.0063		4

References: 1. Venturini et al., 1992; 2. Görlich, 1980; 3. Iwata et al., 1990; 4. Sanchez et al., 1988.

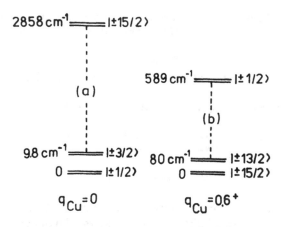

FIGURE 4.39. Crystal field splitting of the erbium $\pm 15/2$ multiplet for $q_{Cu} = 0$ and $q_{Cu} = +0.6$. The energy-level diagram shows the strongest components of the wave function (Yakinthos, 1985).

• $B_2^0 = -0.22$ K, $B_4^0 = 0.0011$ K, and $B_4^4 = -0.002$ K from magnetic data

4.8.3.1.7. ErT_2X_2 Compounds

In all these systems, the magnetic moment localized on the Er ion has been found to be smaller than the free-ion value of 9 μ_B. With the exception of $ErCu_2Ge_2$, the magnetic moment in the other compounds is normal to the tetragonal axis. The case of $ErCu_2Ge_2$ may be explained as the effect of CEF. The level scheme for the Er^{3+} ion was deduced assuming that the charge is localized on Er and is zero on Cu. The level scheme is displayed in Figure 4.39.

TABLE 4.33
Values of the B_n^m Crystal Field Parameters in TmT_2Si_2 Compounds

Compound	$B_2^0(K)$	$B_4^0(K)$	$B_4^4(K)$	$B_6^0(K)$	$B_6^4(K)$	Ref.
$TmFe_2Si_2$	2.53	-0.017	0.049	0	0.002	1
$TmCu_2Si_2$	0.12	-0.0312	-0.049	-0.00014	0.00126	2

References: 1. Umarji et al., 1984; 2. Stewart and Żukrowski, 1982.

The magnetic moment is given by:

$$\mu_z = \mu_B g_J \langle a_+ | J_z | a_- \rangle$$

$$\mu_x = \mu_B g_J \langle a_+ | J_x | a_- \rangle \qquad (4.11)$$

where $|a_+\rangle$ and $|a_-\rangle$ are the two wave functions of the fundamental doublet, $|a_-\rangle$ is the time conjugate of $|a_+\rangle$, g_J is the Lande factor, J_z is the z component of the lanthanide total angular momentum, and is the magnetic moment in the z and x directions (x is normal to z). The wave functions were calculated to be:

$$|a_+\rangle = 0.156| \pm 7/2\rangle + 0.984| \pm 1/2\rangle + 0.081| \pm 9/2\rangle \quad (4.12)$$

Hence, $\mu_z = 0.5$ and $\mu_x = 4.8$ μB. However, this model is not consistent with the experimental data. If a nonzero charge is assumed on the Cu ion, the splitting of the $| \pm 15/2\rangle$ multiplet becomes reversed (see Figure 4.39) and the ground doublet obtains $| \pm 15/2\rangle$ as its strongest component. For $q_{Cu} = 0.6^+$, the calculated wave functions are

$$|b_\pm\rangle = 0.920| \pm 15/2\rangle + 0.249| \pm 9/2\rangle + 0.297| \pm 1/2\rangle \quad (4.13)$$

Consequently, the moments are $\mu_z = 7.9$ μ_B and $\mu_x = 0.4$ μ_B. The direction of the magnetic moment is thus along the tetragonal axis, in agreement with the neutron diffraction results. The CEF parameters obtained from magnetic data are only available for $ErRh_2Si_2$: $B_2^0 = 0.241$ K, $B_4^0 = -0.00135$ K, and $B_4^4 = -0.00246$ K.

4.8.3.1.8. TmT_2X_2 Compounds
The CEF parameters have only been found for $TmFe_2Si_2$ and $TmCu_2Si_2$. They are listed in Table 4.33.

The CEF splitting in $TmRh_2Si_2$, $TmFe_2Si_2$, and $TmCu_2Si_2$ has shown that the ground state of thulium was deduced to be $|0\rangle$, with the first excited state ($| \pm 1\rangle$) located at 49 K in $TmRh_2Si_2$ and 8 K in $TmFe_2Si_2$. Such splitting leads to a reduction of the magnetic moment value on the Tm^{3+} ion, yielding 4.2 μ_B in $TmRh_2Si_2$ and 6.2 μ_B in $TmCu_2Si_2$. Both values are smaller than the free-ion value of 7 μ_B.

4.8.1.9. Summary

A vast amount of CEF data obtained for RT_2X_2 compounds shows that the B_2^0 coefficient plays a dominant role, since the other components are smaller by an order of magnitude. At the site with tetragonal point symmetry, the easy axis of magnetization is aligned along the fourfold axis if B_2^0 is negative, but is normal to it when B_2^0 is positive. The values of the B_2^0 coefficients for a number of RT_2X_2 compounds are collected in Table 4.34, which also includes the orientation of the magnetic moment localized on a lanthanide ion as deduced from neutron diffraction experiments. The correlation between the sign of B_2^0 and moment orientation is clearly visible. For example, in $TbCo_2Si_2$, $DyCo_2Si_2$, and $HoCo_2Si_2$, the B_2^0 is negative and the magnetic moments are parallel to the tetragonal axis, but in $ErCo_2Si_2$ and $TmCo_2Si_2$, for which the B_2^0 coefficients are positive, they are normal to it. In copper-containing compounds, the situation is opposite: B_2^0 was found to be positive for the Tb, Dy, and Ho compounds, and the alignment of magnetic moments was in the basal plane. In $ErCu_2Si_2$, the moments point along the fourfold axis and B_2^0 is positive.

The data collected in Table 4.35 also indicate that the sign of the B_2^0 coefficient is dependent on the number of 4f and 5d electrons. Figure 4.40 displays the plot of B_2 against the R element for a number of RT_2Si_2 compounds, where R = Tb, Dy, Ho, Er, Tm, Yb, and T stands for 3d, 4d, and 5d elements. In all of them, the change of sign of the B_2^0 coefficient occurs for Ho and Er compounds. Different behavior is observed in the case of nd^{10} elements (Cu, Ag, Au). This regularity is not obeyed by the Ni- and Pd-containing systems, for which the magnitudes of the B_2^0 coefficients are small compared to other compounds. Thus, the signs of the CEF higher-order terms may be correlated with those observed in the experimental moment direction.

Table 4.35 lists the B_2^0 coefficients, ground states, and first-excited level energies determined for RFe_2Si_2 by Mössbauer spectroscopy (Noakes et al., 1983).

The effect of B_2^0 on the magnetic transition temperatures is well illustrated by the case of RCu_2Si_2 compounds. T_N do not follow the De Gennes rule (Figure 4.41), but if the CEF Hamiltonian is added to the exchange Hamiltonian, the agreement with the De Gennes function improves rapidly (Noakes and Shenoy, 1982)

$$H = -2(g_J - 1)^2 J_z \langle J_z \rangle + B_2^0 [3J_z^2 - J(J + 1)] \qquad (4.14)$$

The magnetic ordering temperature is given:

$$T_N = 2(g_J - 1)^2 \sum J_z^2 \exp(-3B_2^0 J_z^2/T_N) \left[\sum_{Jz} \exp(-3B_2^0 J_z^2/T_N) \right]^{-1} \qquad (4.15)$$

The deduced values of B_2^0 are as follows: 0.8 K for Tb, 0.57 K for Dy, 0.175 K for Ho, −0.2 K for Er, and −0.79 K for Tm (Budkowski et al., 1987).

TABLE 4.34
The Values of B_2^0 Coefficients and Direction of the Magnetic Moments for RT_2Si_2 Compounds

R/T	Mn	Fe	Co	Ni	Cu	Ru	Rh	Pd	Ag	Os	Ir	Pt	Au
Ce	[a]		-7.08[b]	-3.99	-3.1		[a]	-11.4[a]	-0.5[a]			30.7	-1.08[b]
Pr	[a]		-1.95[a]		3.72	[b]							
Nd		-0.63[b]		[a]	0.99		-0.915[b]						
Tb		-3.0[b]	-2.65[a]	-0.66[b]	1.69[a]	-8.33[b]	-1.09[b]	-0.18[a]	1.77	-8.32[b]	-3.91	2.48	-0.25[a]
Dy	-1.35	-1.8[b]	-1.59[b]	-0.17	0.56[a]	-4.94[b]	-1.93[c]	-0.11[c]	1.05	-4.93	-2.32[b]	1.47	-0.15
Ho		-0.61[c]	-0.63[b]	-0.13[a]	0.18[a]	-1.64[b]	-0.22[c]	0.04[c]	0.35	-1.64[b]	-0.77	0.49	-0.05
Er	[a]	0.67[a]	0.53[a]	0.14[a]	-0.21	1.78[a]	0.24[a]	0.04[c]	-0.38	1.78[a]	0.84	-0.53	0.05
Tm		2.53	2.12[a]	0.55[a]	-0.69	6.89	0.69[a]	0.15	-1.46	6.88	3.24	-2.06	0.21
Yb		10.12[a]	5.58[a]	0.65	-3.23	20.74	8.11	0.45	-4.41	20.71	9.75	-6.19	0.62

[a] Magnetic moment is perpendicular to the c axis.   [b] Magnetic moment is parallel to the c axis.   [c] Magnetic moments form angle φ with the c axis.

TABLE 4.35
Values of B_2 for Trivalent Lanthanide Ions in RFe_2Si_2 Compounds Derived from ^{155}Gd Quadrupole Interaction and Consequent Effects on R Electronic State, if B_2 Dominates

R	B_2^0 (in K)	Ground state (J_z)	First excited level (K)	Overall splitting (K)
Nd	-0.63	9/2	15.0	38
Tb	-3.00	6	100.0	325
Dy	-1.80	15/2	75.0	305
Ho	-0.61	8	27.0	120
Er	0.67	1/2	4.0	110
Tm	2.54	0	7.6	275
Yb	7.65	1/2	46.0	275

Noakes et al., 1983.

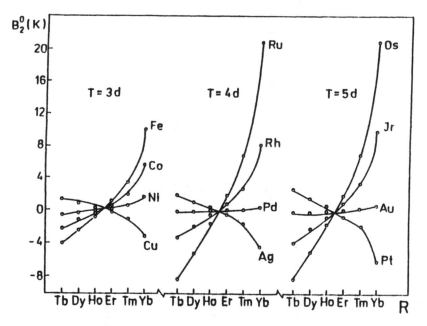

FIGURE 4.40. Dependency of B_2^0 coefficients as a function of the R elements in RT_2Si_2 compounds (R = Tb–Yb, T = 3d, 4d, and 5d metal).

4.8.3.2. RTX$_2$ Compounds

The Néel points of $RNiSi_2$ and $RNiGe_2$ plotted against the R element (Figure 4.42.) show that the De Gennes rule is not obeyed, suggesting that the effect of the CEF is much larger than in the corresponding RNi_2Si_2 and RNi_2Ge_2 compounds. However, to date neither experimental nor calculated data have been published for RTX_2.

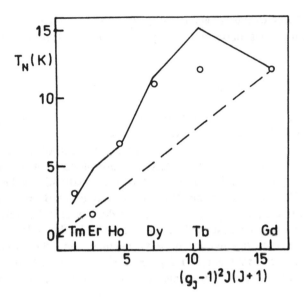

FIGURE 4.41. Comparison of experimental (○) and calculated magnetic transition temperatures T_N for RCu_2Si_2 compounds. The broken smooth line represents the De Gennes rule. The solid line represents the T_N predicted by the B_2^0 model (Budkowski et al., 1987).

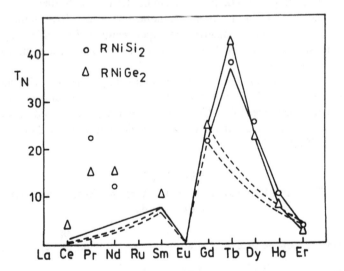

FIGURE 4.42. Magnetic ordering temperatures of $RNiSi_2$ and $RNiGe_2$ compounds as a function of the rare-earth ion. Prediction of the De Gennes rule (dashed line), crystal field model with the B_2^0 term only (solid line) (Gil et al., 1983a).

4.8.3.3. RTX Compounds

Although a large number of the 1:1:1 compounds have been synthetized and examined by magnetometric, X-ray, neutron diffraction, and Mössbauer spectroscopy methods, the CEF data are only available for CeCuSi. Its crystal structure belongs to the hexagonal Ni_2In type. Ce^{3+} ions are located at the sites with 3m point symmetry. Consequently, the CEF Hamiltonian is given as:

$$H_{CEF} = B_2^0 O_2^0 + B_4^0 O_4^0 + B_4^3 O_4^3$$

The values of B_n^m parameters were determined to be as follows: $B_2^0 = 9.14$ K, $B_4^0 = -0.035$ K, $B_4^3 = 6.56$ K. The magnetic moment of the Ce ion lies in the basal plane, which is in agreement with the sign of B_2^0. Under these conditions, the multiplet $J = 5/2$ is split into three doublets which are mainly $\pm |M\rangle$ states ($|M_J\rangle = 1/2, 3/2$, or $5/2$) with a small mixing between the $\pm|1/2\rangle \pm|5/2\rangle$ states due to the B_4^3 term. For the above B_n^m parameters, the ground state in the paramagnetic region is the doublet $\pm|1/2\rangle$, well separated ($\Delta = 90$ K) from the first excited level $\pm|3/2\rangle$. In the ordered state, the basal plane is thus favored as the easy magnetization direction, with an associated magnetic moment of 1.2 μ_B at 0 K. This value turned out to be quite consistent with the experimental data ($\mu_R = 1.25$ μ_B) (Gignoux et al., 1986a).

4.8.3.4. RT$_2$X Compounds

In lanthanide Heusler intermetallic compounds, the ordering temperatures and associated magnetic energies are small. CEF effects therefore play an essential role in determining their magnetic properties. The values of lanthanide magnetic moments determined experimentally for cubic RT_2X compounds are smaller than the free R^{3+} ion values. This result indicates the strong influence of CEF.

In RT_2X compounds the lanthanide ions occupy the site with the cubic point symmetry. The CEF will thus lift the $(2J + 1)$-fold degeneracy of the free-ion state. The values of the CEF parameters were determined for a number of RT_2X compounds using the neutron inelastic scattering method. They are collected in Table 4.36.

TABLE 4.36
Crystal Electric Field Parameters and Ground States for Some RT$_2$X Compounds

Compound	Ground state	W(K)	x	B_4^0 (in 10^{-2}K)	B_4^4 (in 10^{-4}K)
CeAg$_2$In	Γ_7	2.9(5)	1		
PrAg$_2$In	Γ_3	−1.2(2)	0(0.2)		
NdAg$_2$In	Γ_6	1.2(2)	0(0.2)		
TbPd$_2$Sn		0.64	−0.785		
DyPd$_2$Sn		−0.43	−0.509	−0.61	0.38
HoPd$_2$Sn		0.35	0.3	0.32	0.41
ErPd$_2$Sn		−0.54	0.302	−0.39	−0.6
TmPd$_2$Sn	Γ_3	0.92	−0.513	−0.104	1.48
YbPd$_2$Sn	Γ_7	−10.27	−0.7	0.13	−33.0

Note: x represents the relative weight of fourth- and sixth-order terms.

Chapter 5

MAGNETIC MATERIALS BASED ON 3d-RICH TERNARY COMPOUNDS

In this chapter, the magnetic properties of several groups of compounds are presented. A large amount of data concern two groups of compounds: $RT_{12-x}M_x$ and $R_2T_{14}X$. These compounds have been intensively investigated in the last several years, since they constitute the basis for the fabrication of the strongest permanent magnets to date. Other groups of these compounds have been only partially investigated.

5.1. $RT_{12-x}M_x$ PHASES

$RFe_{12-x}M_x$ compounds exhibit the tetragonal $ThMn_{12}$ type of crystal structure (space group I4/mmm). M is Ti, V, Cr, Mn, Mo, W, Al, or Si, and x is in the range $1 < x < 4$ (Yang et al., 1981; De Mooij and Buschow, 1987; Ohashi et al., 1987; Buschow and De Mooij, 1989).

In this type of structure, the lanthanide atoms fully occupy the 8f and 8j sites, while the 8i site is populated by a mixture of Fe and M atoms. An exception occurs when M = Si. In $RFe_{10}Si_2$, the Si atoms share the 8f and 8j positions with the Fe atoms (Buschow, 1988).

The above compounds have been proposed as cheap alternative materials for permanent magnets (De Mooij and Buschow, 1987; Li and Coey, 1991). The results of an intensive and systematic study of these compounds gave the following general features, which emerged from the mass of data collected for the iron series (see also Table 5.1).

1. Curie temperatures range from 260 to 650 K, with the highest values for the Gd compound in each series except for the Mo-containing series, where it occurs for Sm (Figure 5.1).
2. The values of magnetization measured at T = 4.2 K plotted in Figure 5.1 change with the R element. For light rare-earths (J = L − S), the total lanthanide magnetic moment (gJ) is coupled parallel to the Fe moment, in contrast to the heavy rare-earths (J = L + S) where the lanthanide magnetic moment couples antiparallel to the Fe moment. The differences between the magnetization values in different series can be correlated with preferential site occupation by additional atoms.
3. The anisotropy due to the iron sublattice is uniaxial. The magnitude of K_1 (Fe) is a monotonic function of temperature. The 3d sublattice anisotropy favors the c axis for Fe compounds and is planar for Co compounds. The rare-earth sublattice anisotropy is comparatively small in all these compounds except Sm compounds, whose room-temperature anisotropy field is near 10 T (Li and Coey, 1991).

TABLE 5.1
Magnetic Data for $RT_{12-x}M_x$ (T = Fe and Co) Compounds with $ThMn_{12}$-type Structure

Compound	T_c(K)	M_s(μ_B per formula unit)		$\mu_o H_a$(T)		Ref.
		4.2 K	293 K	4.2 K	293 K	
$NdFe_{11}Ti$	547	21.27	16.8			1, 2
$SmFe_{11}Ti$	584	20.09	16.98	15	10.5	1–5
$GdFe_{11}Ti$	607	12.46	12.5	7.81	3.34	1, 2, 6, 7
$TbFe_{11}Ti$	554	9.7	10.6			1, 2
$DyFe_{11}Ti$	534	9.7	11.3		2.3	1, 2
$HoFe_{11}Ti$	530	9.58				1, 2
$ErFe_{11}Ti$	505	9.2	12.4	8.3	2.4	1, 2, 8
$TmFe_{11}Ti$	496					1, 2
$LuFe_{11}Ti$	488		15.7		2.2	1, 2
$YFe_{11}Ti$	524	19.0	16.6	4.0	2.1	1, 6, 8
$GdCo_{11}Ti$	1080	7.8[a]	9.3			9
$HoCo_{11}Ti$	1073	6.0[a]	10.6			9
$ErCo_{11}Ti$	1066	7.5[a]	12.7			9
$YCo_{11}Ti$	943	12.69[a]	12.93			8, 10
$CeFe_{10}V_2$	440					11
$NdFe_{10}V_2$	570	18.07	17.2		1.89	12, 13, 14
$SmFe_{10}V_2$	610		11.55	15.0	5.4	12, 15
$GdFe_{10}V_2$	616	8.93			3.34	7, 12
$TbFe_{10}V_2$	570	7.92				12, 16
$DyFe_{10}V_2$	540	6.66				12, 16
$HoFe_{10}V_2$	525	6.35		10.0		12, 16
$ErFe_{10}V_2$	505	8.37	10.6		1.76	8, 12, 16
$TmFe_{10}V_2$	496	11.17				12
$LuFe_{10}V_2$	483	15.49				12
$YFe_{10}V_2$	532	16.15	14.21	4.0	1.97	8, 12, 17
$YCo_{10}V_2$	611	7.19[a]				18
$NdFe_{10}Si_2$	574	19.8[a]	16.5		1.75	13, 19
$SmFe_{10}Si_2$	606		15.3		5.2	11, 15
$GdFe_{10}Si_2$	610	11.4	11.5		3.3	7, 11, 19
$TbFe_{10}Si_2$	585	9.6	10.7			11, 19
$DyFe_{10}Si_2$	566	9.4	10.9			11, 19
$HoFe_{10}Si_2$	558	9.1	10.6			11, 19
$ErFe_{10}Si_2$	550	10.4	12.2			11, 19
$TmFe_{10}Si_2$	546	10.2	11.9			11, 19
$LuFe_{10}Si_2$	540					11
$YFe_{10}Si_2$	540	14.06[a]	13.05	3.65[a]	2.18	8, 11
$NdFe_{10}Cr_2$	530	19.8[a]	16.2		4.6	11, 20
$SmFe_{10}Cr_2$	565		12.28		5.9	11, 15
$GdFe_{10}Cr_2$	580	11.2[a]	11.3		2.7	11, 20
$TbFe_{10}Cr_2$	525	9.3[a]	9.6			11, 20
$DyFe_{10}Cr_2$	495	9.8[a]	10.0		4.9	11, 20
$HoFe_{10}Cr_2$	485	9.9[a]	10.9		4.0	11, 20
$ErFe_{10}Cr_2$	475	10.2[a]	11.4		2.7	11, 20
$TmFe_{10}Cr_2$	465	10.1[a]	11.0		2.8	11, 20
$LuFe_{10}Cr_2$	450					11

TABLE 5.1 (continued)
Magnetic Data for $RT_{12-x}M_x$ (T = Fe and Co) Compounds
with $ThMn_{12}$-type Structure

Compound	$T_c(K)$	$M_s(\mu_B$ per formula unit)		$\mu_oH_a(T)$		Ref.
		4.2 K	293 K	4.2 K	293 K	
$YFe_{10}Cr_2$	510	16.67	14.9		2.6	11, 17, 20
$CeFe_{10}Mo_2$	260					11
$PrFe_{10}Mo_2$	385					11
$NdFe_{10}Mo_2$	400	16.4	12.6			11, 21
$SmFe_{10}Mo_2$	483	15.2	10.48		3.7	15, 21, 22
$GdFe_{10}Mo_2$	430	9.79	7.04		2.19	7, 11, 21
$TbFe_{10}Mo_2$	390	4.9				21
$DyFe_{10}Mo_2$	365	4.4				21
$HoFe_{10}Mo_2$	345	3.5				21
$ErFe_{10}Mo_2$	310	3.6				21
$TmFe_{10}Mo_2$	290	5.5				21
$LuFe_{10}Mo_2$	260	10.6				21
$YFe_{10}Mo_2$	360	13.0				11, 21, 22
$NdFe_{10}W_2$	547		14.5			23
$SmFe_{10}W_2$	532				2.47	7
$GdFe_{10}W_2$	569		11.29		2.52	7
$YFe_{10}W_2$	500					11
$GdFe_{10}Mn_2$	445		7.81		2.17	7

ᵃ At 77 K.

References: 1. Hu et al., 1989; 2. Yang et al., 1989b; 3. Liu et al., 1988; 4. Yang et al., 1989a; 5. Kaneko et al., 1989; 6. Christides et al., 1988; 7. Chin et al., 1989a; 8. Solzi et al., 1989; 9. Sinha et al., 1989; 10. Yang et al., 1988a; 11. Buschow and De Mooij, 1989; 12. Xing and Ho, 1989; 13. Chin et al., 1989b; 14. Christides et al., 1990; 15. Ohashi et al., 1988; 16. Christides et al., 1989b; 17. Verhoef et al., 1988; 18. Jurczyk and Chistjakov, 1989; 19. Wang et al., 1988; 20. Wrzeciono et al., 1989; 21. Christides, 1990; 22. Yang et al., 1990; 23. Chin et al., 1989b.

4. The series show a variety of spin reorientations, since the anisotropy contributions from the iron sublattice and the rare-earth sublattice are of comparable magnitudes but opposite sign. Figure 5.2 presents the magnetic structure diagram for some of the $RFe_{12-x}M_x$ compounds. Spin reorientations reflect the different thermal variations of crystal-field terms of different order acting on the lanthanide ion.

5. The measured average iron magnetic moments in R ($Fe_{12-x}M_x$) range from 1.35 to 1.93 μ_B. The detailed experimental data concerning the individual iron moments located at the three different crystallographic sites, obtained by neutron diffraction and ^{57}Fe Mössbauer spectroscopy, are given in Table 5.2. These data indicate the different values of the

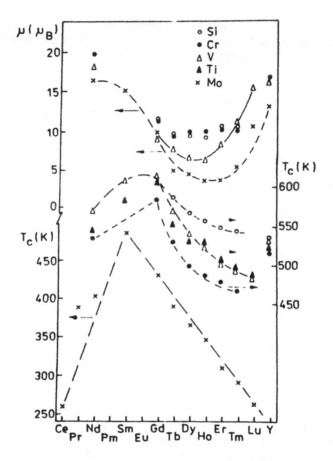

FIGURE 5.1. Plot of the average atomic moment μ at 4.2 K and Curie temperature in $RFe_{10}M_2$ (M = Si, Cr, Mo, V) and $RFe_{11}Ti$.

magnetic moments located in different sites. It is possible to interpret these obtained results on the basis of the following prediction:

- The average Fe–Fe distances follow the relation: $d_{FeFe}(8i) > d_{FeFe}(8j) > d_{FeFe}(8f)$.
- The coordination number of iron in different sites indicates that the (8i) site should exhibit the largest magnetic moment, suggesting that $\mu_{Fe}(8i) > \mu_{Fe}(8j) > \mu_{Fe}(8f)$. Band calculations for YFe_{12} (Coehoorn, 1988, 1990) and $YFe_{10}M_2(M = V$ or $Cr)$ (Jaswal et al., 1990) support this prediction.

The influence of the substitution atoms is observed in $YCo_{12-x}V_x$ compounds (Table 5.3). With an increase of vanadium content, a large decrease of the Curie temperature and magnetization is observed (Jurczyk, 1990).

The substitution of vanadium atoms in the $YFe_{12-x}V_x$ system indicates that compounds with the tetragonal $ThMn_{12}$ structure are formed for x ranging

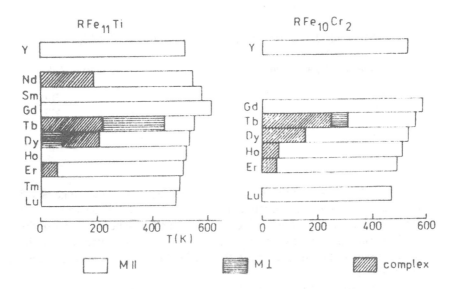

FIGURE 5.2. Temperature dependence of ferromagnetic phases for $RFe_{11}Ti$ (Hu et al., 1989) and $RFe_{10}Cr_2$ (Stefański, 1991) series.

TABLE 5.2
Fe Moment on the Three Different Crystallographic Sites of the $ThMn_{12}$ Structure[a]

Compound	Method	$\mu_{Fe}(\mu_B)$ at 4.2 K			$\mu_{Fe}(\mu_B)$ at RT			Ref.
		8i	8j	8f	8i	8j	8f	
				Experimental Data				
$YFe_{10}V_2$	Neutron	1.83(7)	1.95(13)	1.52(16)	1.53(5)	1.70(9)	1.16(11)	1
$YFe_{11}Ti$	Neutron				1.80	1.92	2.28	2
$YFe_{11}Ti$	Mössbauer	1.50(3)[b]	1.97(3)[b]	1.75(3)[b]	1.32(3)	1.74(3)	1.51(3)	3, 4
$LuFe_{11}Ti$	Mössbauer	1.43(3)[b]	1.90(3)[b]	1.67(3)[b]	1.23(3)	1.67(3)	1.43(3)	3, 4
$YFe_{10}Mo_2$	Mössbauer	1.16	1.92	1.55				5
$YFe_{10}V_2$	Mössbauer	1.25	1.75	1.47				5
$YFe_{10}Cr_2$	Mössbauer	1.20	1.82	1.57				5
				Calculation				
$YFe_{10}V_2$		1.68	2.1	1.99				6
$YFe_{10}Cr_2$		1.77	2.23	2.09				6

[a] Determined by neutron diffraction and ^{57}Fe Mössbauer spectroscopy
[b] At 77 K.

References: 1. Helmholdt et al., 1988; 2. Yang et al., 1988; 3. Hu et al., 1989a; 4. Hu et al., 1989b; 5. Denissen et al., 1990; 6. Jaswal et al., 1990.

TABLE 5.3
Magnetic Data for $YCo_{12-x}V_x$ Compounds, at 77 K

Composition x	$M_s(\mu_B$ per formula unit)	$\mu_{Co}(\mu_B$ per formula unit)	$T_c(K)$
1.6	9.85	0.95	710
2.0	7.19	0.72	610
2.5	4.57	0.48	390
3.5	0.29	0.03	77

Jurczyk, 1990.

from $x = 1.5$ to $x = 3$. In $GdFe_{12-x}V_x$ and $NdFe_{12-x}Mo_x$, the range is 1.5 to 2.5. For all compounds, the value of the Curie temperature decreases with an increase in the concentration of x (De Mooij and Buschow, 1988). In the case of $GdFe_{12-x}M_x$, where M = V, Si, Mo, W, or Ti, a decrease of the magnetic moment at all sites has been observed with an increase of x (Sinnema et al., 1989,a).

The composition dependence of the Curie temperature T_c and magnetization M_s for $RFe_{10-x}Co_xV$ (R = Y, Sm, or Dy) is presented in Figure 5.3. In the range $x < 5$, substitution of Fe by Co produces an average increase in T_c of approximately 50 K per Co atom. This suggests that in the $ThMn_{12}$ structure, cobalt atoms prefer the iron sites involved in negative exchange interactions. This prediction is confirmed by neutron diffraction data, which indicate the strong preference of cobalt for the 8f site (Yang et al., 1990). Among the various sites, the interatomic distance between the 8f–8f site atoms is the shortest. Occupation of the 8f sites by Co reduces the negative exchange between Fe–Fe moments, consequently leading to an increase of T_c.

For R = Y and Dy, the saturation moment values increase with the concentration of x up to $x = 2$, and up to $x = 3$ for R = Sm. At higher cobalt contents, M_s decreases (Figure 5.3). As mentioned earlier, such a dependence is characteristic of a large number of Fe–Co-based alloys and is associated with changes of the exchange interactions and different values of the magnetic moments of iron and cobalt atoms. Such dependencies can be understood in terms of the 3d band structure. According to the Slater-Pauling curve, there is a crossover from weak to strong ferromagnetism in binary $Fe_{1-x}Co_x$ or $Fe_{1-x}Ni_x$ alloys as x increases (Friedel, 1958; Williams et al., 1983).

The substitution also influences other magnetic properties. The magnetic structure phase diagram for $TbFe_{10.5-x}Co_xW_{1.2}$ and $TbFe_{12-x}Co_xTi$ is presented in Figure 5.4 (Jurczyk and Rao, 1991). In both systems, the increase of temperature favors spin orientation in the plane normal to the c axis. The substitution of Co atoms thus stabilizes the plane spin orientation.

Investigations of the magnetic properties of $RTiFe_{11-x}Co_x$ compounds (R = Nd, Pr, Tb, Dy, Ho, and Er, $x = 0$ to 11) indicate that the substitution of Co for Fe causes an increase of the Curie temperature (Figure 5.5). The

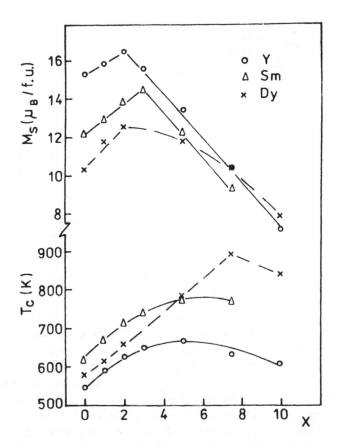

FIGURE 5.3. Concentration dependence of the saturation magnetization and Curie temperature for $RFe_{10-x}Co_xV_2$: (R = Y, Sm, Dy) (Jurczyk and Chistjakov, 1989).

magnetic moment exhibits the maximum value at x = 4. In the $RTiCo_{11}$ system, the magnetic anisotropy of the rare-earth sublattice appears to be planar for the rare earths with a positive second-order Stevens factor (i.e., Er) and axial for the rare earths with a negative second-order Stevens factor (i.e., Tb, Dy, and Ho). This behavior is opposite that observed in $RTiFe_{11}$ compounds. The magnetic anisotropy phase diagrams of the $RTiFe_{11-x}Co_x$ systems were determined (R = Y, Gd, Tb, Dy, Ho, and Er). A complicated diagram is observed for R = Y, Gd, and Er. A change to the conical and planar structure has been found in the composition range $5 < x < 9$. A different situation is observed in Tb, Dy, and Ho compounds (Cheng et al., 1990).

The rare-earth-iron intermetallic compounds of the $RFe_{11}Ti$ family can form hydrides and nitrides by a proper thermal treatment. The interstitial hydrogen or nitrogen atoms exert a large effect on the magnetic properties of the $RFe_{11}TiH_x$ and $RFe_{11}TiN_x$ compounds. In the series with R = Er, Ho,

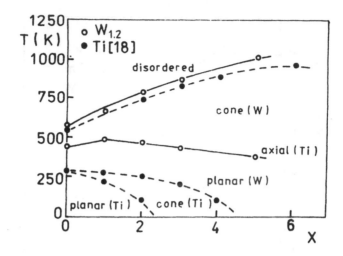

FIGURE 5.4. Diagram of spin configuration types observed in $RFe_{10.8-x}Co_xW_{1.2}$ and $RFe_{11-x}Co_xTi$ (Jurczyk, 1990; Jurczyk and Rao, 1991).

Dy, Tb, and Gd, the magnetic ordering temperatures of the host $RTiFe_{11}$ compounds were found to be between 502 and 603 K. As shown in Table 5.4, upon hydrogenation, the T_c values of the host materials decrease from 63 K for $ErFe_{11}Ti$ to 12 K for $GdFe_{11}Ti$ (Zhang et al., 1990). This decrease is probably related to the expansion of the lattice parameters, since the T_c is mainly determined by the 3d-3d exchange interaction and is very sensitive to the Fe–Fe distance in the binary R–Fe intermetallics (Wallace, 1986). The average saturation magnetization of the Fe sublattice also increases; it is correlated with the electron charge transfer from H to the Fe sublattice. The observed spin reorientation transitions indicate that the influence of hydrogen on the anisotropy of the $RTiFe_{11}$ system has a rather complicated character (Zhang et al., 1990).

In $YFe_{11}Ti$ and $SmFe_{11}Ti$, substitution of hydrogen atoms causes an approximately 60-K increase of the Curie temperature and increases magnetization by 3 to 5%. In the case of $SmFe_{11}TiH_x$, an increase of the anisotropy field from 140 to 170 kOe has been observed with an increase of x (Zhang and Wallace, 1989). Nitrogenation of $YTiFe_{11}$, $NdTiFe_{11}$, and $SmTiFe_{11}$ (Yang et al., 1990) brings about an almost 200-K increase of the Curie temperature, while the iron moment increases about 17%. The magnetic moment is parallel to the c axis in the case of first two compounds, but lies in the basal plane for $RTiFe_{11}N_x$. Interstitial nitrogen atoms thus lead to an increase of both the effective hyperfine field and the isomer shift (Yang et al., 1992). These results indicate that the magnetic moment of iron becomes larger with an increase of nitrogen content.

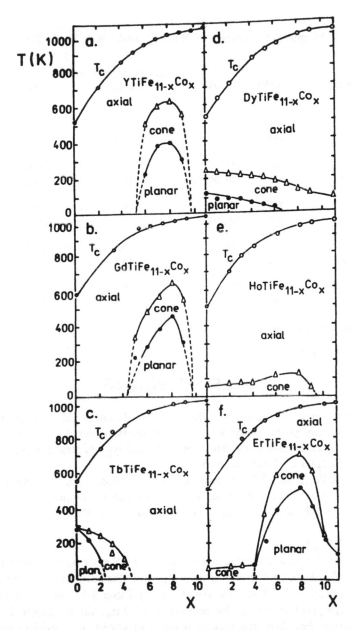

FIGURE 5.5. Magnetic phase diagrams of (a) YTiFe$_{11-x}$Co$_x$, (b) GdTiFe$_{11-x}$Co$_x$, (c) Tb-TiFe$_{11-x}$Co$_x$, (d) DyTiFe$_{11-x}$Co$_x$, (e) HoTiFe$_{11-x}$Co$_x$, and (f) ErTiFe$_{11-x}$Co$_x$ (Cheng et al., 1990).

TABLE 5.4
Magnetic Properties of RTiFe$_{11}$ and Their Hydrides

Compound	T$_c$(K)	T$_{SR}$(K)	μ(μ$_B$ per formula unit) at 4.2 K	DOM[a] 295 K	77 K
GdTiFe$_{11}$					
Host	603	—	11.9	axial	axial
Hydride	615	—	12.3	axial	axial
TbTiFe$_{11}$					
Host	556	327	9.7	planar	conical
Hydride	578	340, 90, 20	10.4	planar	conical
DyTiFe$_{11}$					
Host	539	230, 105	9.1	axial	planar
Hydride	556	210, 95	10.0	axial	planar
HoTiFe$_{11}$					
Host	518	50	9.3	axial	axial
Hydride	562	90	10.8	axial	planar
ErTiFe$_{11}$					
Host	502	50	9.9	axial	axial
Hydride	565	40	11.5	axial	axial

[a]   DOM, direction of magnetization.

Zhang et al., 1990.

5.2. R$_2$T$_{14}$X PHASES

The other group of intermetallic compounds intensively investigated in the last several years are the R$_2$T$_{14}$X compounds, where T = Fe and X = B, C. They crystallize in the tetragonal Nd$_2$F$_{14}$B type of structure (see Figure 2.35 in Chapter 2 of this volume).

The R$_2$Fe$_{14}$B and R$_2$Fe$_{14}$C series are the most complete; they extend from R = La to Lu. On the other hand, compounds with the composition R$_2$Co$_{14}$B are formed only with R elements ranging from La to Tb.

5.2.1. R$_2$Fe$_{14}$B

The fundamental magnetic properties of R$_2$F$_{14}$B systems have been studied by several independent groups of investigators. The bulk magnetic properties of high-purity R$_2$Fe$_{14}$B systems are shown in Table 5.5. The magnetization of Y$_2$Fe$_{14}$B gives an average Fe moment of 2.20 μ$_B$, slightly larger than that in elemental Fe. From the measured magnetizations, it is evident that the light rare-earths Pr, Nd, and Sm are coupled ferromagnetically with the Fe moments, whereas with the heavy lanthanides (Gd–Tm), antiferromagnetic coupling of the R and Fe moments was found. Hence, the aforementioned systems obey the coupling systematics observed in the binary R$_n$T$_m$ compounds (Wallace, 1973; Buschow, 1977; Kirchmayr and Poldy, 1979).

TABLE 5.5
Magnetic Properties of $R_2Fe_{14}B$ Compounds

$R_2Fe_{14}B$	$T_c(K)$	$T_s(K)$	$M_s(\mu_B$ per formula unit)	$K_1(MJ/m^3)$	$K_2(MJ/m^3)$	$K_3(MJ/m^3)$	$\mu_oH_a(T)$	Ref.
R = Y	571		30.7	0.8			1.5	1–7
La	516		30.6				3.0	1–3
Ce	533		29.4	1.8			3.0	1–5
Pr	565		37.6	24.0			7.5	1–6, 11
Nd	588	142	37.6	−16.0	28.0	0.45	7.5	1–11, 16
Sm	618		33.3	−26.0	—	−1.4		1–5, 17
Gd	660		17.79	0.65			2.5	1–5, 16
Tb	629		13.2	6.9			3.3	1, 3, 13
Dy	593		11.3	3.8				1, 3, 5
Ho	574	58	11.2	−1.1	4.4	0.12	9.2	1, 11, 16
Er	557	325	12.7	−1.4	—	−0.29		1, 10, 16
Tm	540	315	18.4	−3.6	—	−0.7		1, 10, 16
Yb	524	115	—					12
Lu	539		28.2				5.0	1, 3, 4

Note: T_c and T_s represent Curie and spin reorientation temperatures, M_s saturation moments at 4.2 K, K_i anisotropy constant and μ_oH_a anisotropy fields at room temperature.

References: 1. Sagawa et al., 1984; 2. Stadelmaier et al., 1985; 3. Sinnema et al., 1984; 4. Abache and Oesterreicher, 1985; 5. Boltich et al., 1985; 6. Burzo et al., 1985; 7. Givord et al., 1984; 8. Buschow et al., 1985; 9. Sagawa et al., 1985; 10. Hirosawa et al., 1986; 11. Grössinger et al., 1986; 12. Burlet et al., 1986; 13. Boge et al., 1985; 14. Tohukara et al., 1985; 15. Yamauchi et al., 1985.

From the known coupling mode and the difference in magnetization of $Y_2Fe_{14}B$ and a compound containing the magnetic rare-earth $R_2Fe_{14}B$, one can estimate the moment carried by the lanthanide ion (assuming, of course, a constant Fe sublattice moment). The moments estimated in this way are in good agreement with the free R^{3+} ion values (Table 5.6).

$R_2T_{14}X$ compounds crystallize in the complicated structure. The rare-earth ions occupy two distinct crystallographic sites (4f and 4g) of the space group $P4_2/mnm$. The iron atoms share six iron-equivalent positions ($16k_1$, $16k_2$, $8j_1$, $8j_2$, 4c, and 4e). Neutron diffraction and ^{57}Fe Mössbauer spectroscopy data gave the information on the values of the magnetic moments at individual sites. The Fe moments obtained by microscopic methods are shown in Table 5.7.

TABLE 5.6
Rare-Earth Moments in $R_2Fe_{14}B$ Systems at $T = 4.2$ K

R	Pr	Nd	Sm	Gd	Dy	Ho	Er	Tm
$\mu(\mu_B)$	3.1	3.2	1.0	6.7	10.1	10.1	9.3	6.7
gJ	3.2	3.27	0.71	7.0	10.0	10.0	9.0	7.0

TABLE 5.7
The Rare-Earth and Iron Moments in the $R_2Fe_{14}B$ and $R_2Fe_{14}C$ Compounds

Compound	T(K)	R_1(4f)	R_2(4g)	Fe_1(4e)	Fe_2(4c)	Fe_3(8j$_1$)	Fe_4(8j$_2$)	Fe_5(16k$_1$)	Fe_6(16k$_2$)	Ref.
				Neutron Diffraction						
$Y_2Fe_{14}B$	4.2	—	—	2.15	1.95	2.40	2.80	2.25	2.25	1
$Nd_2Fe_{14}B$	4.2	2.30	2.25	2.10	2.75	2.30	2.85	2.60	2.60	1
$Nd_2Fe_{14}B$	77	2.3	3.2	1.1	2.2	2.7	3.5	2.4	2.4	2, 3
$Er_2Fe_{14}B$	77	2.6	2.8	1.7	1.7	2.4	3.3	2.6	2.5	3
$Dy_2Fe_{14}B$	77	-8.9	-9.2	2.4	2.5	2.5	3.0	2.6	2.5	3
$Ce_2Fe_{14}B$	77	—	—	2.1	2.4	2.7	3.4	2.7	2.2	4
$Lu_2Fe_{14}B$	77	—	—	1.7	2.2	2.9	3.6	2.8	2.4	4
				Mössbauer Effect Data						
$Nd_2Fe_{14}B$	77	—	—	2.33	2.16	1.87	2.60	2.24	2.55	5
$Er_2Fe_{14}B$	77	-	—	2.41	2.19	1.99	2.65	2.25	2.35	5
$Lu_2Fe_{14}B$	10	—	—	1.83	2.15	2.06	2.38	2.04	2.07	6
$Lu_2Fe_{14}B$	300	—	—	1.46	1.71	1.49	1.90	1.62	1.67	6
$Gd_2Fe_{14}B$	10	—	—	1.92	2.39	2.13	2.53	2.20	2.30	6
$Gd_2Fe_{14}B$	300	—	—	1.77	1.88	1.89	2.33	1.95	2.08	6

									Ref.
Gd₂Fe₁₄C	10	—	1.91	2.06	2.19	2.65	2.19	2.38	6
Gd₂Fe₁₄C	300	—	1.69	1.85	1.90	2.36	1.94	2.11	6
Lu₂Fe₁₄C	10	—	1.73	2.02	1.81	2.26	1.94	1.99	6
Lu₂Fe₁₄C	300	—	1.46	1.71	1.49	1.90	1.62	1.67	6
Nd₂Fe₁₄C	300	—	1.55	1.78	1.80	2.19	1.80	1.97	7
Nd₂Fe₁₄B	300	—	1.76	1.91	1.89	2.31	1.95	2.07	7
Band Calculation									
Y₂Fe₁₄B		—	2.06	2.53	2.07	2.74	2.08	2.15	8
		—	2.40	2.10	2.22	2.51	2.11	2.31	9
		—	1.80	1.60	2.50	3.08	2.16	2.47	10
		—	2.32	2.28	2.16	2.74	2.41	2.11	11
		—	2.11	2.31	2.35	2.61	2.36	2.62	12
Nd₂Fe₁₄B		—	2.42	2.49	2.33	2.61	2.40	2.40	13
		—	2.13	2.59	2.12	2.74	2.15	2.18	8
		—	2.34	3.11	2.39	3.35	2.18	2.34	14
		—	2.28	2.31	2.50	2.80	2.46	2.67	12

References: 1. Givord et al., 1985; 2. Herbst et al., 1984; 3. Herbst et al., 1985; 4. Herbst and Yelon, 1986; 5. Wallace et al., 1985; 6. Denissen et al., 1988a,b; 7. Buschow et al., 1988; 8. Jaswal, 1990; 9. Coehoorn, 1991; 10. Gu and Ching, 1987; 11. Inoue and Shimizu, 1986; 12. Itoh et al., 1987; 13. Szpunar and Szpunar, 1985; 14. Zhong and Ching, 1990.

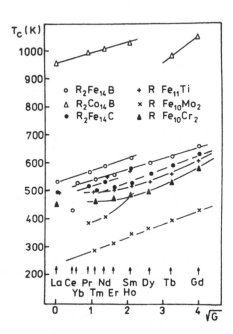

FIGURE 5.6. Curie temperatures of $R_2Fe_{14}B$, $R_2Co_{14}B$, $R_2Fe_{14}C$, $RFe_{11}Ti$, $RFe_{10}Mo_2$, and $RFe_{10}Cr_2$ compounds vs. the root of the De Gennes function.

A considerable variation of the moment, depending on the Fe site, caused by the different environments of the Fe positions is observed. In all cases, the 8j Fe atoms have the largest moments. They also have the largest number of near-neighbor Fe atoms; hence, large moments are expected. As shown earlier in the Fe–Si system, the larger the Fe moment, the larger the number of nearest-neighbor Fe atoms (Stearns, 1971).

The data listed in Table 5.5 show the Curie temperatures of $R_2Fe_{14}B$, which range from 437 K (R = Ce) to 660 K (R = Gd). In Figure 5.6, T_c is plotted against the effective spin of the rare-earth element (the square root of the De Gennes function). It should be noted that linearity is shown in all systems except $Ce_2Fe_{14}B$. The deviation observed in this case is a consequence of the quadripositive nature of Ce in this compound. The linearity confirms that the increase of T_c with the effective spin of the rare-earth element is a consequence of the R–T exchange. It is small, but not negligible compared to the T–T exchange (T-transition metal). The plots in Figure 5.6 show that the T_c values for the light and heavy members of the rare-earth series do not coincide. The T_c of the former are about 50 K higher than those of the latter. This is a consequence of the short Fe–Fe distances in $R_2Fe_{14}B$ systems.

Measurements of polycrystalline materials performed by various authors indicate that several $R_2Fe_{14}B$ compounds change their magnetic structure when cooled down from T_c to 4.2 K (Givord et al., 1984; Hirosawa et al., 1986; Pędziwiatr and Wallace, 1986). Two distinct types of spin reorientation have been evidenced:

1. Transitions from a complex noncolinear ordering to a c axis structure due to the strong temperature dependence of the higher-order crystal-field terms have been found when the temperature is raised. The spin-tilt transition at T_{SR} = 135 K was observed in $Nd_2Fe_{14}B$. Below T_{SR}, the net magnetization begins to tilt away from the c axis, while K_1 crosses through zero. In addition, a change in unit cell volume manifests itself by a small specific-heat anomaly. The tilt angle reaches 32° at 4.2 K (Hirosawa et al., 1986). A similar reorientation below T_{SR} = 65 K occurs in $Ho_2Fe_{14}B$: the tilt angle reaches 22° at 4.2 K.

2. Spin reorientations occur when the net magnetization changes its orientation from the c axis to the basal plane as a function of temperature. These transitions are due to the faster decrease of the 4f sublattice anisotropy with increasing temperature compared to the relatively temperature-independent 3d sublattice anisotropy. They occur in iron compounds with rare-earths which have a positive α_J. Such spin reorientations have been observed in $Er_2Fe_{14}B$ (T_{SR} = 325 K), $Tm_2Fe_{14}B$ (T_{SR} = 315 K), and $Yb_2Fe_{14}B$ (T_{SR} = 115 K). Transitions of this type have also been observed in Co-based compounds, namely, $Nd_2Co_{14}B$, $Pr_2Co_{14}B$, and $Tb_2Co_{14}B$ (Pędziwiatr and Wallace, 1987).

5.2.2. $R_2Fe_{14}C$

The $Nd_2Fe_{14}B$-type structure has been found in $R_2F_{14}C$ (R = Pr, Nd, Sm, Gd, Tb, Dy, Ho, Er, Tm, or Lu) (Gueramian et al., 1987). The magnetic parameters of $R_2Fe_{14}C$ are listed in Table 5.8. The Curie temperatures are systematically 40 to 50 K lower than those of $R_2Fe_{14}B$ (see Figure 5.5). This fact might be attributed to the weakening of the exchange interaction with the shortening of some Fe–Fe distances, which is due to the smaller c/a ratio and cell volume of the carbides. The other reason might be a reduction of the iron moment due to the different electron transfers between iron and the boron and carbon (Cadeville and Daniel, 1966).

The values of the magnetic moments of iron for each individual crystallographic site, determined using ^{57}Fe Mössbauer spectroscopy, are in good agreement with this prediction (Buschow et al., 1988; Denissen et al., 1988; Jacobs et al., 1989) (see Table 5.7). An analysis of the values of the anisotropy constants and fields for isostructural borides and carbides $R_2T_{14}X$ shows that these parameters are larger for carbides than for borides.

5.2.3. $R_2Co_{14}B$

The magnetic properties of $R_2Co_{14}B$ are listed in Table 5.9 (Buschow et al., 1988). When R is nonmagnetic, the magnetic anisotropy is due to the Co sublattice and favors an easy plane magnetization, which agrees with the data obtained for $Y_2Co_{14}B$ (Le Roux et al., 1985; Kapusta et al., 1990; Matsuura et al., 1985).

The temperature dependence of the magnetization for some $R_2Fe_{14}B$ and $R_2Co_{14}B$ compounds is shown in Figure 5.7. In $R_2Co_{14}B$ compounds, the

TABLE 5.8
Magnetic Properties of the $R_2Fe_{14}C$ Compounds

$R_2Fe_{14}C$	$T_c(K)$	$T_s(K)$	$M_s(\mu_B$ per formula unit)	$\mu_o H_a(T)$	Ref.
			T = 4.2 K		
R = La			27.5[a]		1, 4, 5
Ce	345		23.9[b]		8
Pr	513		34.5		1, 9
Nd	530	140	32.5	9.5	3, 5–7, 10, 13
Sm	580		30.2		1, 10, 13
Gd	630		18.1	3.53	1, 10–13
Tb	585		12.0		1, 10, 13
Dy	555		10.5		1, 2, 6, 10, 13
Ho	525	35	10.9		1, 6, 10, 13
Er	510	322	12.3	0.1	1, 2, 10, 13
Tm	500	310	18.4	0.4	1, 10, 13
Lu	495		27.2	3.1	1, 10, 13

[a]   At an applied field up to 7 T.
[b]   Extrapolated values from $R_2Fe_{14-x}Mn_xC$ series.

References: 1. Gueramian et al., 1987; 2. Pędziwiatr et al., 1986; 3. Buschow et al., 1988; 4. Marusin et al., 1985; 5. Denissen et al., 1988; 6. De Boer et al., 1988; 7. Buschow et al., 1988; 8. Jacobs et al., 1989; 9. Buschow et al., 1988; 10. De Boer et al., 1988; 11. Abache and Oesterreicher, 1985; 12. Kou et al., 1990; 13. Grössinger et al., 1990.

Note: T_c and T_s represent Curie and spin reorientation temperatures, M_s saturation moment and μH_a anisotropy field.

TABLE 5.9
Magnetic Properties of $R_2Co_{14}B$ Compounds

$R_2Co_{14}B$	$T_c(K)$	$T_s(K)$	$M_s(\mu_B$ per formula unit)	$K_1(MJ/m^3)$	$K_2(MJ/m^3)$	Ref.
R = Y	1010		19.8	−1.2		1 3
La	955		19.3	−1.3		2, 4
Pr	990	664	24.8			2, 5
Nd	997	36,544	26.3	−1.4	19	1, 4–6
Sm	1029		18.1			2
Gd	1053		5.36	−1.27		2–4
Tb	1033	795				2, 5

Note: T_c and T_s represent Curie temperatures and spin reorientation temperatures, M_s saturation moment and K_i the anisotropy constant refer to 4.2 K.

References: 1. Le Roux et al., 1985; 2. Buschow et al., 1985; 3. Huang et al., 1986; 4. Hirosawa et al., 1987; 5. Pędziwiatr and Wallace, 1986; 6. Abache and Oesterreicher, 1986.

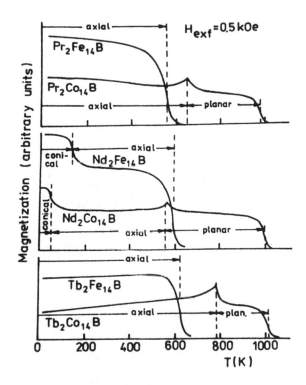

FIGURE 5.7. Magnetization vs. temperature curves for selected $R_2Fe_{14}B$ and $R_2Co_{14}B$ compounds (Pędziwiatr et al., 1988).

Curie temperatures are higher. In addition, a large number of magnetic structure transitions are observed. Spin reorientations at 664 and 795 K have been found in $Pr_2Co_{14}B$ and $Tb_2Co_{14}B$. As the temperature increases, $Nd_2Co_{14}B$ undergoes two spin transitions, one at 34 K and the second at 546 K.

The magnetic phase transitions that occur in the aforementioned three systems are schematically displayed in Figure 5.8. The first two based on the iron atoms, are very similar; the Pr, Gd, Tb, Dy, and Lu compounds have an easy c axis at all temperatures, while the Sm compound is planar. Changes to a centered structure are observed at low temperatures in Nd and Ho compounds. Plane-to-c-axis spin reorientation occurs in the phases with R = Er, Tb, and Yb.

In $R_2Co_{14}B$ compounds, the moments lie in the plane. Only in the case of R = Pr and Tb the reorientation from axis to plane is observed. In $Nd_2Co_{14}B$, two transitions were detected (Pędziwiatr and Wallace, 1986).

These results indicate the strong influence of 3d sublattice anisotropy on the magnetic ordering.

5.2.4. SUBSTITUTION COMPOUNDS

The effect of iron substitution by other metals in $R_2Fe_{14}B$-based compounds was also investigated. The magnetization of $R_2Fe_{14-x}T_xB$, where T

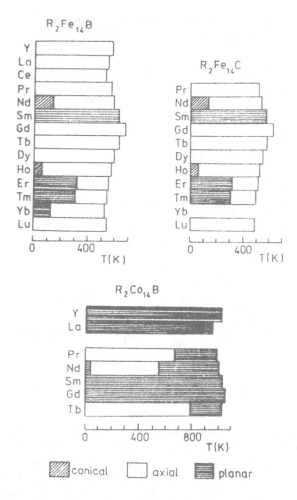

FIGURE 5.8. Temperature variation of the magnetic structure of $R_2Fe_{14}B$, $R_2Fe_{14}C$, and $R_2Co_{14}B$. Data from Coey et al. (1986), Pędziwiatr and Wallace (1986), and De Mooij and Buschow (1988), respectively.

is different from the Fe 3d or nonmagnetic atom, decreases as the concentration of x increases. In the simple dilution model, the magnetization, M_s, may be described by the relation: $M_s = 14M_{Fe} - x(M_{Fe} - M_T)$. This relation turned out to be valid for a large number of compounds (Burzo and Kirchmayr, 1989). A different dependence is observed in $R_2Fe_{14-x}Co_xB$, where an increase, followed by a decrease of magnetization has been observed (Herbst et al., 1984; Matsuura et al., 1985a).

The concentration dependence of the Curie temperature T_c is different. For T = Co-, Ni-, Cu-, Nb-, and Si-containing phases, T_c increases, whereas in other cases (e.g., T = Cr, Mn, Ru, and Al), it decreases. This dependence can be correlated with the preferential occupancy of the substitution atoms.

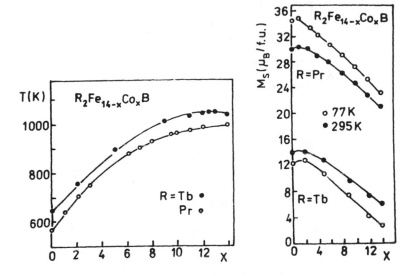

FIGURE 5.9. Composition dependence of the Curie temperature and magnetization of $R_2Fe_{14-x}Co_xB$ (R = Pr and Tb) (Pędziwiatr, 1988).

The interactions between iron atoms in $8j_1$ and $16k_2$ sites are strongly negative, while those between $8j_1$ and $8j_1$ are also negative, but exhibit less intensity. The magnetic interactions between $8j_1$ and those located in the $8j_2$ and $16k_1$ sites are positive and impose a parallel alignment of iron moments.

The increase of T_c suggests preferential occupancy of Fe sites, where the exchange interactions are negative. The decrease of T_c indicates the preferential occupancy sites with positive exchange interactions. Only cobalt has been able to replace iron over the whole concentration range. Figure 5.9 shows the $T_c(x)$ and $M_s(x)$ for $PrFe_{14-x}Co_xB$ and $TbFe_{14-x}Co_xB$, which is representative of the other R systems (Pędziwiatr, 1988). The increase of T_c is accompanied by a depression of the saturation magnetization.

The magnetic phase diagrams for some $R_2Fe_{14-x}Co_xB$ compounds are presented in Figure 5.10. In the case of compounds with R = Y and Gd, the magnetic phase diagram is simple. Close to a concentration of x = 9, a change of the magnetic structure from an axial spin alignment (the axial tendency of the Fe sublattice dominates) to a planar spin arrangement (planar tendency of the Co sublattice is dominant) is observed.

A different spin phase diagram is observed in the case of compounds with R = Pr, Nd, and Tb. The dominant regions in these diagrams are those with axial alignment. This is expected, since these systems are based on strongly axial lanthanide ions whose anisotropy dominates that of the 3d sublattice. However, due to the competing effects of the Co sublattice and to temperature effects, planar and conical types of spin arrangements also occur.

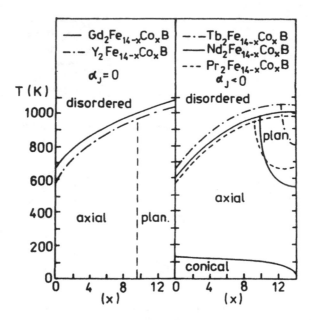

FIGURE 5.10.   Magnetic phase diagrams for $R_2Fe_{14-x}Co_xB$ (R = Y, Gd, Tb, Nd, Pr) (Pęd-ziwiatr, 1988).

The magnetic properties of the $(R_{1-x}R_x')_2Fe_{14}B$-type are also interesting. Suitable substitutions of the R component should lead to changes in the easy magnetization direction, spin reorientation temperature, and Curie temperature. Conclusive information would thus be obtained regarding the exchange interactions and the crystal field interactions.

The magnetic phase diagrams are shown in Figure 5.11. For $Nd_{2-x}Y_xCo_{14}B$, two spin reorientations occur, cone-to-axis and axis-to-plane, while $Pr_{2-x}Y_xCo_{14}B$ exhibits only one change (Pędziwiatr, 1988).

The magnetic phase diagram of $Tm_{2-x}Dy_xFe_{14}B$, is shown in Figure 5.12 (Pourarian et al., 1988). At low temperatures, the changes of magnetic structure from planar to canted and, further, to axial are observed as the concentration of x increases. This dependence shows good agreement with the predictions based on a Hamiltonian which includes the second-order CEF terms (Boltich et al., 1988).

In partially substituted $R_2Fe_{14}B_{1-x}C_x$ samples, the Curie temperature and saturation magnetization were observed to fall with an increase of carbon concentration (Leccabue et al., 1985; Bolzoni et al., 1985; Hirosawa et al., 1987).

5.2.5. HYDRIDES

Many R-3d intermetallic compounds are capable of absorbing considerable amounts of hydrogen (for a review of hydrides based on binary intermetallics, see Buschow, 1984).

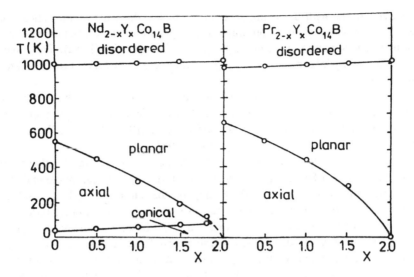

FIGURE 5.11. Magnetic phase diagrams for $R_{2-x}Y_xCo_{14}B$ (R = Pr, Nd) (Pędziwiatr, 1988).

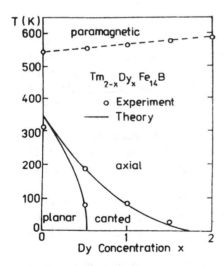

FIGURE 5.12. Magnetic phase diagram for $Tm_{2-x}Dy_xFe_{14}B$ (Pourarian et al., 1988).

Magnetization measurements performed on $Y_2Fe_{14}XH_x$ have established that the average Fe moment increases a few percentage points as a result of hydrogenation (Andreev et al., 1986; Coey et al., 1986). This effect certainly indicates that hydrogen exerts an appreciable influence on the electronic structure.

The influence of hydrogen on the magnetic properties of $R_2Fe_{14}B$ has also been studied (Pourarian et al. 1986; Zhang et al., 1988a). The Curie temperatures T_c increase, while those of the spin reorientation transition T_{sR} and anisotropy field H_Θ decrease upon hydrogenation. The most interesting behavior of the hydrides is the observation that hydrogenation influences the R-3d and 3d-3d exchange couplings. Hydrogenation also causes the changes of magnetic structure, e.g., in $Gd_2Fe_{14}BH_x$ (Zhang et al., 1988a). These data indicate that hydrogen causes a change of interactions and that it favors planar anisotropy.

In $R_2T_{14}X$ compounds, the hydrogenation also influences the spin reorientation phenomena. Axial-planar transitions as functions of hydrogen concentration and temperature are observed in $R_2Fe_{14}BH_x$ (R = Ce, Pr, Y, and Gd) (Pourarian et al., 1986; Zhang et al., 1988a).

$R_2Co_{14}BH_x$ compounds (R = La, Pr, Nd, Sm, Gd, Tb, and Y) also show the tetragonal $Nd_2Fe_{14}B$ type of crystal structure. Hydrogenation expands the unit cell volume 1.5 to 3.0%, depending on the nature of the R ion and the concentration of hydrogen. Magnetization increases for light lanthanides and decreases for heavy lanthanides. For Pr- and Tb-based intermetallics, hydrogenation significantly decreases the anisotropy field H_a, and the spin-reorientation transition temperature T_{SR}. The results indicate that hydrogen makes the compounds magnetically softer. This is attributed to the influence of hydrogen on both the 3d and R sublattices (Zhang et al., 1988).

5.2.6. MAGNETIC MOMENTS

In both groups of compounds, the atoms with the magnetic moments occupy different sites. It is possible to determine the magnetic moment of rare-earth atoms by magnetization measurements using the difference in the magnetization of rare-earth and yttrium compounds. The results obtained for $R_2F_{14}B$ systems (Hirosawa et al., 1986) give good agreement with free R^{3+} ion values (see Table 5.6).

A different situation is observed in the case of 3d metal atoms. The 3d moments at particular sites are closely related to the local environment. In these compounds, the average moment per Fe atom is approximately 2.1 to 2.2 μ_B, i.e., as in Fe metal.

The values of the magnetic moments at different sites can be obtained from Mössbauer spectroscopy and neutron diffraction data, including polarized neutron measurements. The values determined for some compounds are presented in Tables 5.2 and 5.7.

The results reported in Table 5.7 indicate that the magnetic moments of Fe change from 1.95(5) to 2.80(4) μ_B.

The $R_2Fe_{14}B$ structure can be described as an alternate sequence of two different pseudolayers of Fe atoms and a layer of Nd and Fe atoms (see Figure 2.25 in Chapter 2 of this volume).

The values of the Fe moments at 4.2 K in $Y_2Fe_{14}B$ may be related to differences in the respective local environments. Fe moment values for atoms at the Fe_1, Fe_3, Fe_5, and Fe_6 sites range from 2.15 to 2.40 μ_B. The high coordination of Fe atoms at the Fe_4 site is associated with a particularly large magnetic moment of 2.80 μ_B. On the other hand, Fe_2 atoms, which have the

TABLE 5.10
Calculated and Experimental Total Densities of
States at the Fermi Energy, $N(E_F)$, for $Y_2Fe_{14}B$

	$N(E_F)$ (states/eV-atom)	References
Theory	0.86	Jaswal, 1990
	0.8	Coehoorn, 1991
	1.88	Gu and Ching, 1987a
	1.96	Inoue and Shimizu, 1986a
Experiment	2.1	Fujii et al., 1987

largest coordination of Y neighbors, have a magnetic moment of only 1.95 μ_B. The reduced value is related to an electron transfer from the 4d band of yttrium to the 3d band and 4d-3d hybridization, as commonly observed in R-transition metal alloys. According to the magnetic measurements performed on Fe–B alloys, an additional reduction of the 3d moment should be observed for Fe atoms surrounded by boron atoms. In fact, the moment at the Fe_1 site is slightly smaller than that at the similar Fe_3, Fe_5, and Fe_6 sites. This suggests that the 2p-3d electron transfer and/or hybridization is not very effective in these phases, where the d states are essentially determined by interactions between 3d and 4d (or 5d) electrons.

In the $RFe_{12-x}M_x$ systems, the average iron moments range from 1.35 to 1.83 μ_B (see Table 5.2). The observed values of the magnetic moments at different sites are not the same. The following relation is observed: $\mu_{Fe}(8i) > \mu_{Fe}(8j)$. Considerable effort has been devoted to calculating the band structure of $R_2Fe_{14}B$ and $R_2Co_{14}B$ compounds. The calculated values of the magnetic moments are compared with the experimental values in Table 5.2.

The obtained results point out large differences in the behavior of the local density of states (LDOS) associated with the various Fe sites. The calculated values of the total density of states at the Fermi energy level $N(E_F)$ for $Y_2Fe_{14}B$ are listed in Table 5.10. The results obtained by different authors are compared with the experimental specific heat data (Fujii et al., 1987). The low-temperature data for $Y_2Fe_{14}B$ fit the expression $C(T) = \gamma T + \beta T^3$. The electronic contribution to the specific heat equals 86 mJ/K^2 mol formula unit. The calculated density of states at the Fermi energy level $N(E_F)$ agrees well with the results of the above measurements (Inoue and Shimizu, 1986; Gu and Ching, 1987).

The experimental and calculated values of the magnetic moments presented in Table 5.7 show good agreement despite the fact that the crystal structure of these compounds has a relatively low symmetry and that the unit cell contains two different R sites and six different Fe sites.

5.2.7. MODEL DESCRIPTION OF EXCHANGE INTERACTIONS

The experimental results were analyzed using the mean-field theory, in which the strength of the R–T exchange coupling is represented by the intersublattice molecular-field coefficient μ_{RT} and microscopic exchange parameter J_{RT}.

It was shown in the previous section that the coupling of the magnetic moments in $R_2T_{14}X$ is similar to that observed in binary R–Fe intermetallics, i.e., there is an antiferromagnetic coupling between the Fe spins and the R spins. For compounds in which R is a light rare-earth element $(J = L - S)$, this means that the total lanthanide moment (gJ) is coupled parallel to the Fe moments. By contrast, when R is a heavy rare-earth element $(J = L + S)$, the total rare-earth moment is coupled antiparallel to the Fe moment. It was also shown that the M_R values are equal to gJ. This is by no means obvious. Owing to the large contribution of rare-earth ions to the magnetocrystalline anisotropy, one can expect a substantial crystal-field splitting and, associated with it, a crystal-field ground state which does not necessarily lead to the maximum moment for R^{3+}. Observation of the maximum moment indicates that the influence of the exchange field experienced by the 4f moments is much greater than that of the crystal field, although the latter still determines the preferred direction of the 4f moments.

The Curie temperatures are due to the cumulative effect of the magnetic interactions between 3d atoms (J_{3d-3d}), 3d and 4f atoms (J_{3d-R}), and between lanthanide ions (J_{R-R}).

The molecular field model can be used to describe the variation of the Curie temperature in the $R_2Fe_{14}B$ series (Sinnema et al., 1984). In this model, T_c can be written as

$$3kT_c = a_{FeFe} + a_{RR} + [(a_{FeFe} - a_{RR})^2 + 4a_{RFe}a_{FeR}]^{1/2} \quad (5.1)$$

where a_{xy} represents the magnetic interaction energy between the X and Y spins. Again, neglecting the relatively weak R–R exchange energy, this expression reduces to

$$3kT_c = a_{FeFe} + (a_{FeFe} + 4a_{FeR}a_{RFe})^{1/2}$$

where

$$a_{FeFe} = ZJ_{FeFe}S_{Fe}(S_{Fe} + 1)$$

$$a_{RFe}a_{FeR} = Z_1Z_2S_{Fe}(S_{Fe} + 1)(g_J - 1)^2(J + 1)J_{RFe} \quad (5.2)$$

In $R_2Fe_{14}B$ compounds, each R atom has an average number of $Z_1 = 18$ Fe nearest-neighbor atoms. Furthermore, each Fe atom has an average number of $Z_2 = 2.5$ lanthanide nearest-neighbor atoms and an average number of $Z = 10$ iron neighbors. In $RFe_{12-x}M_x$, each R atom has an average number of $Z_1 = 17$ nearest-neighbors, while each Fe atom has an average number of $Z_2 = 2$ lanthanide nearest-neighbors. The values of the J_{FeFe} coupling constants were determined using magnetic data for compounds with R = Y. They were

TABLE 5.11
The Values of the Exchange Interactions in $R_2T_{14}X$
and $RFe_{10}M_2$ Compounds

Compound	$J_{R-T} \times 10^{-22}$ J	$J_{T-T} \times 10^{-22}$ J
	$R_2Fe_{14}B$	
R = La		4.7
Gd	−1.25(2)	
Tb	−1.28(4)	
Dy	−1.19(7)	
Ho	−1.10(8)	
Er	−0.95(2)	
Tm	−1.03(6)	
	$R_2Fe_{14}C$	
R = Gd	−1.32(3)	
Ho	−1.10(2)	
Er	−1.05(2)	
	$R_2Co_{14}B$	
R = Y		17.8
La		15.8
Gd	−2.23	
	$RFe_{10}M_2$	
M = Cr	−1.57	9.41
Si	−2.2	9.45

subsequently used to determine the coupling constant J_{RFe}. The results obtained for a number of different compounds are listed in Table 5.11.

Below T_c, a large, positive, anomalous thermal expansion occurs (Givord et al., 1985) as a result of the minimization of magnetic energy at the expense of elastic energy. The volume anomaly is approximately proportional to the square of the spontaneous magnetization M_s^2, as expected for isotropic exchange interactions (Givord et al., 1986).

A comparison of the J_{R-Fe} for $RFe_{12-x}M_x$ compounds with constant R and different (Table 5.12) shows that coupling constants differ with the transition element M. For example, J_{R-Fe} is equal to 3.20×10^{-22} J, when R = Sm and M = V, 2.74×10^{-22} J for M = Ti, 2.51×10^{-22} J for M = Si, and 2.30×10^{-22} J for M = Cr. This indicates that a change of nearest-neighbors exerts a certain influence on the coupling constants. It is a well-established fact that T atoms have site-occupation preferences V and Ti atoms occupy the 8i positions. Si atoms share the 8j and 8f positions with Fe atoms, and chromium atoms prefer 8i sites (Helmholdt et al., 1988; Helmholdt and Buschow, 1989; De Mooij and Buschow, 1988; Yang et al., 1988).

TABLE 5.12

Calculated Intra- and Intersublattice Exchange Coupling Parameters
J_{Fe-Fe} and J_{R-Fe} (with an accuracy of 0.01)

R	$RFe_{10}V_2$ $J_{R-Fe}(10^{-22}$ J)	$RFe_{10.9}Ti_{1.1}$ $J_{R-Fe}(10^{-22}$ J)	$RFe_{10}Si_2$ $J_{R-Fe}(10^{-22}$ J)	$RFe_{10}Cr_2$ $J_{R-Fe}(10^{-22}$ J)
Nd	3.36	2.77		3.65
Sm	3.30	2.74	2.51	2.30
Gd	1.78	1.64	1.62	1.47
Tb	1.41	1.38	1.58	0.84
Dy	0.77	1.06	1.51	0.93
Ho	0.89	0.75	1.57	1.54
Er	2.27	1.61	1.61	2.35
Tm	3.84		2.0	3.95
J_{Fe-Fe} $(10^{-22}$ J)	10.14	8.79	9.45	8.61

From Stefański, et al., *Phys. Status Solidi* (b), 156, 657, 1989. With permission.

The exchange interactions in iron sublattices for various metalloid atoms are similar, but they have a maximum of J_{FeFe} = 10.14 × 10^{-22} J when T = V and are smallest (J_{FeFe} = 8.61 × 10^{-22} J) when T = Cr. For a constant T, the values of intersublattice exchange interactions are always substantially smaller than the intrasublattice ones.

Although estimating intra- and intersublattice molecular interactions with the mean-field method is probably a rough approximation, it makes it possible to estimate one of the most important parameters describing the magnetism of a lanthanide atom in its intermetallic compounds. The second, more important parameter is the magnetocrystalline anisotropy; investigations of this class of compounds are still in progress.

The magnitude of J_{R-3d} in $RFe_{12-x}M_x$ is comparable to that for $R_2T_{14}X$ compounds.

5.2.8. MAGNETOCRYSTALLINE ANISOTROPY

In $R_2T_{14}X$ and $RFe_{12-x}M_x$ compounds, contributions to magnetocrystalline anisotropy arise from both the 3d and rare-earth sublattices. The former can be determined experimentally from measurements made on compounds with nonmagnetic rare earths or yttrium. The R sublattice contribution can be calculated using the single ion-model theory.

For $R_2T_{14}X$ and $RFe_{12-x}M_x$ which show tetragonal structures, the magnetocrystalline anisotropy may be expressed as

$$E_A = K_1\sin^2\theta + K_2\sin^4\theta + K_3\sin^4\theta\cos^4\varphi \qquad (5.3)$$

where Θ and φ are polar angles of magnetization. The values of the anisotropy constants, as determined from experiments, are presented in Tables 5.1 and 5.7.

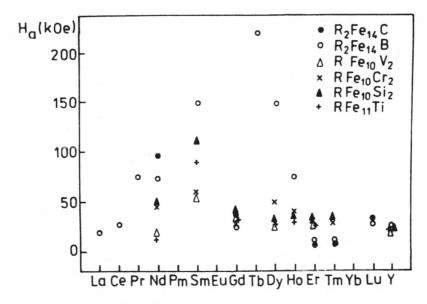

FIGURE 5.13. Anisotropy field H_a in $R_2Fe_{14}X$ (X = B, C), $RFe_{10}M_2$ (M = V, Cr, Si) and $RFe_{11}Ti$.

The values of K_1 determined for some $R_2T_{14}X$ and $RT_{12-x}M_x$ compounds are large relative to that of metallic Fe (cubic $K_1 = 0.06$ MJ/m^3) (Bozorth, 1951), but smaller compared with, e.g., hexagonal YCo_5 ($K_1 \approx 7.4$ MJ/m^3) (Alameda et al., 1981).

The second parameter that characterizes magnetocrystalline anisotropy is the anisotropy field H_a.

The composition dependence of the anisotropy fields H_a in $R_2T_{14}X$ and $RFe_{12-x}M_x$ compounds is displayed in Figure 5.13. In compounds with yttrium or nonmagnetic lanthanides such as La, Ce, and Lu, H_a values are small. Similar behavior was observed in the $Gd_2Fe_{14}X$ and corresponding $RFe_{12-x}M_x$ compounds. In all of them, the anisotropy fields are determined by the iron sublattice. In these cases, magnetization is oriented along the c axis.

The temperature dependence of the anisotropy fields in $R_2Fe_{14}B$ compounds (R = Y, La, Ce, and Lu) (Grosinger et al., 1985, 1986) indicates that they increase to a flat maximum at a characteristic temperature. This suggests that iron atoms located at different sites provide opposite contributions to the bulk anisotropy.

For other compounds, the values of anisotropy field are large. They can be correlated to the influence of the rare-earth sublattice. In this case, the single-ion, crystal field-induced anisotropy of the rare-earth component has been assumed to operate. It arises as a consequence of the interaction between the electrostatic field of the ions surrounding the rare-earth site and the asymmetric charge cloud of the 4f electrons. The charge cloud may be shaped like a cigar or a pancake. The difference in shape is reflected by the difference in the sign of the so-called second-order Stevens factor (α_J), which can be positive or negative. In the case of Gd, the 4f charge cloud has spherical symmetry (L = 0) and α_J = 0. As a consequence, in materials containing Gd, crystal field-induced anisotropy is absent. Rare-earth atoms for which L = O and which have different signs of α_J display changes in the preferred magnetization direction of the corresponding 4f moment when placed a crystallographic site experiencing the same crystal field. In compounds of rare-earth elements with 3d metals, the same magnetization direction is adopted by the 3d sublattice magnetization, owing to the strong coupling between the 3d and 4f moments, which leads to colinear sublattices.

In compounds with hexagonal or tetragonal crystal structures, the easy magnetization direction for a given rare-earth element (α_J fixed) may be oriented parallel or perpendicular to the c axis, depending on the nature of the electrostatic field associated with the point symmetry of the rare-earth site in the particular crystal structure. This specific property of the crystallographic arrangement of atoms surrounding the rare-earth atom is commonly expressed in terms of crystal-field parameters. For many hexagonal and tetragonal structures, the leading term is the second-order field parameter V_2^0. The specific properties of the rare-earth atom (α_J) and its environment (V_2^0) are determined by the sign and magnitude of the uniaxial anisotropy constant K_1 as well as the easy magnetization direction.

In $Nd_2Fe_{14}B$ for each of the two rare-earth sites with mm point symmetry, the 4f crystal-field Hamiltonian contains a number of terms (Hutchings, 1964):

$$H_{cf} = B_2^0O_2^0 + B_2^{2c}O_2^{2c} + B_4^0O_4^0 + B_4^{2c}O_4^{2c} + B_4^{4c}O_4^{4c} + B_6^0O_6^0$$

$$+ B_6^{2c}O_6^{2c} + B_{6c}^4O_6^{4c} + B_6^{6c}O_6^{6c} \tag{5.4}$$

At the single rare-earth site with 4/mmm point symmetry in the $ThMn_{12}$ structure, the crystal-field Hamiltonian is simpler:

$$H_{cf} = B_2^0O_2^0 + B_4^0O_4^0 + B_4^4O_4^4 + B_6^0O_6^0 + B_6^4O_6^4 \tag{5.5}$$

B_n^m are the crystal field parameters for a given lanthanide ion and O_n^m are Stevens operators.

$$B_2^0 = \alpha_J\langle r^2\rangle(1 - \sigma_2)A_2^0 \tag{5.6}$$

B_2^0 expresses the contribution of the second-order crystal-field term, α_J is the Stevens coefficient of the rare-earth, and A_2^0 is the electric field gradient at the R site.

Crystal-field parameters for $Nd_2Fe_{14}B$ were calculated using the point-charge model (Cadogan and Coey, 1984; Cadogan et al., 1988; Honma et al., 1985; Adam et al., 1986). Different sets of ionic charges for Nd, Fe,

TABLE 5.13

Crystal-Field Parameters in $Nd_2Fe_{14}B$, $TbFe_{10}Cr_2$, $DyFe_{10}Cr_2$, and $ErFe_{10}Cr_2$ Obtained from Point-Charge Calculations

Term	$Nd_2Fe_{14}B$[a] 4f	$Nd_2Fe_{14}B$[a] 4g	$TbFe_{10}Cr_2$[b]	$DyFe_{10}Cr_2$[b]	$ErFe_{10}Cr_2$[b]
$B_2^0(K)$	-29.6	-28.34	2.9	2.4	0.97
$B_2^2(K)$	32.61	-40.96			
$B_4^0(K)$	1.85×10^{-2}	1.72×10^{-2}	-1.15×10^{-2}	1.07×10^{-2}	-1.08×10^{-2}
$B_4^2(K)$	-4.69×10^{-2}	3.64×10^{-2}			
$B_4^4(K)$	-9.09×10^{-2}	6.20×10^{-2}	4.23×10^{-3}	-6.3×10^{-4}	3.9×10^{-4}
$B_6^0(K)$	-1.05×10^{-3}	-9.23×10^{-4}	1.79×10^{-7}	-1.47×10^{-7}	-2.2×10^{-7}
$B_6^6(K)$	6.88×10^{-5}	-6.10×10^{-5}	2.20×10^{-8}	1.8×10^{-8}	2.7×10^{-8}

[a] From Cadogen and Coey, 1984.
[b] From Stefański, 1991.

and B have been used. A charge of $+3$ was taken for Nd in all cases, while the charge attributed to Fe varied from $-9/14$ to $+1.8$ and that for B from 0 to $+3$. The obtained data are listed in Table 5.13. Similar calculations were performed for other $RFe_{10}M_2$ compounds (Table 5.13). The results indicate that at and above room temperature, the second-order interaction is dominant.

This effect is certainly true for $R_2Fe_{14}B$ (all R except Gd and Yb), where the sign of the total magnetocrystalline anisotropy at room temperature is determined by the sign of B_2^0 ($B_2^0 = \alpha_J < r_2 > A_2^0$). Its sign depends on the sign of the second-order Stevens coefficient α_J, and the electric field gradient at the rare-earth site. A_2^0 and α_J are positive for Sm, Er, Tm, and Yb, but negative for the other lanthanides with $L = 0$. In the $Nd_2Fe_{14}B$ structure, A_2^0 is positive at both rare-earth sites; hence, the rare-earth contribution to magnetocrystalline anisotropy favors the easy axis for Nd and other lanthanides with a negative α_J. The planar rare-earth contribution to the anisotropy of $R_2Fe_{14}B$ alloys where α_J is positive is sufficient to overcome the axial iron contribution at room temperature in all cases except R = Yb.

The situation in $RFe_{10}V_2$ is different. In contrast to the $R_2Fe_{14}B$ system, where the easy magnetization direction along the c axis is expected for R elements with a negative second-order factor ($\alpha_J < 0$), in $RFe_{10}V_2$ compounds, R elements with $\alpha_J > 0$ prefer the c axis. In the R-3d systems, the temperature dependence of the rare-earth and transition-metal sublattice anisotropies are completely different. Generally, the anisotropy due to the 4f ion prevails at low temperatures, but declines rapidly with an increase of temperature. The anisotropy due to Fe is less sensitive to temperature and has a positive sign in the whole range of temperatures. Whenever the 4f sublattice and 3d anisotropies favor different directions, spin reorientation is likely to appear. A

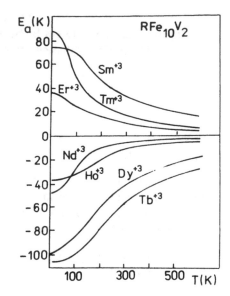

FIGURE 5.14. Calculated temperature dependence of the magnetocrystalline anisotropy of trivalent rare-earth ions in $RFe_{10}V_2$ compounds (Stefański and Wrzeciono, 1989).

series of complex spin-reorientation transitions are found in the Nd, Tb, and Dy compounds due to the interplay of terms of different orders in the crystal-field Hamiltonian.

The local magnetocrystalline anisotropy of the 4f moment is defined as $E_{anis} = E_\perp - E_\parallel$, where $E_\perp$ and $E_\parallel$ are the ground-state energy of the Hamiltonians. The $H_\perp = B_2^0 O_2^0 \, g_J \, \mu_B H m J^x$ and $H_\parallel = B_2^0 O_2^0 \, g_J \, \mu_B H m J^z$ have been calculated for $RFe_{10}V_2$ (Stefański and Wrzeciono, 1989). The results are presented in Figure 5.14. The E_{anis} show positive values for Sm^{3+}, Tm^{3+}, and Er^{3+}, and negative values for Nd^{3+}, Ho^{3+}, Tb^{3+}, and Dy^{3+}. The R sublattice anisotropy predominates at low temperatures, but decreases much more with temperature than does the 3d sublattice anisotropy. Magnetization measurements made on $YFe_{10}V_2$ have shown that the 3d sublattice has a preferred magnetization direction parallel to the c axis (De Boer et al., 1987).

The experimentally determined values of A_2^0 coefficient for a large number of $GdFe_{12-x}M_x$ compounds are listed in Table 5.14. They are smaller than those found in hard magnetic materials of the $R_2T_{14}X$ type (Buschow et al., 1988). The aforementioned results also show that the rare-earth contribution to the magnetocrystalline anisotropy in $RFe_{12-x}M_x$ compounds is almost the same when M = V, Mo, W, and Si, but is significantly smaller in the titanium compounds (Dirken et al., 1989). There will be no such effect when R = Sm, Er, and Tm.

A theoretical analysis of spin reorientation has been done using the crystal-field theory (Boltich and Wallace, 1985). It assumes that the fourth-order CEF term in the Hamiltonian expression is sufficiently large to explain the occurrence of spin reorientation in $Nd_2Fe_{14}B$. In this analysis, the sixth-order and nondiagonal CEF terms were ignored, and the value of B_4^0/B_2^0 was two

TABLE 5.14
Values of A_2^0 Parameters for $GdFe_{12-x}M_x$ Compounds Derived from Fits of ^{155}Gd Mössbauer Spectra[a]

Compound	A_2^0 (Ka_0^{-2})	3d Sublattice anisotropy
$GdFe_{10.8}Ti_{1.2}$	-79	easy axis
$GdFe_{10}V_2$	-144	easy axis
$GdFe_{10}Mo_2$	-118	easy axis
$GdFe_{10.8}W_{1.2}$	-140	easy axis
$GdFe_{10}Si_2$	-123	easy axis

[a] From Dirken et al., 1989.

to three orders of magnitude larger than that estimated by the point-charge model.

Spin reorientations have been observed experimentally for R = Nd, Dy, Ho, and Er in $RFe_{10}V_2$ systems (Christides et al., 1989) and for R = Nd, Tb, Dy, and Er in the $RFe_{11}Ti$ system (Hu et al., 1989). Good agreement between experimental and calculated values of spin reorientation temperatures was obtained for both systems. A Hamiltonian which included five crystal-field coefficients was used (Christides et al., 1989; Hu et al., 1989).

Fourth- and sixth-order CEF terms with appropriate positive values explain the spin reorientation in the case of $ErFe_{10}Cr_2$ (Stefański et al., 1991).

When comparing the magnetocrystalline anisotropy properties of $R_2T_{14}X$ and $RFe_{12-x}M_x$ compounds, the following points should be noted:

1. The magnitudes of A_2^0. For example, in $RFe_{11}Ti$, the A_2^0 value is about 1/3 to 1/5 times smaller than that of $Nd_2Fe_{14}B$.
2. The sign of A_2^0. $A_2^0 < 0$ in $RFe_{11}Ti$ and $A_2^0 > 0$ in $R_2Fe_{14}B$.
3. The influence of high-order CEF terms. Below room temperature, the contribution of A_2^0 in $R_2Fe_{14}B$ is dominant, but the influence of higher-order crystal fields in $RTiFe_{11}$ cannot be ignored.

5.3. RT_6X_6 PHASES

RMn_6Sn_6 compounds (R = Y, Sc, Lu, and Tm–Gd) (Malaman et al., 1988) crystallize in the hexagonal $HfFe_6Sn_6$-type structure (space group P6/mmm) (Olenitch et al., 1981).

The compounds with diamagnetic R elements (Sc, Y, and Lu) and Tm are antiferromagnetic, while those with R = Gd, Tb, Dy, and Ho are ferromagnetic. $ErMn_6Sn_6$ exhibits an intermediate behavior, since it is antiferromagnetic down to 75 K and ferromagnetic at lower temperatures. Magnetic ordering temperatures range from 449 K for Gd to 337 K for Tm (Malaman et al., 1988; Venturini et al., 1991).

The ferromagnetic compounds (Tb–Er) are characterized by antiferromagnetically coupled ferromagnetic Mn and R (001) planes. Neutron diffraction studies indicate that both the Mn and R = Tb, Dy, and Ho sublattices order simultaneously at T_C. Above room temperature, the rare-earth and Mn moments lie in the (001) plane, while a spin reorientation to the c axis occurs at low temperatures. Some of these compounds (Tb–Er) exhibit large coercive fields (11 kOe in $TbMn_6Sn_6$ at 4.2 K) due to uniaxial anisotropy (Venturini et al., 1991; Chafik El Idrissi et al., 1991).

The ^{155}Gd Mössbauer spectrum of $GdMn_6Sn_6$ indicates a magnetic ordering at low temperatures and provides information on the second-order crystal-field parameter A_2^0 associated with the Gd atom. The obtained value of A_2^0 ($+92$ K/A_2^0) is fairly small compared to the A_2^0 values found in highly anisotropic materials (Dirken et al., 1991). The small A_2^0 values suggest that the anisotropic behavior of RMn_6Sn_6 compounds will be determined only to a limited extent by the second-order parameter, leaving an important role for the parameters of the higher-than-A_2^0 orders.

Neutron diffraction studies of YMn_6Sn_6 and $ScMn_6Sn_6$ show complex helical structures built of ferromagnetic Mn (001) planes rotating by various angles along the (001) direction. $LuMn_6Sn_6$ is a colinear antiferromagnet consisting of the same ferromagnetic sheets found after the stacking sequence along the c axis: $+ + - - + + - -$ (Venturini et al., 1991).

RFe_6Ge_6 compounds show different types of crystal structures (see Chapter 2 of this volume). All are antiferromagnets with Néel temperatures between 452 K for R = Lu and 482 for R = Gd (Venturini et al., 1992).

The relevant magnetic data are presented in Table 5.15.

5.4. $RT_{4+x}Al_{8-x}$ PHASES

The structure of the $ThMn_{12}$-type compounds has been known for a long time. However, binary compounds with f-electron metals exist only as RMn_{12} and RZn_{12} (Wang and Gilfrich, 1966; Kirchmayer, 1969; Stewart and Coles, 1974). In RMn_{12}, two ordering temperatures have been observed: the first, T_N, close to that of YM_{12}, corresponds to the antiferromagnetic ordering of the Mn moments. The second, T_C, is lower and corresponds to the ferromagnetic ordering of the lanthanide moments (Table 5.16) (Deportes et al., 1977). In YMn_{12}, the Mn moments order antiferromagnetically. Below 120 K, they cancel each other in the same sublattice. The amplitudes of the Mn moments have been reported to be 0.42 μ_B for the 8i and 8j sites and 0.14 μ_B for the 8f site (Deportes and Givord, 1976). Lanthanide R = Gd–Er magnetic moments couple ferromagnetically below T_C (about 5 K), and Mn moments couple antiferromagnetically below T_N (about 100 K) (Deportes et al., 1977) (Table 5.17).

A neutron diffraction study indicates that the alignment of the Mn moments is the same as that found in YMn_{12} and that the direction of the lanthanide moments is along the (100) axis (Deportes et al., 1977). For RMn_{12}

TABLE 5.15
Magnetic Data for RT$_6$X$_6$ Compounds

Compound	Type of magnetic ordering	$T_{C,N}(K)$	$\Theta_p(K)$	Mn $\mu_{eff}(\mu_B)$	R $\mu_{eff}(\mu_B)$	Ref.
		RMn$_6$Sn$_6$				
R = Sc	AF	384	439	3.48	0	1
Y	AF	333	394	3.61	0	1
Gd	F	435	139	3.80	7.97	1
Tb	F	423	125	3.58	9.7	1, 2
Dy	F	393	138	3.79	10.6	1
Ho	F	376	157	3.64	10.6	1, 2
Er	AF	352	241	3.52	9.6	1
Tm	AF	347	304	3.48	7.6	1
Lu	AF	353	406	3.47	0	1
		RMn$_6$Ge$_6$				
R = Sc	AF	516	500	3.05		3
Y	AF	473	483	3.0		3
Nd	F	417	418	3.16		3
Sm	F	441	439	3.09		3
Gd	F	463	NCW[a]			3
Tb	AF	427	NCW			3
Dy	AF	423	NCW			3
Ho	AF	466	NCW			3
Er	AF	475	NCW			3
Tm	AF	482	NCW			3
Yb	AF	480	449	2.75		3
Lu	AF	509	485	3.00		3
		RFe$_6$Ge$_6$				
R = Sc	AF	473	159			3
Y	AF	477	108			3
Gd	AF	482	5			3
Tb	AF	480	5			3
Dy	AF	475	62			3
Ho	AF	473	45			3
Er	AF	470	105			3
Tm	AF	465	121			3
Yb	AF	475	103			3
Lu	AF	452	112			3

[a] NCW, non-Curie-Weiss.

References: 1. Venturini et al., 1991; 2. Chafik el Idrissi, 1991; 3. Venturini et al., 1992.

TABLE 5.16
Magnetic Properties of RMn$_{12}$ Compounds

R	T$_N$(K)	T$_C$(K)	μ$_s$(μ$_B$)	$T < T_N$		$T > T_N$		Ref.
				Θ$_p$(K)	μ$_{eff}$(μ$_B$)	Θ$_p$(K)	μ$_{eff}$(μ$_B$)	
Y	120							1
Gd	160	5.0	4.2	10	4.3	−80	7.92	2
Tb	108	4.7	4.4	4	9.3	−7	9.74	2
Dy	100	2.2	5.0	2	10.0	−11	10.5	2
Ho	90	1.7	6.4	−9	10.1	−8	10.4	2
Er	87	1.9	6.5				9.8	1
Tm		<1.2					7.6	1

References: 1. Deportes et al., 1977; 2. Okamoto et al., 1987.

TABLE 5.17
Magnetic Data for RT$_x$Al$_{12-x}$ (T = Cr, Mn, Fe, or Cu, x = 4, 5, and 6) Compounds with ThMn$_{12}$-Type Structure

Compound	T$_{C,N}$(K)	Θ$_p$(K)	μ$_{eff}$(μ$_B$)	μ$_s$(μ$_B$ per formula unit)	Ref.
			RFe$_4$Al$_8$		
R = La	135.4	−68	4.4	0.36	1–3
Ce	159.7	23	4.2	0.26	1–3
Pr	107	35	4.2	1.06	1–3
Nd	142	44	4.4	0.18	1–3
Sm	108	68	4.0	0.46	1–3
Eu	137	−148	4.5	0.27	1, 2, 4, 5
Gd	172.3	−151	4.5	2.4	1–3, 6
Tb	165.3	−105	4.4	2.1	1–3
Dy	122	−80	4.2	2.3	1–3
Ho	137	−66	4.4	1.65	1–3
Er	183	−38	4.3	2.8	1–3
Tm	186.6	7	4.3	2.0	1–3
Yb	103	62	4.6	2.8	1–3
Lu	197.3	79	4.3	0.6	1–3
Y	184.7	−16	4.6	0.28	1–3, 6
			RCu$_4$Al$_8$		
R = Ce		−17			1, 7
Pr	15	−15			1, 7
Nd	20	−18			7
Sm	25	−40			7
Gd	32	−16			1, 2, 7
Tb	22	−14			7
Dy	17	−5			7
Ho	7	−7			7
Er	6	−10			1, 7
Tm	5	−6			7

TABLE 5.17 (continued)
Magnetic Data for RT_xAl_{12-x} (T = Cr, Mn, Fe, or Cu, x = 4, 5, and 6) Compounds with $ThMn_{12}$-Type Structure

Compound	$R_{C,N}(K)$	$\Theta_p(K)$	$\mu_{eff}(\mu_B)$	$\mu_s(\mu_B$ per formula unit)	Ref.
			RCr_4Al_8		
R = Pr	6	−16	0.16	0.38	1, 7
Nd		−11	0.12		7
Sm	19	−24	0.61	0.03	7
Gd	8	−9	0.15	1.6	1, 7
Tb	12	−5	0.08	3.3	7
Dy	16	−5	0.12	4.2	7
Ho	8	−5	4.1	4.1	7
Er	14	−4	0.08	3.8	1, 7
Tm	11	−11	0.07	2.4	7
			RMn_4Al_8		
R = Ce		8	0.9		1, 7
Pr	11	−1	1.6	0.38	1, 7
Nd	7	−8	1.7	0.48	7
Sm	12	−18	1.4	0.14	7
Eu	20			0.46	7
Gd	28	4	1.6	3.1	1, 7
Tb	21	7	1.7	4.3	7
Dy	19	−26	1.9	4.0	7
Ho	14	−2	1.5	4.8	7
Er	15	−2	1.0	4.1	1, 7
Tm	13	5	1.5	3.7	7
Lu		10	0.9		7
Y		5	0.8		1, 7
			RFe_5Al_7		
R = Sm	220	265	3.9	5.1	8
Gd	268	241	3.8	0.4	8
Tb	248	174	2.5	0.6	8
Dy	227	177	3.4	1.8	8
Ho	227	186	3.5	2.3	8
Er	218	192	3.6	1.7	8
Tm		109			8
Yb	207	224	3.9	3.6	8
Lu	212	224	4.2	5.4	8
Y	215	230	4.4	5.8	8
			RFe_6Al_6		
R = Eu	135				9–11
Gd	345			2.11	9–12
Tb	335				9–11
Dy	325				9–11
Ho	310				9–11

TABLE 5.17 (continued)
Magnetic Data for RT_xAl_{12-x} (T = Cr, Mn, Fe, or Cu, x = 4, 5, and 6) Compounds with $ThMn_{12}$-Type Structure

Compound	$R_{C,N}(K)$	$\Theta_p(K)$	$\mu_{eff}(\mu_B)$	$\mu_s(\mu_B$ per formula unit)	Ref.
			RFe_6Al_6		
Er	320				9–11
Tm	320				9–11
Yb	210				9–11
Y	308				9–11
			RCu_6Al_6		
R = Gd	21	−22			10, 13
Tb	33.4	−54			10, 13
Dy	3.9	−14			10, 13
Ho	1.9	−11			10, 13
Er	2.6	−3			10, 13
Tm	3.9	−5			10, 13
			RMn_6Al_6		
R = Gd	15	−9			10, 13
Tb	8	−26			10, 13
Dy	6	−7			10, 13
Er	2.6	−4	1.4		10, 13
Tm	2	−18	1.2		10, 13
Yb		−23	1.2		10, 13
Y		−44	1.4		10, 13
			RCr_6Al_6		
R = Gd	170	158			10, 13
Dy	20	−5			10, 13
Er	15	−1	1.6		10, 13

References: 1. Buschow et al., 1976; 2. Felner and Nowik, 1978a; 3. Buschow and Van der Kraan, 1978; 4. Felner and Nowik, 1987b; 5. Darshan and Padalia, 1984; 6. Van der Kraan and Buschow, 1977; 7. Felner and Nowik, 1979; 8. Felner et al., 1983; 9. Chełkowski et al., 1988; 10. Felner, 1980; 11. Felner et al., 1981a; 12. Wang et al., 1988; 13. Felner et al., 1981b.

compounds, the effective moments μ_{eff} for $T_C < T < T_N$ are smaller than those for $T > T_N$. The μ_{eff} for $T > T_N$ are mainly due to the R ions; the contribution of the Mn atoms was found to be very small (Okomoto et al., 1987). The difference between μ_{eff} ($T_C < T < T_N$) and μ_{eff} ($T > T_N$) is due to the residual interaction of the R moments with the Mn moments above T_C.

The addition of a third element stabilizes the $ThMn_{12}$-type phase. The structure and magnetic properties of RT_xAl_{12-x} compounds with $4 < x < 6$ have been investigated since the 1970s. Detailed studies of the site occupancy of 3d atoms (Fe, Cu) in rare-earth and actinide $RT_{4+x}Al_{8-x}$ compounds by neutron diffraction and ^{57}Fe Mössbauer spectroscopy indicate a strong preferential site occupancy (Moze et al., 1990; Schafer et al., 1989). The 2a position is occupied by a lanthanide atom. For x = 4, the 8f position is, in principle, occupied by a transition element, whereas the other two positions (8i and 8j) are occupied by aluminum atoms. When x > 4 transition metal atoms also occupy the 8j positions.

Magnetic measurements indicate that $GdCr_6Al_6$ is ferromagnetic (T_C = 170 K), whereas $DyCr_6Al_6$ and $ErCr_6Al_6$ are antiferromagnetic with T_N equal to 20 and 15 K, respectively. $LuCr_6Al_6$ is diamagnetic (Felner et al., 1981).

The magnetic properties of RMn_6Al_6 are not fully known. The rare-earth moments order antiferromagnetically at low temperatures (see Table 5.17) (Felner et al., 1981). The ordering of the Mn sublattice has not been investigated.

RCu_6Al_6 (R = Yb and Lu) are diamagnetic, whereas those with R = Gd, Tb, Dy, Ho, Er, and Tm order antiferromagnetically at low temperatures (Felner et al., 1981).

In all these compounds, Cr and Cu do not carry a magnetic moment; in the case of Mn-containing samples, the localization of the magnetic moment on the Mn ions is not certain.

Different magnetic properties have been observed in RFe_4Al_8 compounds. Magnetic susceptibility measurements of YFe_4Al_8 indicate a single, magnetic ordering temperature, whereas for R = Pr, Nd, Sm, Gd, Tb, Dy, Ho, Er, or Tb, two magnetic transitions are observed. The iron sublattice orders independently of the rare-earth sublattice at about 130 K, while the R sublattice orders at about 20 K (Felner and Nowik, 1978; Buschow and van der Kraan, 1978). In $TbFe_4Al_8$ (Felner and Nowik, 1988), $HoFe_4Al_8$ (Gal et al., 1989), and $AnFe_4Al_8$ (An = actinide element) (Gal et al., 1990), magnetic measurements indicate spin glass-like behavior at low temperatures. A neutron diffraction study of $HoFe_4Al_8$ at low temperatures established the coexistence of a long-range magnetic ordering of the Fe sublattice, together with a spin-glass state of the Ho sublattice. Neutron diffraction measurements made on $TbFe_4Al_8$ show that the rare-earth sublattice orders in a complicated antiferromagnetic scheme with a ferromagnetic component (Shaked et al., 1979).

The Curie temperatures of RFe_5Al_7 range from 200 to 268 K, with a maximum value for the Gd compound (Felner et al., 1983). Iron magnetization is 5.4 μ_B per formula unit in $LaFe_5Al_7$ and 5.8 μ_B per formula unit in YFe_5Al_7. The rare-earth moments lie antiparallel to the 8j site iron moments and parallel to a ferromagnetic component of the canted antiferromagnetic structure of iron in the 8f sites. Iron ions located in the 8i sites are nonmagnetic. The population of Fe atoms over three sites determined by ^{57}Fe Mössbauer spectroscopy is 58, 6, and 36% for 8f, 8i, and 8j, respectively (Felner et al., 1983). RFe_5Al_7 compounds display a variety of unusual magnetic effects, including a huge thermal hysteresis of the magnetization in as-cast samples. For example, Figure 5.15 shows the hysteresis loop of $HoFe_5Al_7$ at 4.2. K.

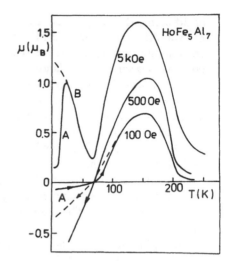

FIGURE 5.15. Temperature dependence of the magnetization of HoFe$_5$Al$_7$. The A curves were obtained when sample was cooled to 4.1 K in 30 Oe. Curve B was obtained when sample was cooled in 5 kOe (Felner et al., 1983).

The magnetic anisotropy in RFe$_5$Al$_7$ originates in part from the rare-earth crystal field and, to a larger extent, from the iron sublattice. The relatively high Curie temperature compared to the RF$_4$Al$_8$ series is attributed to the intrasublattice exchange interactions at the 8j sites. The temperature dependence of the magnetization of the Sm, Tb, and Lu compounds shows a negative magnetization value when the samples are cooled down quickly in the presence of a relatively weak field.

A different situation is observed in RFe$_6$Al$_6$ compounds. The magnetic and Mössbauer spectroscopy data indicate that the magnetic moment is localized on the rare-earth and iron sublattice. In RFe$_6$Al$_6$ systems, all magnetic sublattices order at relatively high temperatures. The magnetization curves start with small values at low temperatures, rise to very high values at T$_{max}$ (230 K), and then drop to 0 at T$_C$ (330 K). Such behavior can be explained by the following model:

- The rare-earth sublattice is coupled ferromagnetically.
- The magnetic moments of Fe atoms order ferromagnetically at the 8j sublattice and form a canted structure in the 8f sublattice (Felner, et al., 1981).

A comparison of the values calculated using this model and the experimental values of the magnetic moments as a function of temperature is presented in Figure 5.16. Although the agreement in absolute values is not perfect, the general behavior, and, in particular, the maximum points on the M(T) curves agree with the experimental data. Large differences among the magnetic properties of the three groups (RFe$_4$Al$_8$, RFe$_5$Al$_7$, and RFe$_6$Al$_6$) are observed. The j–j and a–j exchange interactions in RFe$_6$Al$_6$ play a dominant role, and raise the Curie temperature from 140 K in RFe$_4$Al$_8$ and 240 K in

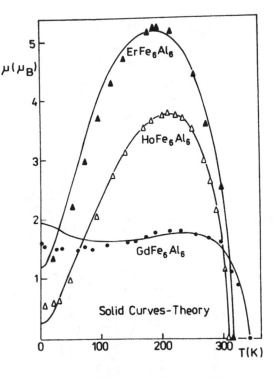

FIGURE 5.16. Theoretical magnetization curves for RFe$_6$Al$_6$ systems calculated from the suggested molecular field model. The experimental points were obtained from the experimental curves at high fields (Felner et al., 1981).

RFe$_5$Al$_7$ to 330 K in RFe$_6$Al$_6$. The magnetic data of RT$_{4+x}$Al$_{8-x}$ systems are summarized in Table 5.17.

5.5. La(T$_{1-x}$X$_x$)$_{13}$ (T = Fe, Co OR Ni, X = Al, Si) PHASES

La(T$_{1-x}$X$_x$)$_{13}$ compounds crystallize in the cubic NaZn$_{13}$ structure (space group Fm3c), have La atoms at the 8a sites, and T and Al atoms at the 8b and 96i sites (see Chapter 2 of this volume).

The only lanthanide binary compound exhibiting this type of structure is LaCo$_{13}$, which is ferromagnetic with μ_{Co} = 1.58 μ_B and Curie temperature T$_c$ = 1290 K (Buschow and Velge, 1977). In the case of Fe- and Ni-based compounds, this type of crystal structure is stabilized by substituting a portion of the T atoms with Si or Al atoms.

The La(Fe$_{1-x}$Si$_x$)$_{13}$ phases were formed for 0.12 < x < 0.19. Magnetic studies showed that T$_c$ increases with x, from 198 K for x = 0.12 to 262 K for x = 0.19, whereas the magnetic moment decreases from 2.08 μ_B for x = 0.12 to 1.85 μ_B for x = 0.19 (Palstra et al., 1983).

La(Fe$_{1-x}$Al$_x$)$_{13}$ compounds exist for x between 0.08 and 0.54 (Palstra et al., 1985). The experimental phase diagram of La(Fe$_x$Al$_{1-x}$)$_{13}$ is displayed in Figure 5.17. For 0.08 < x < 0.14, an antiferromagnetic order appears

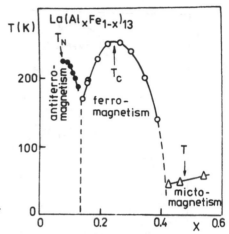

FIGURE 5.17. Magnetic phase diagram of La(Fe$_{1-x}$Al$_x$)$_{13}$ after (Palstra et al., 1985).

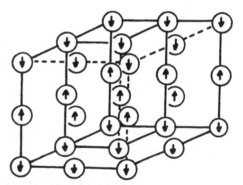

FIGURE 5.18. Model of the magnetic structure of La(Al$_x$Fe$_{1-x}$)$_{13}$. Each spin represents the total spin of the cluster of 13 atoms (Palstra, 1986).

along with a sharp metamagnetic transition in the external field of 4T (Palstra et al., 1984). With a decrease of the Al concentration ($0.14 < x < 0.38$), a soft ferromagnetic phase was found, which, at lower temperatures, shows anisotropy effects related to re-entrant mictomagnetic behavior. At larger values of x ($0.38 < x < 0.54$), a mictomagnetic region occurs with a distinct cusp on the ac susceptibility curve at about 50 K. The large positive Curie-Weiss temperature $\Theta_p = +110$ K indicates the presence of predominantly ferromagnetic exchange interactions. A linear decrease of the saturation magnetic moment with a decrease of iron concentration from 2.14 μ_B per Fe atom for x = 0.08 to 1.35 μ_B per Fe atom for x = 0.35 is observed.

The magnetic structures were determined for ferromagnetic (x = 0.32) and antiferromagnetic (x = 0.09) samples. For the x = 0.32 compound, the magnetic moment is $\mu = 1.41(8)$ μ_B per Fe atom. The magnetic structure of the x = 0.91 sample is presented in Figure 5.18. This compound crystallizes in a complicated structure (see Chapter 2 of this volume). Each icosahedron of 12 Fe^{II} atoms together with the central Fe^{I} atom is considered as one cluster.

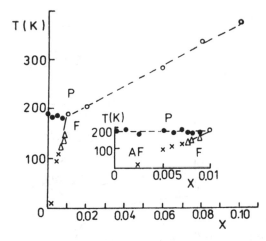

FIGURE 5.19. Magnetic phase diagram La(Fe$_{0.88-x}$Co$_x$Al$_{0.12}$)$_{13}$ system, o, Curie temperature, o, Néel temperature, x, temperature phase transition F–AF (Yermolenko et al., 1988).

Magnetic moments of 2.14(3) μ_B per Fe atom for the Fe^{II} moment and 1.10(7) μ_B per Fe atom for the Fe^I moment have been obtained.

In the La (Fe$_{0.86}$Al$_{0.14}$)$_{13}$ compound, a ferro- to antiferromagnetic phase transition induced by pressure as low as p < 0.1 GPa has been detected (Abd-Elmequid et al., 1987). A ^{57}Fe Mössbauer high-pressure experiment performed on La (Fe$_{0.88}$Al$_{0.12}$)$_{13}$ shows an abrupt drop of the magnetic ordering temperature. The average magnetic hyperfine field B$_{eff}$ at critical pressure p$_c$ = 45 kbar, corresponding to an average Fe distance d$_c$ = 2.53(5) Å was also found (Ludorf et al., 1989).

Studies on La(Co$_{1-x}$Al$_x$)$_{13}$ with x up to 0.3, La(Co$_{1-x-y}$Fe$_x$Al$_y$)$_{13}$ with y = 0.15 and x up to 0.3 (La$_{0.7}$Nd$_{0.3}$) (Co$_{0.7}$Fe$_{0.3}$)$_{13}$, and La(Co$_{1-x}$Fe$_x$)$_{13}$ with x up to 0.6 have been reported (Ido et al., 1990). Fe substitution raises the magnetization but lowers the Curie temperature, while Al substitution decreases the average magnetic moment and the Curie temperature.

In a La(Fe$_{0.88-x}$Co$_x$Al$_{0.12}$)$_{13}$ sample, the ferromagnetic-paramagnetic transitions for x > 0.01 and ferromagnetic-antiferromagnetic-paramagnetic transitions for x < 0.01 were observed under hydrostatic pressures (Medvedeva et al., 1992).

The magnetic phase diagram of the La(Fe$_{0.88-x}$Co$_x$Al$_{0.12}$)$_{13}$ system is shown in Figure 5.19 (Yermolenko et al., 1988). At low temperatures, alloys with a cobalt concentration x < 0.0075 are in the antiferromagnetic state. The Néel temperatures of these alloys are practically insensitive to x. For alloys with 0.0075 < x < 0.01, there are two transitions at T$_t$ and T$_N$. The temperature of the first-order ferro- to antiferromagnetic transition T$_t$ is strongly dependent on x, dT$_t$/dx = 3 10^2 K (at %)$^{-1}$. One can obtain a magnetic field-induced ferromagnetic state at 4.2 K in alloys with 0.0025 < x < 0.0075. When x > 0.01, second-order ferro- to paramagnetic transitions are observed. The Curie temperature depends on x, dT$_c$/dx = 20 K (at %)$^{-1}$

The temperature dependence of the total resistivity of La(Fe$_{1-x}$Al$_x$)$_{13}$ in examples typical for each of the three regions is displayed in Figure 5.20.

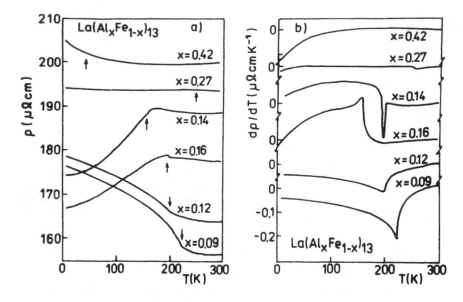

FIGURE 5.20. (a) Electrical resistivity vs. temperature for $La(Al_xFe_{1-x})_{13}$. Arrows indicate magnetic ordering temperatures. (b) Temperature derivative of electrical resistivity $d\rho/dT$ vs. temperature for $La(Al_xFe_{1-x})_{13}$ (Palstra, 1986).

The general trend is that room-temperature resistivity increases from 157 $\mu\Omega$ cm for $x = 0.09$ to 200 $\mu\Omega$ cm for $x = 0.42$. The $d\rho/dT$ changes with the concentration of x, from negative $(x < 0.12)$ to cross positive $(x = 0.15)$ to negative $(x > 0.27)$. Large anomalies in the resistance are observed around the magnetic-ordering temperatures. To explain these anomalies, $d\rho/dT$ vs. T has been plotted (see Figure 5.20) (Palstra, 1986).

Pressure produces a decrease of the Curie and Néel temperatures $(dT_c/dp = -0.8$ K/kbar^{-1}, $dT_N/dp = -0.3$ K/kbar$^{-1})$.

In $La(Fe_{0.88-x}Mn_xAl_{0.12})_{13}$, a phase transition from the antiferro- to the micromagnetic state is observed for $x > 0.045$ (Scherbakova et al., 1989).

The magnetic properties of these compounds are strongly dependent on the surroundings of the magnetic ions and the interatomic distances. For the $La(Fe_{0.88}Al_{0.12})_{13}$ compound, the average distance between iron atoms is $d = 2.544$ Å, which is close to the critical value of d_e, for which the exchange interactions reverse sign. Substitution or pressure induces the F–AF transition, with large values of dT_c/dp. $LaNi_{11}Si_2$ is a Pauli paramagnet (Palstra et al., 1983).

5.6. RT_9Si_2

$NdCo_9Si_2$ crystallizes in a tetragonal structure (space group $I4_1/amd$) derived from the $BaCd_{11}$ type of structure (Bodak and Gladyshevskii, 1969). The cobalt atoms occupy two different sites, 32i and 4b, while the neodymium

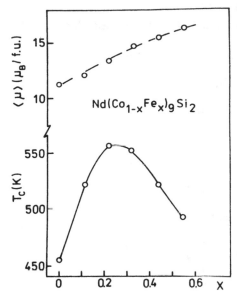

FIGURE 5.21. Iron concentration dependence of Curie temperature and average 3d moment in Nd(Co$_{1-x}$Fe$_x$)$_9$Si$_2$ (Berthier et al., 1988).

and silicon atoms are located in the 4a and 8d sites, respectively. Each cobalt atom has eight cobalt nearest-neighbors, but the average Co distances are significantly different for 32i sites (2.537 Å) compared with 4b sites (2.623 Å).

NdCo$_9$Si$_2$ is ferromagnetic, with a Curie temperature of 454 K (Malik et al., 1989). The magnetization at 4.2 K is 11.1 μ_B per formula unit, from which the value of 0.9 μ_B was deduced for the Co moment. This is smaller than the value in Co metal, RCo$_5$, R$_2$Co$_{17}$ (Wallace, 1986), or R$_2$Co$_{14}$B (Buschow et al., 1988), and reflects the relatively large silicon content. The small moment accounts for the comparatively low T_c of NdCo$_9$Si$_2$. The anisotropy field deduced from the magnetization curves of an oriented powder sample of NdCo$_9$Si$_2$ was 1.2 T at 295 K and more than 2.0 T at 77 K (Malik et al., 1989). A spin reorientation is observed below 77 K in SmCo$_9$Si$_2$ (Pourarian et al., 1989). There is no such effect in NdCo$_9$Si$_2$, which indicates that the contributions to the anisotropy of the cobalt and Sm sublattices are opposite in sign.

Nd(Co$_{1-x}$Fe$_x$)$_9$Si$_2$ alloys are isostructural, with NdCo$_9$Si$_2$ for $0 < x < 0.55$ (Berthier et al., 1988; Sanchez, 1989). Figure 5.21 shows the concentration dependence of the average 3d magnetic moment and the Curie temperature for the Nd(Co$_{1-x}$Fe$_x$)$_9$Si$_2$ series. The average 3d moment increases monotonically with x, whereas the Curie temperature passes through a maximum of 556 K at $x = 0.22$, and then drops. The Fe atoms preferentially occupy the larger 4b sites, and the iron moment attains a maximum (1.74 μ_B) for $x = 0.33$, whereas the cobalt moment increases uniformly with x. It is suggested that the Co moments are canted away from the ferromagnetic axis. The magnetic data for these compounds are listed in Table 5.19.

TABLE 5.18
Magnetic Data for La(T$_{1-x}$X$_x$)$_{13}$ Compounds (T = Fe, Co and X=Si, Al)

Compound	T$_C$(K)	$\mu_T(\mu_B)$ at 4.2 K	Ref.
La(Fe$_{1-x}$Si$_x$)$_{13}$			
x = 0.115	198	2.08	1
0.139	211		1
0.146	219		1
0.154	230	1.95	1
0.162	234		1
0.169	245		1
0.193	262	1.85	1
La(Co$_{1-x}$Si$_x$)$_{13}$			
x = 0	1290	1.58	1, 2
0.115		1.29	1
0.154		1.14	1
0.193		0.88	1
La(Fe$_{1-x}$Al$_x$)$_{13}$			
x = 0.08	224	2.14	3
0.12	202	2.02	3
0.14	158	1.97	3
0.16	196	1.91	3
0.27	250	1.59	3
0.35	201	1.35	3

References: 1. Palstra et al., 1983; 2. Buschow and Velge, 1977; 3. Palstra, 1986.

5.7. RT$_{12}$B$_6$ SERIES

The RCo$_{12}$B$_6$ compounds (R = Y, La, Ce, Pr, Nd, Sm, Gd, and Dy) crystallize in a rhombohedral crystal structure of the SrNi$_{12}$B$_6$ type. They order ferromagnetically, with the T$_c$ ranging from 154 to 177 K. Magnetic data for this series are given in Table 5.20. Figure 5.22 shows the magnetic isotherms and M vs. T curves for selected alloys. Assuming that the YCo$_{12}$B$_6$ moment represents the Co moments in RCo$_{12}$B$_6$ systems, the R–Co coupling schematics can be directly inferred from these results. These compounds were found to be ferromagnetic for R = Y–Sm and ferromagnetic for Gd and Dy, thus following the general coupling scheme observed in all R-based compounds. As expected for ferromagnets, a spin compensation effect is observed in GdCo$_{12}$B$_6$ and DyCo$_{12}$B$_6$ (Figure 5.23) due to the different temperature dependencies of the R and Co sublattice magnetizations.

The average cobalt moment in RCo$_{12}$B$_6$ changes only slightly with the rare-earth element; it is 0.44 μ_B in YCo$_{12}$B$_6$ and 0.43 μ_B in GdCo$_{12}$B$_6$ (Ro-

TABLE 5.19
Magnetic Data for RCo$_9$Si$_2$ and R(Co$_{1-x}$Fe$_x$)$_9$Si$_2$ Compounds

Compound	T$_C$(K)	M(μ_B/per formula unit)	Ref.
RCo$_9$Si$_2$			
R = Nd	454	11.1	1
Sm	473	8.6	1
Gd	483	1.8	1
Tb	463	2.0	1
Er	415	2.0	1
Nd(Co$_{1-x}$Fe$_x$)$_9$Si$_2$			
x = 0.11	521	11.97	2, 3
0.22	556	13.32	2, 3
0.33	551	14.69	2, 5
0.44	521	15.31	2, 3
0.55	492	16.24	2, 3

References: 1. Porarian, 1990; 2. Chevalier et al., 1987; 3. Berthier et al., 1988.

TABLE 5.20
Structural and Magnetic Data for RT$_{12}$B$_6$ (T = Fe or Co) Compounds

Compound	T$_c$(K)	M$_s$(μ_B per formula unit) 4.2 K	77 K	T$_{comp.}$ (K)	Ref.
NdFe$_{12}$B$_6$	230	19.5			1
LaCo$_{12}$B$_6$	160				2, 3
CeCo$_{12}$B$_6$	154	5.4	4.5		2–4
PrCo$_{12}$B$_6$	177	8.4	5.6		2, 3
NdCo$_{12}$B$_6$	174	8.8	6.0		2, 3
SmCo$_{12}$B$_6$	172	6.8	5.5		2, 3, 5
GdCo$_{12}$B$_6$	169	2.1	1.5	47.8	2, 3, 6–8
DyCo$_{12}$B$_6$	165	5.9	1.9	72.0	2, 3
YCo$_{12}$B$_6$	151	6.5	5.5		3, 6, 8

References: 1. Buschow et al., 1986; 2. Mittag et al., 1989; 3. Jurczyk et al., 1987a; 4. Niihara and Yajima, 1972; 5. El Masry et al., 1983; 6. Rosenberg et al., 1988; 7. Chaban and Kuźma, 1977; 8. Rosenberg et al., 1989.

senberg et al., 1988). The small difference in the cobalt moment or Curie temperature between YCo$_{12}$B$_6$ and GdCo$_{12}$B$_6$ reflects the weakness of the Gd–Co exchange interactions compared to the Co–Co exchange interactions, which largely determine the magnitude of T$_c$. The ineffectiveness of the rare-earths in influencing the magnetic properties is probably due to the small concentration of rare-earth atoms in these structures and the fact that they are

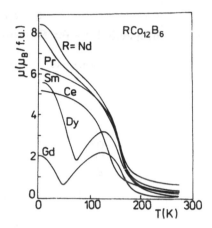

FIGURE 5.22. Temperature dependence of the magnetization of $RCo_{12}B_6$ compounds in an applied field of 2 T (Jurczyk et al., 1987a).

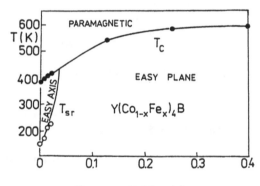

FIGURE 5.23. Magnetic phase diagram of the $Y(Co_{1-x}Fe_x)_4B$ system (Thuy et al., 1990).

relatively isolated by a surrounding cage of boron atoms (Kuźma et al., 1981). The exchange integral $J_{GdCo} = 5.54$ K in $GdCo_{12}B_6$ is four times smaller than that in $Gd_2Co_{14}B$ (Rosenberg et al., 1988), whereas the magnitudes of the J_{CoCo} integrals in both compounds are similar.

Figure 5.22 shows the temperature dependence of the magnetization of $RCo_{12}B_6$ for R = Ce, Nd, Pr, Sm, Gd, or Dy in a field of 2 T (Jurczyk et al., 1987a). The R–Co coupling follows the usual trend, with R and Co moments coupling ferromagnetically in compounds containing the light rare-earths and antiferromagnetically in those containing the heavy rare-earths. A ferromagnetic spin compensation phenomenon has been detected in $GdCo_{12}B_6$ and $DyCo_{12}B_6$ at 46 and 72 K, respectively.

A metastable compound in the Nd–Fe–B system with the rhombohedral $SrNi_{12}B_6$ structure was also found (Buschow et al., 1986). $NdFe_{12}B_6$ exhibits a Curie temperature at 230 K. The magnetization at 4.2 K is 19.5 μ_B per formula unit.

TABLE 5.21
Magnetic Data for RT$_4$B (T = Co, Fe) Compounds

Compound	T_C(K)	T_{comp}(K)	M_s(μ_B per formula unit) at 4.2 K	K_{eff}(10^7 J/m^3)	Ref.
CeCo$_4$B	297		1.1		1
PrCo$_4$B	455		5.1		2, 3
NdCo$_4$B	458		5.8		2, 4
SmCo$_4$B	510		3.8		3, 5
GdCo$_4$B	505	410	3.25	0.05	2, 3, 6, 7
TbCo$_4$B	455		5.17		1
DyCo$_4$B	427	340	6.2	−1.65	1, 7
HoCo$_4$B	396	240	6.1	−1.35	1, 7
ErCo$_4$B	386	163	5.1	0.55	1, 7
LuCo$_4$B	396				2, 14
YCo$_4$B	382		2.65		2, 3, 8, 9
ErFe$_4$B	620		2.08		10
TmFe$_4$B	610		1.16		10
LuFe$_4$B	573		5.22		3, 5, 9, 10

References: 1. Burzo et al., 1989a; 2. Pędziwiatr et al., 1987; 3. Spada et al., 1984; 4. Ido et al., 1990; 5. Oesterreicher et al., 1984; 6. Drzazga et al., 1990; 7. Drzazga et al., 1989; 8. Dung et al., 1988; 9. Burzo et al., 1989b; 10. Van Noort et al., 1985.

5.8. RT$_4$B

The RT$_4$B compounds crystallize in the CeCo$_4$B type of structure. This structure can be derived from the CaCu$_5$ type by an ordered replacement of cobalt atoms with boron atoms at the 2c sites (Kuzma and Bilonizhko, 1973). Only RFe$_4$B compounds (R = Er, Tm, and Lu) were studied (Spada et al., 1984; Van Noort et al., 1985; Vaishnana et al., 1985; Aly and Hadjipanayis, 1987). As can be seen from Table 5.21, the Curie temperatures are around 600 K. The anisotropy fields at room temperature are H$_a$ = 5.4, 2 and 1.6 Tesla for R = Er, Tm, and Lu, respectively. The magnetization in ErFe$_4$B and TmFe$_4$B was found to be rather small, owing to the antiparallel coupling of the R and Fe sublattices.

RCo$_4$B compounds exist for R = Y, Pr, Nd, Gd, and Er. The thermal variation of the magnetization of compounds with R = Y, Pr, and Nd indicates a ferromagnetic order, whereas in the case of R = Gd and Er, ferromagnetic ordering is observed. For R = Y-, Nd- and Gd-based compounds, the easy direction of magnetization is the c axis. PrCo$_4$B evidenced basal plane anisotropy. Above the Curie temperature, the reciprocal susceptibilities obey a hyperbolic dependence of the Néel type (Néel, 1948):

$$\chi^{-1} = \chi_0^{-1} + TC^{-1} - \sigma(T - \theta)^{-1} \qquad (5.7)$$

The χ_0, σ, and θ parameters are corrected with the molecular field coefficient J_{ij} (i, j = R, Co), characterizing the exchange interactions inside and among

TABLE 5.22
The Molecular Field Coefficients (in K) for RCo$_4$B

	GdCo$_4$B			ErCo$_4$B		
	J_{Co-Co}	J_{Gd-Co}	J_{Gd-Gd}	J_{Co-Co}	J_{Er-Co}	J_{Er-Er}
From thermal variation of spontaneous magnetization	352	-34	20	320	-18	5
From paramagnetic data	165	-26	43	120	-21	19

Pędziwiatr, 1988.

the magnetic sublattices. They were obtained by two-sublattice ferromagnet approximation (Néel, 1948; Pędziwiatr, 1988). The J_{ij} values determined with the paramagnetic data and spontaneous magnetization are listed in Table 5.22.

An interesting feature of the RT$_4$B series is the huge rare-earth anisotropy found in these systems, e.g., 90.6 T for the uniaxial anisotropy field observed in SmCo$_4$B (Oesterreicher et al., 1984). ^{155}Gd Mössabuer and magnetic measurements carried out for GdCo$_4$B, Gd$_3$Co$_{11}$B$_4$, Gd$_2$Co$_7$B$_3$, and GdCo$_3$B$_2$ show that the second-order crystal-field coefficient A$_2^0$ increases with increasing B substitution (Smit et al., 1988). Averaged over the rare-earth sites, A$_2^0$ reaches -1000 Ka$_0^{-2}$ in RCo$_4$B compared with -700 Ka$_0^{-2}$ in RCo$_5$ and 670 Ka$_0^{-2}$ in R$_2$Fe$_{14}$B. The negative sign of A$_2^0$ means that the rare earths having a positive second-order Stevens coefficient α_J (Sm^{3+}, Er^{3+}, Tm^{3+}, and Yb^{3+}) show uniaxial anisotropy similar to that of RCo$_5$, but unlike that of R$_2$Fe$_{14}$B. A ^{57}Fe Mössabuer study of oriented ErFe$_4$B and TmFe$_4$B absorbers confirmed the uniaxial anisotropy (Gros et al., 1988a). There is no indication of spin reorientation (Zouganelis et al., 1990a). It has been reported that NdCo$_4$B has the easy c axis and NdCo$_3$FeB the easy c plane, which indicates that Fe atoms occupy mainly 2c sites. Thus, the transition metal has a crucial effect on uniaxial anisotropy (Gros et al., 1989).

The R(Co$_{1-x}$Fe$_x$)$_4$B solid solution covers the whole concentration range, the samples are either ferromagnets or ferrimagnets. In Lu(Co$_{1-x}$Fe$_x$)$_4$B, the Curie temperatures and average magnetic moments increase with x. An anomalous dependence is observed for the anisotropy constant K$_1$, similar to that observed in the Y(Co$_{1-x}$Fe$_x$)$_4$B system. The magnetic phase diagram for this system is presented in Figure 5.23.

K$_1$ is negative in LuCo$_4$B, and correlates with planar anisotropy. With an increase of the iron concentration, a K$_1$ minimum is observed for x = 0.3. In the sample with x = 0.8, a change of sign was detected. This dependence is quite different from that observed in 1:5 compounds in which a K$_1$ maximum has been found (Thuy and Franse, 1986).

A relatively small concentration of Fe in Gd(Co$_{4-x}$Fe$_x$)B (x = 0.15) causes the easy direction of magnetization to shift from the c axis to the basal plane (Drzazga et al., 1990).

Magnetic coupling in $(R_{1-x}Lu_x)Fe_4B$ (R = Ho, Er, or Tm) was studied using the temperature dependence of the magnetization and its high-field dependence (up to 30 T). The value of the coupling constant (J_{FeFe}) was 75 K in all the compounds, but J_{RFe} ranged from -9.7 to -11.0 K (Buschow et al., 1990a).

Large intrinsic coercivities were found in RFe_2Ni_2B (R = Sm or Er). (Strzeszewski et al., 1988). In $SmFe_2Ni_2B$, the coercivity is larger than 8 T at 4.2 K. By contrast, $NdFe_2Ni_2B$ is quite soft because of its planar anisotropy. $RCo_{4-n}Fe_nB$ alloys with R = Nd, Sm, or Er have been examined by magnetometry and X-ray diffraction (Aly ct al., 1988). The $CeCo_4B$-type phase has been found for all n in the case of Er, for n < 3 in Sm, and for n < 2 in Nd compounds. Maximum coercivities, $\mu_0H_c > 1.7$ T at room temperature, were obtained in a crystallized $SmFe_2Co_2B$ sample. The unusually high achievable ratios of extrinsic coercivity to anisotropy field in these nonmetal-stabilized materials have been explained by their chemically relatively inert layer structure, which appears to lead to less corrugated surface structures, and thus is responsible for the characteristic domain-wall nucleation process (Oesterreicher et al., 1984). The $R(Co_{1-x}Fe_x)_4B$ systems have been studied by Spada et al. (1984) (Sm), Hong et al. (1988) (Lu), Burzo et al. (1989b) (Y), and Gros et al. (1988b) (Pr, Nd, Sm, and Er).

5.9. $R_{1+\epsilon}T_4B_4$

The $R_{1+\epsilon}T_4B_4$ compounds crystallize in a tetragonal structure with a complicated arrangement of atoms (see Chapter 2 of this volume).

The lanthanide-iron borides exhibit very low Curie temperatures, with a maximum T_c of 37 K for R = Sm, (Rechenberg and Sanchez, 1987).

[57]Fe Mössbauer measurements (Rechenberg et al., 1986) have indicated that all compounds show no trace of magnetic ordering at temperatures as low as 1.5 K. This is a clear indication that the Fe atoms are nonmagnetic; magnetic ordering is observed only in the rare-earth sublattice. At 4.2 K, the spontaneous magnetization in $Gd_{1+\epsilon}Fe_4B_4$ ($\epsilon = 1.143$) is 8 μ_B per formula unit, i.e., 7 μ_B per Gd atom, which implies a ferromagnetic colinear structure of the rare-earth sublattice. High-field magnetization data on a single crystal of $Nd_{1.1}Fe_4B_4$ (Givord et al., 1985, 1986) indicate that the direction of easy magnetization is in the basal plane. The ferromagnetic Nd moment at 4.2 K 2.2 μ_B per Nd atom is smaller than the free-ion value ($gJ = 3.27 \mu_B$). This reduction is due to the crystal-field effect. [155]Gd and [161]Dy Mössabuer studies performed on $Gd_{1.14}Fe_4B_4$ and $Dy_{1.2}Fe_4B_4$ compounds made it possible to determine the second-order crystal-field coefficient $A_2^0 = -2450$ Ka_0^2 (Rechenberg et al., 1987; Bogé et al., 1989), which is the largest for any rare-earth intermetallic compound. This implies that in $R_{1+\epsilon}T_4B_4$ compounds, the crystal-field interactions are comparable to or stronger than the exchange interactions.

TABLE 5.23
Magnetic Data for $R_{1+\epsilon}Fe_4B_4$ and RCo_4B_4 Compounds

Compound	$T_C(K)$	$M_s(\mu_B)$	Ref.
$Pr_{1+\epsilon}Fe_4B_4$	7.5	1.5	1, 2
$Nd_{1+\epsilon}Fe_4B_4$	14	2.4	1–3
$Sm_{1+\epsilon}Fe_4B_4$	37		1, 2, 4
$Gd_{1+\epsilon}Fe_4B_4$	20	6.1	1, 2
$Tb_{1+\epsilon}Fe_4B_4$	16	3.5	1, 2
$Dy_{1+\epsilon}Fe_4B_4$	11.5	5.7	1
$Ho_{1+\epsilon}Fe_4B_4$	6.5	5.5	1
$PrCo_4B_4$	171	1.7	5
$NdCo_4B_4$	182	1.9	5
$SmCo_4B_4$	176	0.2	5

References: 1. Niarchos et al., 1986; 2. Bezinge et al., 1985; 3. Givord et al., 1985; 4. Rechenberg and Sanchez, 1987; 5. Jurczyk et al., 1987b.

The orientation of the magnetic moments are in agreement with the signs of the second-order Stevens coefficient: α_J is positive for Sm (the magnetic moment is parallel to the c axis) and negative for Nd and Dy (the magnetic moments lie in the c plane).

Magnetic measurements of RCo_4B_4 (R = Pr, Nd, and Sm) indicate a ferromagnetic ordering with Curie temperatures ranging from 171 to 182 K. Magnetization data suggest a complicated magnetic ordering scheme (Jurczyk et al., 1987b; El Masary and Stadelmeier, 1984).

The magnetic data for $R_{1+\epsilon}$ T_4B_4 compounds are given in Table 5.23.

Chapter 6

CONCLUDING REMARKS

The analysis of the structural and magnetic properties of ternary phases containing lanthanide-transition metal-nonmetal (Si, Ge, Sn, Ga) elements makes it possible to draw a number of general conclusions. The first characteristic feature is the existence of many phases in each of these ternary systems. For example, in the Ce–Ni–Si system, 21 phases, each exhibiting different magnetic properties and a characteristic crystal structure, have been detected (Bodak et al., 1973). In other systems, usually from few to more than ten different phases have been found. However, it should be mentioned that the number of lanthanide ternary systems which were fully examined is still rather small. The hitherto most complete account of phase equilibria and phase diagrams was published a few years ago (Gladyshevsky et al., 1990). Explaining the existence of so many phases in each system turned out to be a difficult task. For binary lanthanide compounds, a theory has been proposed (Miedema, 1976) which postulates that the heat of formation of a binary phase H involves contributions bound to the interactions among the neighboring atoms in the particular crystal structure. Three kinds of contributions were distinguished: the difference in electronegativity of the two component elements ($\Delta\Phi^*$); the difference in component atom electron density at the Wigner-Seitz cell boundary (Δn_{ws}), the hybridization energy when a lanthanide atom becomes bound to a polyvalent transition metal atom. The heat of formation has been expressed as

$$\Delta H = f(c)[Pe(\Delta\Phi^*)^2 + Q_o(\Delta n_{ws}^{1/3})^2 - R]$$

In the case where the metallic radii of both components do not differ much, the quantity $f(c)$ can be expressed by the following relation:

$$f(c) = c(1 - c)[1 + 8c^2(1 - c^2)]$$

c is the concentration of the lanthanide element in a binary compound, e-elementary charge, P, Q_o, R-constants. This theorem satisfactorily represents the stability condition of a binary compound, but, unfortunately, it does not properly work in the case of ternary phases. Nevertheless, it has been used to calculate the magnitudes of observed isomer shifts in GdT_2Si_2 compounds (De Vries et al., 1985; Łątka, 1989) and a reasonable agreement with the experiment was obtained.

Based on the Miedema theorem, it has been suggested that the ternary compounds with composition $A_xB_yC_z$ ($x > y > z$; $x + y + z = 1$) are stable when $x < 0.85$ and $z > 0.05$. Only in a few cases was this suggestion found to be true; for example: $RFe_{11}Ti = R_{0.077}Fe_{0.846}Ti_{0.077}$ or $R_2F_{14}B = R_{0.118}Fe_{0.823}B_{0.059}$.

The crystal structure data collected up to now for the lanthanide ternary phases indicate that they crystallize in a large number of different structure types. An attempt to explain this phenomenon turned out to be difficult, since the stability of a particular structure is dependent on many factors, such as the geometrical, the electron, the electrochemical, and the chemical bond system. An analysis of interatomic distances observed in the RT_2X_2 ($ThCr_2Si_2$-type structure) shows that ionic, covalent, and metallic bonding systems simultaneously operate in these phases. Making a proposal for reasonable systematics thus becomes rather difficult. Moreover, a lack of dominant factors has been indicated — the structure tends to reach the highest symmetry, the largest space filling and the largest number of bonds among the atoms. The more quantitative factors are represented by the size of constituent atoms, the coordination number, and by the interatomic distances observed in the experiment (Iandelli and Palenzona, 1979).

A discussion of unit-cell dimensions of a large number of RT_2X_2 and $ThMn_{12}$-type phases lead to a conclusion that only selected interatomic constants control the unit-cell dimensions (Pearson and Villars, 1984). For example, in the RFe_2X_2 ($X = Si$, Ge) phases it has found that the R–X bonds control the unit-cell dimensions (Bara et al., 1990). In the binary compounds with the crystal structure of $ThMn_{12}$ type, the unit cell dimensions depend solely on the Tb–Mn(8i) and Th–Mn(8f) constants (Pearson, 1984).

When discussing their magnetic properties, two groups of lanthanide ternaries can be distinguished. The phases showing a large content of the transition metals (particularly the 3-d metals) are associated with the first group. Their magnetic behavior is mainly due to the 3d-transition metal ions, so that their ordering temperatures are usually higher than the room temperature, similar to transition metal alloys. However, the distinct influence of lanthanide ions is usually observed at low temperatures. The dominating influences are the magnetic interactions between the moments localized on 3d-transition metals. The interactions between the latter and the moments carried by rare-earth ions are considerably weaker, similar to lanthanide-lanthanide interactions.

The latter are usually responsible for moment reorientations. Compounds with a large content of lanthanide elements belong to the second group. Their magnetism arises from the interactions of the magnetic moments localized on lanthanide ions. The d transition ions usually do not carry magnetic moments. The only exceptions are the RMn_2X_2 and $RFe_{12-x}Al_x$ phases. Two ordering temperatures have been observed — at temperatures larger than 300 K associated with the Mn(or Fe) magnetic sublattice and at low temperatures (mostly below 78 K) connected with the rare-earth sublattice. This effect suggests that the coupling between the two sublattices is rather weak. The type of magnetic ordering in the Mn or Fe sublattice has been found to bear relation to a critical Mn–Mn (or Fe–Fe) contact length. In RMn_2X_2 it amounts to 2.857 Å; it is 2.60 Å in $RFe_{12-x}Al_x$ compounds. Antiferromagnetic coupling has been observed to operate when the Mn–Mn critical distance is smaller than 2.85 Å. It becomes ferromagnetic when d_{Mn-Mn} exceeds this value. The same regularity and critical values have been detected earlier in binary lanthanide-3d transition metal compounds (Forrer, 1952).

In most phases constituting this group, only lanthanide ions carry localized magnetic moments, but the magnetic interactions show a different character, since the R–R distances are usually longer than 3.5 Å. The discovery of helicoidal magnetic structures with fairly long periodicities and metallic-like character of electrical conductivity suggests that the magnetic interactions operate on long distances in a crystal.

In many cases, these observations have been satisfactorily explained in terms of the RKKY model, which postulates the magnetic interactions proceeding via the conducton electrons. The stabilities of a number of observed experiments in magnetic ordering schemes have been successfully accounted for by this model.

The Curie and Néel temperatures determined for many families of ternary phases only in part follow the De Gennes scaling. This effect was shown to be due to the action of the CEF, since addition of a corresponding term to the Hamiltonian describing the RKKY interactions has explained, in many cases, the observed deviations from the De Gennes function.

A conclusion may thus be drawn that the CEF plays a significant role in stabilizing the particular magnetic ordering schemes. The determination of the CEF parameters is, therefore, of importance for understanding the nature of magnetic properties of lanthanide ternaries. The respective investigations have been, up to now, carried out only for a limited number of phases. The most complete data so far have been collected for the RT_2X_2 compounds, so that more general conclusions can already be drawn:

- The B_n^m parameters depend on the number of f and d electrons. In the isostructural RT_2X_2 series, the compounds with T = Mn–Ni, Ru, Rh, Pd, Os, show a change of the sign of the B_2^0 parameter from plus to minus when passing from R = Ho to Er. On the other hand, in compounds with T = Cu, Pt, and Au, the change is from minus in HoT_2X_2 to plus in ErT_2X_2.

- The sign of the B_2^0 parameter decides the orientation of the ordered moment localized on R^{3+} ion. When B_2^0 is negative, the moment is observed to align along the fourfold axis (the c axis), but in the cases when $B_2^0 > 0$ the moment is normal or makes an angle with it. This effect clearly follows from a vast amount of observations made in the RT_2X_2 phases. Exceptions were detected in some $RNi_2 X_2$ and $RPd_2 X_2$ compounds; however, the determined for them the B_2^0 values are small. Therefore, the influence of the higher-order CEF parameters cannot be neglected.

The other physical properties of lanthanide ternaries have not been as exhaustively studied, until recently. However, the discovery of the super-conductive and, in particular, the heavy fermion effect in a number of them has attracted considerable attention to the lanthanide ternaries. An increasing flow of new and more detailed data can be expected. Moreover, single crystals of many phases become gradually available, so that experiments can now be performed with better accuracy.

A number of lanthanide ternaries have already been found to show properties being of interest from the practical point of view. One can thus expect that, in the course of time, these new materials will be produced on an industrial scale. This constitutes an important factor which will stimulate rapid progress in the physics and chemistry of ternary lanthanide intermetallic systems.

REFERENCES

1. Abache, C. and Oesterreicher, H., *J. Appl. Phys.*, 57, 4112 1985.
2. Abd-Elmequid, M. M., Schleede, B., Michlitz, M., Palstra, T. T. M., Nieuwenhuys, G. J., and Buschow, K. H. J., *Solid State Commun.*, 43, 177, 1987.
3. Adam, S., Adam, Gh., and Burzo, E., *J. Magn. Magn. Mater.*, 61, 260, 1986.
4. Adam, A., Sakurai, J., Yamaguchi, Y., Fujiwara, Y., Mibu, K., and Shinjo, T., *J. Magn. Magn. Mater.*, 90–91, 544, 1990.
5. Adroja, D. T., Malik, S. K., Nagarajan, R., Padalia, B. D., and Vijayaraghavan, R., *Conf. Abstr. Int. Conf. Magnetism*, Paris, 1988, 1988b, 156, 5P G–11.
6. Adroja, D. T., Malik, S. K., Nagarajan, R., Padalia, B. D., and Vijayaraghavan, R., *Phys. Rev.*, B39, 4831, 1989.
7. Adroja, D. T. and Malik, S. K., *Phys. Rev.*, B45, 779, 1992.
8. Adroja, D. T., Nagarajan, R., Malik, S. K., Padalia, B. D., and Vijayaraghavan, R., *Solid State Commun.*, 66, 1201, 1988a.
9. Adroja, D. T. and Rainford, B. D., *Int. Conf. Strongly Correlated Electron Systems*, Sendai, 1992, 214, 10P–64.
10. Alameda, J. M., Givord, D., Lemaire, R., and Lu, Q., *J. Appl. Phys.*, 52, 2079, 1981.
11. Aliev, F. G., Moshchalkov, V. V., Kozyrkov, V. V., Zalyalyutdinov, M. K., Pryadun, V. V., and Skolozdra, R. V., *J. Magn. Mater.*, 76–77, 295, 1988.
12. Allain, Y., Bouree-Vigneron, F., Oleś, A., and Szytuła, A., *J. Magn. Magn. Mater.*, 75, 303, 1988.
13. Allemand, J., Bertrand, C., Le Roy, J., Moreau, J. M., Paccard, D., Paccard, D., Fremy, M. A., and Givord, D., *Concerted European Action on Magnets (CEAM)*, Mitchell, I. V., Coey, J. M. D., Givord, D., Harris, I. R., and Hanitsch, R., Eds., Elsevier, London, 1989, 98.
14. Aly, S. H. and Hadjipanayis, G. C., *J. Appl. Phys.*, 61, 3757, 1987.
15. Aly, S. H., Singleton, E. W., Hadjipanayis, G. C., Sellmye, D. J., and Zhao, Z. R., *J. Appl. Phys.*, 63, 3704, 1988.
16. Ammarguellat, C., Escorne, M., Mauger, A., Beaurepaire, E., Ravet, M. F., Krill, G., Lapierre, F., Haen, F., and Godart, C., *Physica Status Solidi (b)*, 143, 159, 1987.
17. Anderson, O. K., *Phys. Rev.*, B12, 3060, 1975.
18. André, G., Bażela, W., Oleś, A., and Szytuła, A., *J. Magn. Magn. Mater.*, 109, 34, 1992b.
19. André, G., Bourée, F., Bombik, A., Kolenda, M., Oleś, A., Sikora, W., and Szytuła, A., *J. Magn. Magn. Mater.*, 125, 1993a.
20. André, G., Bourée, F., Kolenda, F., Oleś, A., Sikora, W., and Szytuła, A., *J. Magn. Magn. Mater.*, 125, 1993b.
20a. André, G., Bourée, F., Bombik, A., Oleś, A., Sikora, W., Kolenda, M., Szytuła, A., Pacyna, A., and Zygmunt, A., *Acta Physica Polonica*, 1993c.
22. André, G., Thuery, P., Pinot, M., Oleś, A., and Szytuła, A., *Solid State Commun.*, 80, 239, 1991.
23. Andreev, A. V., Deryagin, A. V., Kudrevatykh, N. V., Mushnikov, N. V., Reimer, V. A., and Terentev, S. V., *Sov. Phys. JETP*, 63, 608, 1986.
24. Andress, K. R. and Alberti, E., *Z. Metallkd.*, 27, 12, 1935.
25. Bak, P. and von Boehm, J., *Phys. Rev.*, B21, 5297, 1980.
26. Balcar, E. and Lovesey, S. W., *Phys. Lett.*, A31, 67, 1970.
27. Ban, Z., Matkovic-Calogovic, D., and Szytuła, A., *Proc. 5th Italian-Yugoslav Crystalogr. Congress*, Padova, June 3 to 6, 1986, C55.
28. Ban, Z. and Sikirica, M., *Acta Crystallogr.*, 18, 594, 1965.
29. Bara, J. J., Hrynkiewicz, H. U., Miłoś, A., and Szytuła, A., *J. Less-Common Met.*, 161, 185, 1990.

30. Bara, J. J., Pędziwiatr, A. T., Zarek, W., Konopka, D., and Gacek, U., *J. Magn. Magn. Mater.*, 27, 159, 1982.

31. Barandiaran, J. M., Gignoux, D., Schmitt, S., Gomez-Sal, J. C., and Rodrigez Fernandez, J., *J. Magn. Magn. Mater.*, 69, 61, 1987.

32. Barandiaran, J. M., Gignoux, D., Schmitt, S., Gomez-Sal, J. C., Rodrigez Fernandez, J., Chieux, P., and Schweizer, J., *J. Magn. Magn. Mater.*, 73, 233, 1988.

33. Barandiaran, J. M., Gignoux, D., Schmitt, S., and Gomez-Sal, J. C., *Solid State Commun.*, 57, 941, 1986b.

34. Barandiaran, J. M., Gignoux, D., Schmitt, S., and Gomez-Sal, J. C., *Solid State Commun.*, 59, 223, 1986a.

35. Bauminger, E. R., Froindlich, D., Nowik, I., Ofer, S., Felner, I., and Mayer, I., *Phys. Rev. Lett.*, 30, 1053, 1973.

36. Bażela, W., *J. Less-Common Met.*, 133, 193, 1987.

37. Bażela, W., Leciejewicz, J., Małetka, K., and Szytuła, A., *J. Magn. Magn. Mater.*, 117, L1, 1992.

38. Bażela, W., Leciejewicz, J., Małetka, K., Szytuła, A., and Zygmunt, A., *Acta Phys. Polon.*, 1993.

39. Bażela, W., Leciejewicz, J., Małetka, K., Szytuła, A., and Zygmunt, A., *J. Magn. Magn. Mater.*, 96, 114, 1991.

40. Bażela, W., Leciejewicz, J., and Szytuła, A., *J. Magn. Magn. Mater.*, 50, 19, 1985a.

41. Bażela, W. and Szytuła, A., *J. Less-Common Met.*, 153, 327, 1989.

42. Bażela, W., Szytuła, A., and Leciejewicz, J., *Solid State Commun.*, 56, 1043, 1985b.

43. Bażela, W., Leciejewicz, J., Ptasiewicz-Bąk, H., and Szytuła, A., *J. Magn. Magn. Mater.*, 72, 85, 1988.

44. Beaudry, B. J. and Gschneidner, K. A., Jr., *Handbook on the Physics and Chemistry of Rare Earths*, Vol. I, Gschneidner, K. A., Jr. and Eyring, L., Eds., North-Holland, Amsterdam, 1978, chap. 2, 173.

45. Bertaut, E. F., *J. Phys. Chem. Solids*, 21, 295, 1961.

46. Bertaut, E. F., *Acta Cryst.*, A24, 217, 1968.

47. Berthier, Y., Chevalier, B., Etourneau, J., Rechenberg, H. R., and Sanchez, J. P., *J. Magn. Magn. Mater.*, 75, 19, 1988.

48. Besnus, M. J., Braghta, A., Hamdaomi, N., and Meyer, A., *J. Magn. Magn. Mater.*, 104–107, 1385, 1992.

49. Besnus, M. J., Braghta, A., Kappler, J. P., and Meyer, A., *Proc. Int. Conf. Valence Fluctuations*, Bangalore, India, 1987a, A27.

50. Besnus, M. J., Braghta, A., and Meyer, A., *Z. Phys. B — Condensed Matter*, 83, 207, 1991.

51. Besnus, M. J., Kappler, J. P., Lehmann, P., and Meyer, A., *Solid State Commun.*, 55, 779, 1985a.

52. Besnus, M. J., Kappler, J. P., Meyer, A., Serebus, J., Siaud, E., Lahiouel, R., and Pierre, J., *Physics*, 130B, 240, 1985a.

53. Besnus, M. J., Kappler, J. P., Ravet, M. F., Meyer, A., Lahiouel, R., Pierre, J., Siaud, E., Nieva, G., and Sereni, J., *J. Less-Common Met.*, 120, 101, 1986.

54. Besnus, M. J., Lehmann, P., and Meyer, A., *J. Magn. Magn. Mater.*, 63–64, 323, 1987.

55. Beyermann, W. P., Hundley, M. F., Canfield, P. C., Godart, C., Selsane, M., Fish, Z., Smith, J. S., and Tompson, J. D., *Physica B*, 171, 373, 1991b.

56. Beyermann, W. P., Hundley, M. F., Canfield, P. C., Tompson, J. D., Fish, Z., Smith, J. S., Selsane, M., Godart, C., and Latroche, M., *Phys. Rev. Lett.*, 66, 3289, 1991a.

57. Bezinge, A., Braun, H. R., Muller, J., and Yvon, K., *Solid State Commun.*, 55, 131, 1985.

58. Birgenau, R. J., *J. Phys. Chem. Solids,* 33, 59, 1972.
59. Blanco, J. A., Gignoux, D., Gomez-Sal, J. C., and Schmitt, D., *J. Magn. Magn. Mater.,* 104–107, 1273, 1992.
60. Blanco, J. A., Gignoux, D., Schmitt, D., and Vetties, C., *J. Magn. Magn. Mater.,* 97, 4, 1991.
61. Bleaney, B. I. and Bleaney, B., *Electric and Magnetism,* Clarendon Press, Oxford, 1965, 621.
62. Bodak, O. I. and Gladyshevskii, E. J., *Dopov. Akad. Nauk Ukr.,* RSR, Ser. A5, 452, 1969.
63. Bodak, O. I. and Gladyshevskii, E. J., *Dopov. Akad. Nauk Ukr.,* RSR, Ser. A5, 42, 1969.
64. Bodak, O. J. and Gladyshevskii, E. J., *Sov. Phys. Cryst.,* 14, 859, 1970.
65. Bodak, O. J. and Gladyshevskii, E. J., *Trojnyje sistiemy riedkoziemlienih mietallov,* Visha Shol. Izdav. Lvov, *Sov. Phys. Cryst.,* 14, 859, 1985.
66. Bodak, O. I., Gladyshevskii, E. J., and Kripyakevych, P. I., *Zh. Struct. Khim,* 11, 283, 1970.
67. Bodak, O. J., Gladyshevskii, E. I., Levin, E. M., and Lutsiv, R. V. *Dopov. Akad. Nauk. Ukr.,* RSR, Ser. A12, 1129, 1977a.
68. Bodak, O. I., Kotur, B. Y., Yarovets, V. I., and Gladyshevskii, E. I., *Kristallografya,* 22, 385, 1977b.
69. Bodak, O. J., Miskin, M. G., Tyvanchuk, A. T., Kharchenko, O. J., and Gladyshevskii, E. J., *Inorg. Mater.,* 9, 777, 1973.
70. Bogé, M., Coey, J. M. D., Czjzek, G., Givord, D., Jeandey, C., Li, H. S., and Oddou, J. L., *Solid State Commun.,* 55, 295, 1985.
71. Bogé, M., Jeandey, C., Oddou, J. L., and Yakinthos, J. K., *J. Phys. (Paris),* 49, C8–443, 1989.
72. Boltich, E. B., Oswald, E., Huang, M. Q., Hirosawa, S., Wallace, W. E., and Burzo, E., *J. Appl. Phys.,* 57, 4106, 1985.
73. Boltich, E. B., Pourarian, F., Obermyer, R. T., Sankar, S. G., and Wallace, W. E., *J. Appl. Phys.,* 63, 3964, 1988.
74. Boltich, E. B. and Wallace, W. E., *Solid State Commun.,* 55, 529, 1985.
75. Bolzoni, F., Leccabue, F., Pareti, L., and Sanchez, J. L., *J. Phys. (Paris),* 46, C6–305, 1985.
76. Bourée-Vigneron, F., Point, M., Gołab, M., Szytuła, A., and Oleś, A., *J. Magn. Magn. Mater.,* 86, 383, 1990.
77. Bozorth, R. M., *Ferromagnetism,* Van Nostrand, Princeton, 1951, 568.
78. Brabers, J. H. V. J., de Boer, F. R., and Buschow, K. H. J., *J. Appl. Comp.,* 179, 227, 1992.
79. Braun, H. F., *J. Less-Common Met.,* 100, 105, 1984.
80. Braun, H. F., Engel, N., and Parthe, E., *Phys. Rev.,* B28, 1389, 1983.
81. Braun, H. F., Jorlborg, T., and Junod, A., *Physica,* 135B, 397, 1985.
82. Braun, H. F., Segre, C. U., Acker, F., Rosenberg, M., Dey, S., and Deppe, P., *J. Magn. Magn. Mater.,* 25, 117, 1981.
83. Bucholz, W. and Schuster, H. M., *Z. Anorg. Allg. Chem.,* 482, 40, 1981.
84. Budkowski, A., Leciejewicz, J., and Szytuła, A., *J. Magn. Magn. Mater.,* 67, 316, 1987.
85. Burlet, P., Coey, J. M. D., Barigan, J. P., Givord, D., and Meyer, A., *Solid State Commun.,* 60, 723, 1986.
86. Burzo, E., Boltich, E. B., Huang, M. Q., and Wallace, W. E., *Proc. 4th Int. Symp. Magnetic Anisotropy and Coercivity in Rare Earth — Transition Metal Alloys, Strant,* K. J. Ed., University of Dayton, OH, 1985, 771.
87. Burzo, G. and Kirchmayr, H. R., *Handbook on the Physics and Chemistry of Rare Earths,* Vol. 12, Gschneidner, K. A., Jr. and Eyring, L., Eds., North-Holland, Amsterdam, 1980, 71.

250   Crystal Structures and Magnetic Properties of Rare Earth Intermetallics

88. Burzo, E., Plugaru, N., Creanga, I., and Ursu, M., *J. Less-Common Met.*, 155, 281, 1989a.
89. Burzo, E., Plugaru, N., Pop, V., and Creanga, I., *Physica Status Solidi* (a), 113, K253, 1989b.
90. Buschow, K. H. J., *Physica Status Solidi* (a) 7, 198, 1971.
91. Buschow, K. H. J., *J. Less-Common Met.*, 33, 239, 1973.
92. Buschow, K. H. J., *J. Less-Common Met.*, 39, 185, 1975.
93. Buschow, K. H. J., *Rep. Progr. Phys.*, 40, 1179, 1977.
94. Buschow, K. H. J., *Rep. Progr. Phys.*, 42, 185, 1979.
95. Buschow, K. H. J., *Ferromagnetic Materials*, Vol. I, Wohlfarth, E. P., Ed., North-Holland, Amsterdam, 1980, chap. 4, p. 297.
96. Buschow, K. H. J., *Handbook on the Physics and Chemistry of Rare Earths*, Vol. 6, Gschneidner, K. A., Jr. and Eyring, L., Eds., North-Holland, Amsterdam, 1984, 1.
97. Buschow, K. H. J., *Handbook on the Physics and Chemistry of Rare Earths*, Vol. 4, Gschneidner, K. A., Jr. and Eyring, L., Eds., North-Holland, Amsterdam, 1988, 1.
98. Buschow, K. H. J., *J. Magn. Magn. Mater.*, 100, 79, 1991.
99. Buschow, K. H. J. and van der Kraan, A. M., *J. Phys. F*, 8, 92, 1978.
100. Buschow, K. H. J. and Velge, W. A. J. J., *J. Less-Common Met.*, 13, 11, 1977.
101. Buschow, K. H. J., Campagna, M. and Wertheim, G. K., *Solid State Commun.*, 24, 253, 1977.
102. Buschow, K. H. J., De Mooij, B. D., Denissen, and C. J. M., *J. Less-Common Met.*, 142, L13, 1988b.
103. Buschow, K. H. J. and De Mooij, B. D., *Philips J. Res.*, 41, 55, 1986.
104. Buschow, K. H. J., De Mooij, D. B., and Denissen, C. J. M., *J. Less-Common Met.*, 115, 357, 1988.
105. Buschow, K. H. J. and De Mooij, D. B., *Concerted European Action on Magnets (CEAM)*, Mitchell, I. V., Coey, J. M. D., Givord, D., Harris, I. R., and Hanitsch, R., Eds., Elsevier, London, 1989, 63.
106. Buschow, K. H. J., De Mooij, D. B., Deams, J. L. C., and van Noort, H. M., *J. Less-Common Met.*, 125, 135, 1986.
107. Buschow, K. H. J., De Mooij, D. B., Zhoung, X.-P., and de Boer, F. R., *Physica B*, 162, 83, 1990a.
108. Buschow, K. H. J., Mooij, D. B., and Denissen, C. J. M., *J. Less-Common Met.*, 142, L13, 1986b.
109. Buschow, K. H. J., De Mooij, D. B., Sinnema, S., Radwański, R. J., and Franse, J. J. M., *J. Magn. Magn. Mater.*, 51, 211, 1985.
110. Buschow, K. H. J., van Diepen, A. M., and de Wijn, H. W., *J. Appl. Phys.*, 42, 4315, 1971.
111. Buschow, K. H. J., van Diepen, A. M., and de Wijn, H. W., *Solid State Commun.*, 12, 417, 1973.
112. Buschow, K. H. J., van Vucht, J. H. N., and van den Hoogenhof, W. W., *J. Less-Common Met.*, 50, 145, 1976.
113. Buschow, K. H. J., De Mooij, D. B., Zhoung, X.-P., and de Boer, F. R., *Physica*, 162, 83, 1990a.
114. Cable, J. M. and Koechler, W. C., *AIP Conf. Proc.*, 5, 1381, 1972.
115. Cadeville, M. C. and Daniel, E., *J. de Phys.* (Paris), 27, 449, 1966.
116. Cadogan, J. M. and Coey, J. M. D., *Phys. Rev.*, B30, 7326, 1984.
117. Cadogan, J. M., Garigan, J. P., Givord, D., and Li, H. S., *J. Phys. F*, 18, 779, 1988.
118. Calemczuk, R., Bonjour, E., Rossat-Mignod, J., and Chevalier, B., *J. Magn. Magn. Mater.*, 90–91, 477, 1990.
119. Campbell, I. A., *J. Phys. F*, 2, L47, 1972.
120. Cattano, E. and Wohlleben, D., *J. Magn. Magn. Mater.*, 24, 197, 1981.

121. **Cenzual, K., Gladyshevskii, R. R., and Parthé, E.**, *Acta Crystallogr.*, C48, 225, 1992.
122. **Chaban, N. F. and Kuzma, Yu. B.**, Izv. Akad. Nauk SSSR, *Neorg. Mater.*, 13, 923, 1977.
123. **Chafik El Idrissi, B., Venturini, G., and Malaman, B.**, *J. Alloys Comp.*, 175, 143, 1991a.
124. **Chafik El Idrissi, B., Venturini, G., and Malaman, B.**, *Mat. Res. Bull.*, 26, 431, 1991.
125. **Chazalviel, J. H., Campagna, M., Wertheim, G. K., and Schimdt, P.**, *Solid State Commun.*, 19, 725, 1984.
126. **Chełkowska, G., Chełkowski, A., and Winiarska, A.**, *J. Less-Common Met.*, 143, L7, 1988.
127. **Cheng, S. F., Demczyk, B. G., Laughlin, D. E., and Wallace, W. E.**, *J. Magn. Magn. Mater.*, 84, 162, 1990.
128. **Chernyak, G. V., Chaban, N. F., and Kuźma, Yu. B.**, *Plenum Publishing Group*, 1982, 590.
129. **Chevalier, B., Coey, J. M. D., Lloret, B., and Etourneau, J.**, *J. Phys. C: Solid State Phys.*, 19, 4521, 1986.
130. **Chevalier, B., Cole, A., Lejay, P., Vlasse, M., Etourneau, J., and Hagenmuller, P.**, *Mat. Res. Bull.*, 17, 215, 1982b.
131. **Chevalier, B., Etourneau, J., Greedan, J. E., Coey, J. M. D., and Moaroufi, A.**, *J. Less-Common Met.*, 111, 171, 1985b.
132. **Chevalier, B., Etourneau, J., Hagenmuller, P., Quezel, P. H., and Rossat-Mignod, J.**, *J. Less-Common Met.*, 111, 161, 1985a.
133. **Chevalier, B., Lejay, P., Cole, A., Vlasse, M., and Etourneau, J.**, *Solid State Commun.*, 41, 801, 1982a.
134. **Chevalier, B., Lejay, P., Etourneau, J., and Hagenmuller, P.**, *Mat. Res. Bull.*, 18, 315, 1983.
135. **Chevalier, B., Lejay, P., Etourneau, J., and Hagenmuller, P.**, *Solid State Commun.*, 49, 753, 1984.
136. **Chevalier, B., Lejay, P.,Lloret, B., Zhang, W.-X., Etourneau, J., and Hagenmuller, P.**, *Ann. Chim. F*, 9, 987, 1984b.
137. **Chevalier, B., Rogl, P., Hill, E. K., Tuilier, M. H., Dordor, P., and Etourneau, J.**, *Z. Phys. B. — Condensed Matter*, 84, 205, 1991.
138. **Chin, T.-S., Chang, W.-C., Ku, H.-C., Weng, C.-C., Lee, H.-T., and Hung, M.-P.**, *IEEE Trans. Magn.*, MAG–25, 3300, 1989a.
139. **Chin, T.-S., Weng, C.-C., Hung, M.-P., Chang, W.-C., and Lin, C.-H.**, *Proc. 10th Int. Workshop Rare Earth Magnets*, Vol. I, Kyoto, 1989b, 201.
140. **Christides, C.**, private information, 1990.
141. **Christides, C., Kostikas, A., Niarchos, D., and Simpoulos, A.**, *J. Phys.* (Paris), 49, C8, 1988.
142. **Christides, C., Kostikas, A., Simpoulos, A., Niarchos, D., and Zouganelis, G.**, *J. Magn. Magn. Mater.*, 86, 367, 1990.
143. **Christides, C. A., Niarchos, D., Kostikas, C. A., Li, H.-S., Hu, B.-P., and Coey, J. M. D.**, *Solid State Commun.*, 72, 839, 1989.
144. **Coehoorn, R.**, *J. Phys.* (Paris), 49, C8, 1988.
145. **Coehoorn, R.**, *Phys. Rev.*, B41, 11790, 1990; *J. Phys.* (Paris), 49, C8–301, 1988.
146. **Coehoorn, R.**, *Proc. of the NATO Advanced Study Inst. on Supermagnets, Hard Magnetic Materials*, G. J. Long, and Grandjean, F., Eds., Kluwer Academic Press, Dordrecht The Netherlands, 1991, 133.
147. **Coehoorn, R., Denissen, C. J. M., and Eppenga, R.**, *J. Appl. Phys.*, 69, 6222, 1991.
148. **Coey, J. M. D., Yaouonc, A., and Fruchart, D.**, *Solid State Commun.*, 58, 413, 1986.
149. **Coqblin, B.**, *J. Magn. Magn. Mater.*, 29, 1, and references therein, 1982.
150. **Corliss, L. M. and Hastings, J. M.**, *J. Appl. Phys.*, 35, 1051, 1964.

151. **Cotton, F. A.**, *Chemical Applications of Group Theory,* Interscience-John Wiley, New York, 1983.
152. **Czjzek, G., Oestreich, V., Schmidt, H., Łątka, K., and Tomala, K.,** *J. Magn. Magn. Mater.,* 79, 42, 1989.
153. **Dakun, S., Rapson, G., and Rainford, B. D.,** *J. Magn. Magn. Mater.,* 108, 117, 1992.
154. **Danan, J., de Novion, Ch., and Lallement, R.,** *Solid State Commun.,* 7, 103, 1969.
155. **Darshan, B. and Padalia, B. D.,** *J. Phys. C.,* 17, L281, 1984.
156. **Darshan, B., Padalia, D. B., Nagarajan, R., and Vijayarghavan, R.,** *J. Phys. C: Solid State Phys.,* 17, L445, 1984.
157. **Das, I. and Sampathkamaran, E. V.,** *Phys. Rev.,* B44, 9711, 1991.
158. **Das, I., Sampathkamaran, E. V., and Vijayaraghavan, R.,** *J. Magn. Magn. Mater.,* 104–107, 874, 1992a.
159. **De Boer, F. R., Klaasse, J. C. P., Veenhuizen, P. A., Bohm, A., Bredl, C. D., Gottwick, U., Mayer, H. M., Pawlak, L., Rauchschwalbe, U., Spille, H., and Steglich, F.,** *J. Magn. Magn. Mater.,* 63–64, 91, 1987.
160. **De Boer, F. R., Huang, Y.-K., De Mooij, D. B., and Buschow, K. H. J.,** *J. Less-Common Met.,* 135, 199, 1987a.
161. **De Boer, F. R., Verhoef, R., Zhang, Z.-D., De Mooij, D. B., and Buschow, K. H. J.,** *J. Magn. Magn. Mater.,* 73, 263, 1988.
162. **De Gennes, P. G.,** *J. Phys. Radium.,* 23, 510, 1962.
163. **De Mooij, D. B. and Buschow, K. H. J.,** *Philips J. Res.,* 41, 400, 1987.
164. **De Mooij, D. B. and Buschow, K. H. J.,** *J. Less-Common Met.,* 136, 207, 1988.
165. **De Vries, J. W. C., Thiel, R. C., and Buschow, K. H. J.,** *J. Less-Common Met.,* 111, 313, 1985.
166. **Denissen, C. J. M., Coehoorn, R., and Buschow, K. H. J.,** *J. Magn. Magn. Mater.,* 87, 51, 1990.
167. **Denissen, C. J. M., De Mooij, D. B., and Buschow, K. H. J.,** *J. Less-Common Met.,* 142, 349, 1988.
168. **Deportes, J. and Givord, D.,** *Solid State Commun.,* 19, 845, 1976.
169. **Deportes, J., Givord, D., Lemaire, R., Nagi,** *Physica B.,* 88–89, 69, 1977.
170. **Dhar, S. K., Sampathkumaran, E. V., Vijayarghavan, R., and Kuentzler, R.,** *Solid State Commun.,* 61, 479, 1987.
171. **Dhar, S. K., Nambudripad, N., and Vijayarghavan, R.,** *J. Phys. F Metals Phys.,* 18, L41, 1988.
172. **Dhar, S. K., Pattalwar, S. M., and Vijayaraghavan, R.,** *J. Magn. Magn. Mater.,* 104–107, 1303, 1992.
173. **Dirken, M. W., Thiel, R. C., and Buschow, K. H. J.,** *J. Less-Common Met.,* 147, 97, 1989.
174. **Dirken, M. W., Thiel, R. C., Brabers, J. H. V. J., de Boer, F. R., and Buschow, K. H. J.,** *J. Alloys Comp.,* 177, L11, 1991.
175. **Dirken, M. W., Thiel, R. C., de Jongh, L. J., Jacobs, T. H., and Buschow, K. H. J.,** *J. Less-Common Met.,* 155, 339, 1989.
176. **Dommann, F., Hulliger, F., and Baerlocher, Ch.,** *J. Less-Common Met.,* 138, 113, 1988.
177. **Donaberger, R. L. and Stager, C. V.,** *J. Less-Common Met.,* 127, 93, 1987.
178. **Doniach, S.,** *Physica,* B91, 231, 1977.
179. **Dorrscheidt, W. and Schafer, H.,** *J. Less-Common Mat.,* 58, 209, 1978.
180. **Doukouré, M. and Gignoux, D.,** *J. Magn. Magn. Mater.,* 30, 111, 1982.
181. **Doukouré, M., Gignoux, D., and Sayetat, F.,** *Solid State Commun.,* 58, 713, 1986.
182. **Drzazga, Z., Winiarska, A., Borgieł, K., and Mydlarz, T.,** *J. Magn. Magn. Mater.,* 83, 155, 1990.
183. **Drzazga, Z., Winiarska, A., and Stein, B. F.,** *J. Less-Common Met.,* 153, L21, 1989.
184. **Duc, N. H.,** *Physica Status Solidi* (b), 164, 545, 1991.

185. Duc, N. H., Hien, T. D., and Givord, D., *J. Magn. Magn. Mater.*, 104–107, 1344, 1992.
186. Dung, T. T., Thuy, N. P., Hong, N. M., and Hien, T. D., *Physica Status Solidi* (a), 106, 201, 1988.
187. Duraj, M. and Szytuła, A., *Acta Phys. Polon.*, A76, 243, 1989.
188. Duraj, M., Duraj, R., Budkowski, A., and Szytuła, A., *Acta Magnetica*, Suppl. 87, 201, 1987.
189. Duraj, M., Duraj, R., and Szytuła, A., *J. Magn. Magn. Mater.*, 79, 61, 1989.
190. Duraj, M., Duraj, R., and Szytuła, A., *J. Magn. Magn. Mater.*, 82, 319, 1989a.
191. Duraj, M., Duraj, R., Szytuła, A., and Tomkowicz, Z., *J. Magn. Magn. Mater.*, 73, 240, 1988.
192. Duthie, J. C. and Heine, V., *J. Phys. F*, 9, 1349, 1979.
193. Dwight, A. E., Proc. 12th Rare Earth Research Conf., Vail, Colorado, July 18–22, 1976, Denver Research Institute, 1976, 480.
194. Dwight, A. E., Mueller, M. H., Conner, R. H., Jr., Downey, J. W., and Knott, H., *Trans. Met. Soc. AIME*, 247, 2075, 1968.
195. Eisenmann, B., May, N., Muller, W., and Schafer, H., *Z. Naturrforsch*, B27, 1155, 1973.
196. Ekino, T., Kadowaki, H., Fujii, H., Nagasawa, M., Takahatake, T., Motoya, K., and Yoshizawa, H., *Int. Conf. Strongly Correlated Electron Systems*, Sendai, 1992, 129, 9P-7.
197. El Masary, N. A. and Stadelmeier, H. H., *J. Less-Common Met.*, 96, 65, 1984.
198. El Masary, N. A., Stadelmeier, H. H., Shahawan, C. J., and Jordan, L. T., *J. Metallkol.*, 74, 33, 1983.
199. Endstra, T., thesis, University of Leiden, 1992.
200. Engel, N., Braun, M. F., and Parthé, E., *J. Less-Common Met.*, 95, 309, 1983.
201. Felner, I., *J. Phys. Chem. Solids*, 36, 1063, 1975.
202. Felner, I., *J. Less-Common Met.*, 72, 241, 1980.
203. Felner, I., *Solid State Commun.*, 52, 195, 1984.
204. Felner, I., *Solid State Commun.*, 56, 315, 1985.
205. Felner, I. and Nowik, I., *J. Phys. Chem. Solids*, 39, 763, 1978.
206. Felner, I. and Nowik, I., *J. Phys. Chem. Solids*, 39, 951, 1978a.
207. Felner, I. and Nowik, I., *Solid State Commun.*, 28, 67, 1978b.
208. Felner, I. and Nowik, I., *J. Phys. Chem. Solids*, 40, 1035, 1979.
209. Felner, I. and Nowik, I., *Solid State Commun.*, 47, 831, 1983.
210. Felner, I. and Nowik, I., *J. Phys. Chem. Solids*, 45, 419, 1984.
211. Felner, I. and Nowik, I., *J. Phys. Chem. Solids*, 46, 681, 1985.
212. Felner, I. and Nowik, I., *J. Magn. Magn. Mater.*, 58, 169, 1986.
213. Felner, I. and Nowik, I., *J. Magn. Magn. Mater.*, 74, 31, 1988.
214. Felner, I. and Schieber, M., *Solid State Commun.*, 13, 457, 1973.
215. Felner, I., Seh, M., and Nowik, I., *J. Phys. Chem. Solids*, 42, 1091, 1981b.
216. Felner, I., Seh, M., Rakavy, M., and Nowik, I., *J. Phys. Chem. Solids*, 42, 369, 1981a.
217. Felner, I., Nowik, I., and Seh, M., *J. Magn. Magn. Mater.*, 38, 172, 1983.
218. Felner, I., Mayer, I., Grill, A., and Schieber, M., *Solid State Commun.*, 11, 1231, 1972.
219. Forrer, R., *Ann. Phys.*, 7, 605, 1952.
220. François, M., thesis, Université de Nancy I, Nancy, 1986.
221. François, M., Venturini, G., Malaman, B., and Roques, B., *J. Less-Common Met.*, 160, 197, 1990.
222. Freeman, A. J., *Magnetic Properties of Rare Earth Metals*, Elliot, R. J., Ed., Plenum Press, London, 1972, chap. 6, p. 245.
223. Freeman, A. J. and Watson, R. E., *Phys. Rev.*, 127, 2058, 1962.

224. Freeman, A. J. and Desclaux, J. P., *J. Magn. Magn. Mater.*, 12, 11, 1979.
225. Friedel, J., *Nouvo Cimento*, 7 (Suppl. 2), 287, 1958.
226. Fuerst, C. D., Herbst, J. F., and Alson, E. A., *J. Magn. Magn. Mater.*, 54–57, 567, 1986.
227. Fujii, H., Inoda, M., Okamoto, T., Shigeoka, T., and Iwata, N., *J. Magn. Magn. Mater.*, 54–57, 1345, 1986.
228. Fujii, H., Nagata, H., Uwatoko, Y., Okamoto, T., Yamamoto, H., and Sagawa, M., *J. Magn. Magn. Mater.*, 70, 331, 1987.
229. Fujii, H., Okamoto, T., Shigeoka, T., and Iwata, N., *Solid State Commun.*, 53, 715, 1985.
230. Fujii, H., Ueda, E., Uwatoko, Y., and Shigeoka, T., *J. Magn. Magn. Mater.*, 76–77, 179, 1988.
231. Fujii, H., Uwatoko, Y., Akayam, M., Satch, K., Maeno, Y., Fujita, T., Sakurai, J., Kamimura, H., and Okamoto, T., *Jpn. J. Appl. Phys.*, 26, 549, 1987.
232. Fujii, M., Takabatake, T., and Andoh, Y., *J. Alloys Comp.*, 181, 111, 1992.
233. Fulde, P., Crystal fields, in *Handbook on the Physics and Chemistry of Rare Earths*, Vol. II, Gschneidner, K. A., Jr. and Eyring, L., Eds., North-Holland, Amsterdam, 1979, chap. 17, p. 295.
234. Gal, J., Yaar, I., Arbaboff, E., Etedgi, H., Litterst, F. J., Aggarwal, K., Pereda, J. A., Kalvius, G. M., Will, G., and Schafer, W., *Phys. Rev.*, B40, 745, 1989.
235. Gal, J., Yaar, I., Reger, D., Fredo, S., Shani, G., Arbaboff, E., Potzel, W., Aggarwal, K., Pereda, J. A., Kalvius, G. M., Litterst, F. J., Schafer, W., and Will, G., *Phys. Rev.*, 42, 8507, 1990.
236. Galera, R. M., Murani, A. P., and Pierre, J., *Crystalline Electric Field Effects in f-electrons Magnetism*, Quertin, R. P., Suski, W., and Żołnierek, Z., Eds., Plenum Press, New York, 1982a, 423; *J. Phys. F: Metals Phys.*, 12, 993, 1982a.
237. Galera, R. M., Pierre, J., and Pannetier, J., *J. Phys. F: Metals Phys.*, 12, 993, 1982b.
238. Galera, R. M., Pierre, J., Siaud, E., and Murani, A. P., *J. Less-Common Met.*, 97, 151, 1984.
239. Geibel, C., Kammerer, C., Seidel, B., Bredl, C. D., Grauel, A., and Steglich, F., *J. Magn. Magn. Mater.*, 108, 207, 1992.
240. Ghadraoui, E. H., Pivan, J. Y., Guerin, R., Pena, O., Padiou, J., and Sergent, M., *Mat. Res. Bull.*, 23, 1345, 1988.
241. Gignoux, P., Morin, P., and Schmitt, D., *J. Magn. Magn. Mater.*, 102, 33, 1991.
242. Gignoux, D., Gomez-Sal, J. C., and Paccard, D., *Solid State Commun.*, 49, 75, 1984.
243. Gignoux, D., Schmitt, S., Zerguine, M., Ayache, C., and Bonjour, E., *Phys. Lett.*, A1117, 145, 1986.
244. Gignoux, D., Schmitt, S., Zerguine, M., Bauer, E., Pillmaye, N., Henry, J. Y., Nguyen, V. N., and Rossat-Mignod, J., *J. Magn. Magn. Mater.*, 74, 1, 1988a.
245. Gignoux, D., Schmitt, S., and Zerguine, M., *J. Phys.*, 49, C8–433, 1988b.
246. Gignoux, D., Schmitt, S., and Zerguine, M., *Solid State Commun.*, 58, 559, 1986a.
247. Gil, A., Leciejewicz, J., Małetka, K., Szytuła, A., Tomkowicz, Z., and Wojciechowski, K., *J. Magn. Magn. Mater.*, in press, 1993a.
248. Gil, A., Szytuła, A., Tomkowicz, Z., Wojciechowski, K., and Zygmunt, A., *J. Magn. Magn. Mater.*, in press, 1993b.
249. Girgis, K. and Fischer, P., *J. Phys.*, 40, C5–159, 1979.
250. Givord, D., Li, H. S., Moreau, J. M., Penier de la Bothie, R., and du Tremolet de Lacheisserie, E., *Physica, B.*, 130, 323.
251. Givord, D., Li, H. S., and Moreau, J. M., *Solid State Commun.*, 50, 497, 1984.
252. Givord, D., Li, H. S., Moreau, J. M., and Tenaud, P., *J. Magn. Magn. Mater.*, 54–57, 445, 1986.
253. Givord, D., Moreau, J. M., and Tenaud, P., *J. Magn. Magn. Mater.*, 54–57, 445, 1985.

254. Givord, D., Moreau, J. M., and Tenaud, P., *Solid State Commun.*, 55, 303, 1985.
255. Gladyshevskii, E. J., Bodak, O. I., Yarovetz, V. I., Gorelenko, K., and Skolozdra, R. V., *Fiz. Magnit. Plenok* (Irkutsk), 10, 182, 1977.
256. Gladyshevskii, E. J., Bodak, O. I., Yarovetz, V. I., Gorelenko, K., and Skolozdra, R. V., *Ukr. Fiz. Zh.*, 23, 77, 1978.
257. Gladyshevskii, E. J., Grin, J. N., and Jarmoluk, J. P., *Visn. Lvivsk Derzh. Univ. Ser. Khim.*, 23, 26, 1981.
258. Godart, C., Gupta, L. C., and Ravet-Krill, M. F., *J. Less-Common Met.*, 94, 187, 1983.
259. Godart, C., Umarjii, A. M., Gupta, L. C., and Vajayarghavan, R., *J. Magn. Magn. Mater.*, 63–64, 326, 1987.
260. Goodenough, J. B., *Magnetism and Chemical Bond*, John Wiley & Sons, New York, 1963, 240.
261. Gorelenko, Yu. K., Shcherba, I. D., and Bodak, O. I., *Ukr. Fiz. Zh.*, 30, 301, 1985.
262. Goremychkin, E. A., Muzychka, A. Yu., and Osborn, R., *Physica B*, 179, 184, 1992.
263. Görlich, E. A., Application of ^{116}Sn Mössbauer spectroscopy to investigations of RET_2X_2-type phases, in Proc. XXVI Zakopane School on Physics, Stanek, J. and Pędziwiatr, A. T., Eds., *World Scientific*, Singapore, 1991, 180.
264. Görlich, E. A., Hrynkiewicz, A. Z., Kmieć, R., and Tomala, K., *J. Phys. C: Solid State Phys.*, 15, 6049, 1982.
265. Gratz, E. and Zuckermann, M. J., *Handbook on the Physics and Chemistry of Rare Earths*, Vol. 5, Gschneidner, K. A., Jr. and Eyring, L. R., Eds., North-Holland, Amsterdam, 1982, 117.
266. Grier, B. H., Lawrance, J. M., Murgai, V., and Park, R. D., *Phys. Rev.*, B29, 266, 1984.
267. Gros, Y., Hartmann-Boutron, F., and Meyer, C., *Concerted European Action on Magnes (CEAM)*, Mitchell, I. V., Coey, J. M. D., Givord, D., Harris, I. R., and Hanitsch, R., Eds., Elsevier, London, 1989, 288.
268. Gros, Y., Hartmann-Boutron, F., Meyer, C., Fremy, M. A., Tenaud, P., and Auric, P., *J. Phys.* (Paris), 49, C8–547, 1988a.
269. Gros, Y., Hartmann-Boutron, F., Meyer, C., Fremy, M. A., and Tenaud, P., *J. Magn. Magn. Mater.*, 74, 319, 1988b.
270. Groshev, M. N., Komerlin, M. D., Levin, E. M., Luciv, R. V., Miftahoy, N. M., Smirnov, J. P., Sovjestnov, A. E., Tjunis, A. V., Shaburov, V. A., Yasnickii, R. I., Kuzmina, S. M., Petrova, V. I., and Tjukavina, V. A., *Fiz. Tv. Tela*, 28, 2711, 1987.
271. Grossinger, R., Krewenka, R., Kou, X. C., and Buschow, K. H. J., *J. Magn. Magn. Mater.*, 83, 130, 1990.
272. Grossinger, R., Sun, X. K., Eibler, R., Buschow, K. H. J., and Kirchmayr, H. R., *J. Phys.* (Paris), 46, C6–221, 1985.
273. Grossinger, R., Sun, X. K., Eibler, R., Buschow, K. H. J., and Kirchmayr, H. R., *J. Magn. Magn. Mater.*, 58, 55, 1986.
273a. Gruttner, A. and Yvon, K., *Acta Crystallogr.*, B35, 451, 1979.
274. Gu, Z.-Q. and Ching, W. Y., *Phys. Rev.*, B36, 8530, 1987.
275. Gueramian, M., Bezinge, A., Yvon, K., and Muller, J., *Solid State Commun.*, 64, 639, 1987.
276. Gupta, L. C., Mac Laughlin, D. E., Tien Cheng, Edwards, M. A., and Parks, R. D., *Phys. Rev.*, B28, 3673, 1983.
277. Gyorgy, E. M., Batlogg, B., Remeika, J. P., van Dover, R. B., Fleming, R. M., Bair, H. E., Espinosa, G. P., Cooper, A. S., and Maines, R. G., *J. Appl. Phys.*, 61, 4237, 1987.
278. Haen, P., Lejay, P., Chevalier, B., Lloret, B., Etourneau, J., and Sera, M., *J. Less-Common Met.*, 100, 321, 1985.
279. Hatwar, T. K., *Solid State Commun.*, 39, 61, 1981.

280. Helmholdt, R. B., Vleggear, J. J. M., and Buschow, K. H. J., *J. Less-Common Met.*, 138, L11, 1988; 144, 209, 1988.
281. Helmholdt, R. B. and Buschow, K. H. J., *J. Less-Common Met.*, 155, 15, 1989.
282. Herbst, J. F. and Yelon, W. B., *J. Magn. Magn. Mater.*, 54–57, 570, 1986.
283. Herbst, J. F., Croat, J. J., Pinkerton, F. E., and Yelon, W. B., *Phys. Rev.*, B29, 4176, 1984.
284. Herbst, J. F., Croat, J. J., and Yelon, W. B., *J. Appl. Phys.*, 57, 4086, 1985.
285. Heusler, O., *Ann. Phys.*, 19, 155, 1934.
286. Hiebl, K. and Rogl, P., *J. Magn. Magn. Mater.*, 50, 39, 1985.
287. Hiebl, K., Horvath, C., Rogl, P., and Sienko, M. J., *J. Magn. Magn. Mater.*, 37, 287, 1983a.
288. Hiebl, K., Horvath, C., Rogl, P., and Sienko, M. J., *Solid State Commun.*, 48, 211, 1983b.
289. Hiebl, K., Horvath, C., Rogl, P., and Sienko, M. J., *Z. Phys.*, B56, 201, 1984.
290. Hiebl, K., Horvath, C., and Rogl, P., *J. Less-Common Met.*, 117, 375, 1986.
291. Hirjak, M., Chevalier, B., Etourneau, J., and Hagenmuller, P., *Mat. Res. Bull.*, 19, 727, 1984.
292. Hirosawa, S., Tokuhara, K., Matsuura, Y., Yamamoto, H., Fujimura, S., and Sagawa, M., *J. Magn. Magn. Mater.*, 6, 363, 1986.
293. Hirosawa, S., Tokuhara, K., Yamamoto, H., Fujimura, S., Sagawa, M., and Yamamuchi, H., *J. Appl. Phys.*, 61, 3571, 1987.
294. Hodges, J. A., *Europhys. Lett.*, 4, 749, 1987.
295. Hodges, J. A. and Jéhanno, G., *J. Phys.*, 49, C8-387, 1988.
296. Hong, N. M., Franse, J. J. M., Thuy, N. P., and Hien, T. H., *J. Phys.* (Paris), 49, C8-545, 1988.
297. Honma, H., Shimotomai, H., and Doyama, M., *J. Magn. Magn. Mater.*, 52, 399, 1985.
298. Horn, S., Holland-Moritz, E., Loewenhaupt, M., Steglich, F., Schener, H., Benoit, A., and Flouquet, J., *Phys. Rev.*, B23, 3171, 1981.
299. Horvath, C. and Rogl, P., *Mat. Res. Bull.*, 18, 443, 1983.
300. Horvath, C. and Rogl, P., *Mater. Res. Bull.*, 20, 1273, 1985.
301. Hovestreydt, E., Engel, N., Klepp, K., Chabot, B., and Parthe, E., *J. Less-Common Met.*, 85, 247, 1982.
302. Hu, B.-P., Li, H.-S., and Coey, J. M. D., *Hyperfine Interactions*, 45, 233, 1986b.
303. Hu, B.-P., Li, H.-S., Gavigan, J. P., and Coey, J. M. D., *J. Phys. Condensed Matter*, 1, 755, 1989a.
304. Huang, M. Q., Boltich, E. B., Wallace, W. E., and Oswald, E., *J. Magn. Magn. Mater.*, 60, 270, 1986.
305. Hull, G. W., Wernick, J. N., Geballe, T. H., Waszczak, J. V., and Bernardini, J. E., *Phys. Rev.*, B24, 6715, 1981.
306. Hulliger, F. and Ott, M. R., *Z. Physik*, B29, 47, 1978.
307. Hutchings, M. T., *Solid State Physics*, Vol. 16, Seits, F., and Turnbull, D., Eds., Academic Press, New York, 1964, 227.
308. Iandelli, A., *J. Less-Common Met.*, 90, 121, 1983.
309. Iandelli, A. and Palenzona, A., *J. Less-Common Met.*, 12, 333, 1969.
310. Iandelli, A. and Palenzona, A., *Handbook on the Physics and Chemistry of Rare Earths*, Vol. 2, Gschneidner, K. A., Jr. and Eyring, L. Eds., North-Holland, Amsterdam, 1979, chap. 13, p. 1.
311. Ido, H., Sohn, J. G., Pourarian, F., Cheng, S. F., and Wallace, W. E., *J. Appl. Phys.*, 67, 4978, 1990.
312. Inoue, J. and Shimizu, M., *J. Phys.*, F16, 1051, 1986.
313. Ishida, S., Asano, S., and Ishida, J., *J. Phys. Soc. Jpn.*, 55, 93, 1986.
314. Itoh, T., Hikosaka, K., Takahashi, H., Ukai, T., and Mori, N., *J. Appl. Phys.*, 61, 3430, 1987.

315. Ivanov, V., Vinokurova, L., and Szytuła, A., *Acta Phys. Polon.*, A81, 693, 1992a.
316. Ivanov, V., Vinokurova, L., and Szytuła, A., *J. Magn. Magn. Mater.*, 114, L225, 1992b.
317. Ivanov, V., Vinokurova, L., and Szytuła, A., *J. Magn. Magn. Mater.*, 110, L259, 1992c.
318. Ivanov, V., Vinokurova, L., Szytuła, A., and Zygmunt, A., *J. Alloys Comp.*, 191, 159, 1993.
319. Ivanov, V., Vinokurova, L., and Szytuła, A., *J. Alloys Comp.*, in press.
320. Iwata, M., Hattori, K., and Shigeoka, T., *J. Magn. Magn. Mater.*, 53, 318, 1986.
321. Iwata, M., Iheda, T., Shigeoka, T., Fujii, H., and Okamoto, T., *J. Magn. Magn. Mater.*, 54–57, 481, 1986a.
322. Iwata, N., Honda, K., Shigeoka, T., Mashimoto, Y., and Fujii, H., *J. Magn. Magn. Mater.*, 90–91, 63, 1990.
323. Jacobs, T. H., Denissen, C. J. M., and Buschow, K. H. J., *J. Less-Common Met.*, 153, L5, 1989.
324. Jarlborg, T. and Arbman, G., *J. Phys. F*, 7, 1635, 1977.
325. Jarlborg, T., Braun, H. F., and Peter, M., *Z. Phys.*, B52, 295, 1983.
326. Jaswal, S. S., *Phys. Rev.*, B41, 9697, 1990.
327. Jaswal, S. S., Ren, Y. G., and Sellmyer, *J. Appl. Phys.*, 67, 4564, 1990.
328. Jaworska, T. and Szytuła, A., *Solid State Commun.*, 57, 813, 1987.
329. Jaworska, T., Szytuła, A., Ptasiewicz-Bąk, H., and Leciejewicz, J., *Solid State Commun.*, 65, 261, 1988.
330. Jaworska, T., Leciejewicz, J., Ptasiewicz-Bąk, H., and Szytuła, A., *J. Magn. Magn. Mater.*, 82, 313, 1989.
331. Jeitschko, W. and Gerss, M. H., *J. Less-Common Met.*, 116, 147, 1986.
332. Jeitschko, W. and Reehuis, M., *J. Phys. Chem. Solids*, 48, 667, 1987.
333. Johnson, V., Du Pont de Nemours Co., *U.S. Patent 3, 963, 829*, 1974, 1976.
334. Jorda, J. l., Ishikawa, M., and Hovestreydt, E., *J. Less-Common Met.*, 92, 255, 1983.
335. Jurczyk, M., *J. Magn. Magn. Mater.*, 87, 1, 1990.
336. Jurczyk, M. and Chistyakov, O. D., *Physica Status Solidi* (a) 115, K229, 1989.
337. Jurczyk, M. and Rao, K. K., *J. Alloys Comp.*, 177, 259, 1991.
338. Jurczyk, M., Pędziwiatr, A. T., Sankar, S. G., and Wallace, W. E., *J. Magn. Magn. Mater.*, 68, 237, 1987.
339. Jurczyk, M., Pędziwiatr, A. T., and Wallace, W. E., *J. Magn. Magn. Mater.*, 67, L1, 1987a.
340. Kaneko, T., Yasui, H., Kanomata, T., Kobayashi, H., and Ondera, H., *J. Phys.*, 49, C8–441, 1988.
341. Kaneko, T., Yamada, M., Ohashi, K., Tawana, Y., Osugi, R., Yoshida, H., Kido, G., and Nakagawa, Y., *Proc. 10th Int. Workshop Rare Earth Magnets, Vol. I, Kyoto*, 1989, 191.
342. Kaneko, T., Yasni, H., Kanomata, T., and Suzuki, T., *J. Magn. Magn. Mater.*, 104–107, 1951, 1992.
343. Kanomata, T., Kawashima, T., Kaneko, T., Takakashi, H., and Mari, T., *Physica Status Solidi* (a), 120, K117, 1990.
344. Kapusta, C., Figiel, H., and Kąkol, Z., *J. Magn. Magn. Mater.*, 83, 151, 1990.
345. Kasuya, T., *Prog. Theor. Phys.* (Kyoto), 16, 45, 1956.
346. Kawashima, T., Kanomata, T., Yoshida, H., and Kaneko, T., *J. Magn. Magn. Mater.*, 90–91, 721, 1990.
347. Kido, H., Shimada, M., and Koizumi, M., *Physica Status Solidi* (a), 70, K23, 1982.
348. Kido, H., Hoshikawa, T., Shimada, M., and Koizumi, M., *Physica Status Solidi* (a), 80, 601, 1983a.

258   *Crystal Structures and Magnetic Properties of Rare Earth Intermetallics*

2

349. **Kido, H., Hoshikawa, T., Shimada, M., and Koizumi, M.,** *Physica Status Solidi* (a), 77, K121, 1983b.
350. **Kido, H., Hoshikawa, T., Koizumi, M., and Shimada, M.,** *J. Less-Common Met.,* 99, 151, 1984a.
351. **Kido, H., Hoshikawa, T., Shimada, M., and Koizumi, M.,** *Physica Status Solidi* (a), 83, 561, 1984b.
352. **Kido, H., Hoshikawa, T., Shimada, M., and Koizumi, M.,** *Physica Status Solidi* (a), 87, K61, 1985a.
353. **Kido, H., Hoshikawa, T., Shimada, M., and Koizumi, M.,** *Physica Status Solidi* (a), 88, K39, 1985b.
354. **Kido, H., Hoshikawa, T., Shimada, M., and Koizumi, M.,** *Physica Status Solidi* (a) 87, 237, 1985c.
355. **Kierstead, H. A., Dunlap, B. D., Malik, S. K., Umarji, A. M., and Shenoy, G. K.,** *Phys. Rev.,* B32, 135, 1985.
356. **Kirchmayer, H. R.,** *Z. Angew. Phys.,* 27, 18, 1969.
357. **Kirchmayer, H. R. and Poldy, C. A.,** *Handbook on the Physics and Chemistry of Rare Earths,* Vol. 2, Gschneidner, K. A., Jr. and Eyring, L., Eds., North-Holland, Amsterdam, 1979, chap. 14, p. 55.
358. **Kirchmeyer, H. R. and Poldy, C. A.,** *J. Magn. Magn. Mater.,* 8, 1, 1978.
359. **Klemm, W. and Bommer, W.,** *Z. Anorg. Allg. Chem.,* 231, 138, 1937.
360. **Klepp, K. and Parthé, E.,** *Acta Cryst.,* B38, 1105, 1982.
361. **Knopp, G., Loidl, A., Knorr, K., Pawlak, L., Duczmal, M., Caspry, R., Gottwick, U., Spille, H., and Steglich, F.,** *J. Magn. Magn. Mater.,* 74, 341, 1988.
362. **Knopp, G., Loidl, A., Knorr, K., Pawlak, L., Duczmal, M., Caspary, R., Gottwick, U., Spille, H., Steglich, F., and Murani, A. P.,** *Z. Phys. B: Cond. Mat.,* 77, 95, 1989.
363. **Kobayashi, H., Onodera, H., and Yamamoto, H.,** *J. Magn. Magn. Mater.,* 79, 76, 1989.
364. **Kobayashi, H., Onodera, H., and Yamamoto, H.,** *J. Magn. Magn. Mater.,* 109, 17, 1992.
365. **Kolenda, M. and Szytuła, A.,** *J. Magn. Magn. Mater.,* 79, 57, 1989.
366. **Kolenda, M., Szytuła, A., and Zygmunt, A.,** *Crystalline Electric Field Effects in f-electron Magnetism,* Guertin, R. P., Suski, W., and Zołnierek, Z., Eds., Plenum Press, New York, 1982, 309.
367. **Kolenda, M., Szytuła, A., and Leciejewicz, J.,** *J. Magn. Magn. Mater.,* 49, 250, 1985.
368. **Koterlin, M. P. and Luciv, R. D.,** *Fiz. Electron,* 17, 18, 1978.
369. **Kotsanidis, P. A. and Yakinthos, J. K.,** *Solid State Commun.,* 40, 1041, 1981.
370. **Kotsanidis, P. A. and Yakinthos, J. K.,** *J. Magn. Magn. Mater.,* 81, 158, 1989.
371. **Kotsanidis, P. A. and Yakinthos, J. K.,** *Physica Status Solidi* (a), 68, K135, 1989.
372. **Kotsanidis, P. A., Yakinthos, J.K., and Roudaut, E.,** *Solid State Commun.,* 50, 413, 1984.
373. **Kotsanidis, R. A., Yakinthos, J. K., and Gamari-Seale, E.,** *J. Less-Common Met.,* 152, 287, 1989.
374. **Kotsanidis, P. A., Yakinthos, J. K., and Gamari-Seale, E.,** *J. Magn. Magn. Mater.,* 87, 199, 1990a.
375. **Kotsanidis, P. A., Yakinthos, J. K., and Gamari-Seale, E.,** *J. Less-Common Met.,* 157, 285, 1990b.
376. **Kotsanidis, P. A., Semitelou, I., Yakinthos, J. K., and Roudaut, E.,** *J. Magn. Magn. Mater.,* 102, 67, 1991.
377. **Kotur, B. Ya., Yarovets, V. I., and Gladyshevskii, E. J.,** *Sov. Phys. Cryst.,* 22, 217, 1977a.
378. **Kou, X. C., Sun, X. K., Chuang, Y. C., Zhao, T. S., Grossinger, R., and Kirchmayr, H. R.,** *Solid State Commun.,* 73, 87, 1990; *J. Less-Common Met.,* 160, 109, 1990.

379. Kozlowski, A., thesis, Institute of Physics of the Polish Academy of Sciences, Warsaw, 1986.

380. Kozlowski, A., Maksymowicz, A., Tarnawski, Z., Lewicki, A., Zukowski, J., and Aniola-Jędrzejek, L., *J. Magn. Magn. Mater.*, 68, 95, 1987.

381. Kripyakevich, P. I., Zevechnyuk, O. S., Gladyshevsky, E. I., and Bodak, O. I., Z. *Anorg. Chem.*, 358, 90, 1968.

382. Ku, H. C. and Shelton, R. N., *Solid State Commun.*, 40, 237, 1981.

383. Kulatov, E., Veselago, V., and Vinkurova, L., *Acta Phys. Pol.*, A77, 709, 1990.

384. Kuźma, Yu. B. and Bilonizhko, N. S. *Kristallografya*, 18, 710, 1973.

385. Kuźma, Yu. B., Bilonizhko, N. S., Mykhalenko, S. I., Stepanchikova, G. F., and Chaban, N. F., *J. Less-Common Met.*, 67, 51, 1979.

386. Kuźma, Yu. B., Chernyak, G. V., and Chaban, N. F., *Dopov. Akad. Nauk Ukr. RSR*, Ser. A, 12, 80, 1981.

387. Kwapulińska, E., Kaczmarska, K., and Szytuła, A., *J. Magn. Magn. Mater.*, 73, 65, 1988.

388. Lahiouel, R., Pierre, J., Siaud, E., and Murani, A. P., *J. Magn. Magn. Mater.*, 63–64, 104, 1987.

389. Łątka, K., *Magnetism and Hyperfine Interactions in GdT₂Si₂ Systems*, Report No. 1443/ PS, Institute of Nuclear Physics, Cracow, 1989.

390. Łątka, K., Schmidt, H., Oestreich, V., Gotz, P., and Czjzek, J. G., *Progress Report KfK 2881* (IAK, Kerforschungszentrum Karlsruhe), 1979, 79.

391. Le Roux, D., Vincent, H., L'Hevitier, P., and Fruchart, R., *J. Phys.* (Paris), 46, C6–243, 1985.

392. Le Roy, J., Moreau, J. M., Bertrand, C., and Fremy, M. A., *J. Less-Common Met.*, 136, 19, 1987.

393. Lebech, B. and Bak, P., *J. Phys.* (Paris), 40, C5–14, 1979.

394. Leccabue, F., Sanchez, J. P., Pereti, L., Bolzoni, F., and Panizzlen, R., *Physica Status Solidi* (a), 91, K63, 1985.

395. Leciejewicz, J., *Crystalline Electric Field Effects in f-electron Magnetism*, Guertin, R. P., Suski, W., and Zolnierek, Z., Eds., Plenum Press, New York, 1982, 279.

396. Leciejewicz, J. and Szytuła, A., *Solid State Commun.*, 48, 55, 1983.

397. Leciejewicz, J. and Szytuła, A., *Solid State Commun.*, 49, 361, 1984.

398. Leciejewicz, J. and Szytuła, A., *Solid State Commun.*, 56, 1051, 1985.

399. Leciejewicz, J. and Szytuła, A., *J. Magn. Magn. Mater.*, 49, 117, 1985a.

400. Leciejewicz, J. and Szytuła, A., *J. Magn. Magn. Mater.*, 63–64, 190, 1987.

401. Leciejewicz, J., Kolenda, M., and Szytuła, A., *Solid State Commun.*, 45, 143, 1983.

402. Leciejewicz, J., Szytuła, A., and Zygmunt, A., *Solid State Commun.*, 45, 149, 1983a.

403. Leciejewicz, J., Szytuła, A., Ślaski, M., and Zygmunt, A., *Solid State Commun.*, 52, 475, 1984.

404. Leciejewicz, J., Szytuła, A., Ślaski, M., and Zygmunt, A., *Solid State Commun.*, 52, 475, 1984a.

405. Leciejewicz, J., Siek, S., and Szytuła, A., *J. Magn. Magn. Mater.*, 40, 265, 1984b.

406. Leciejewicz, J., Kolenda, M., and Szytuła, A., *J. Magn. Magn. Mater.*, 53, 309, 1986.

407. Lee, W. H. and Shelton, R. N., *Phys. Rev.*, B35, 5369, 1987.

408. Lee, W. H. and Shelton, R. N., *Solid State Commun.*, 68, 443, 1988.

409. Lee, W. H., Kwan, K. S., Klavius, P., and Shelton, R. N., *Phys. Rev.*, B42, 6542, 1990.

410. Legvold, S., *Ferromagnetic Materials*, Wohlfarth, E. P., Ed., North-Holland, Amsterdam, 1980, chap. 3, p. 183.

411. **Leon, B. and Wallace, W. E.**, *J. Less-Common Met.*, 22, 1, 1970.
412. **Lewis, W. B.**, *Magnetic Resonance and Related Phenomena*, Proc. 16th Conf., Urzu, Ampere I., Ed., House of the Academy of Romania, Bucharest 1971, 717.
413. **Li, H.-S. and Coey, J. M. D.**, *Handbook of Magnetic Materials*, Vol. 6, Buschow, K. H. K., Ed., Elsevier, Amsterdam, 1991. chap. 6, p. 1.
414. **Li, W. H., Lynn, J. W., Stanley, H. B., Udovic, T. J., Shelton, R. N., and Klavius, P.**, *Phys. Rev.*, B39, 4119, 1989.
415. **Liang, G. and Croft, M.**, *Phys. Rev.*, B40, 361, 1989.
416. **Liang, G., Croft, M., Johnston, D. C., Anbalagan, N., and Mihalisin, T.**, *Phys. Rev.*, B38, 5302, 1988.
417. **Lin, C., Sun, Y. X., Liu, Z. X., and Li, G.**, *J. Appl. Phys.*, 69, 5554, 1991.
418. **Liu, N. C., Kamprath, N., Wickvamasekava, L., Cadieu, F. J., and Stadelmaier, H. H.**, *J. Appl. Phys.*, 63, 3589, 1988.
419. **Lloret, B., Chevalier, B., Buffat, B., Etourneau, J., Quezel, S., Lamharrar, A., Rossat-Mignod, J., Calemczuk, R., and Bonjour, E.**, *J. Magn. Magn. Mater.*, 63–64, 85, 1987.
420. **Loidl, A., Knopp, G., Spille, H., Steglich, F., and Murani, A. P.**, *Physica B*, 156–157, 794, 1989.
420a. **Loidl, A., Knorr, K., Knopp, G., Krimmel, A., Caspary, R., Böhm, A., Sparn, G., Geibel, C., and Steglich, F.**, *Phys. Rev. B*, 46, 9341, 1992.
421. **Ludorf, W., Abd-Elmequid, M. M., and Micklitz, H.**, *J. Magn. Magn. Mater.*, 78, 171, 1989.
422. **Madar, R., Chaudouet, P., Boursier, D., Senateur, J. P., and Lambert, B.**, *Extended Abstr. VIII Int. Conf. Solid Compounds of Transition Elements*, Vienna, 1985, P4, A13.
423. **Malaman, B., Venturini, G., Meot, M., and Roques, B.**, *VII Int. Conf. Solid Compounds of Trans. Met.*, Vienna, 1985, P3, B8.
424. **Malaman, B., Venturini, G., Pontonnier, L., and Fruchart, D.**, *Abstr. Int. Conf. Magnetism*, Paris, 1988, 5P G–19, 341.
425. **Malaman, B., Venturini, G., and Roques, B.**, *Mater. Res. Bull.*, 23, 1629, 1988.
426. **Malaman, B., Venturini, G., Pontonnier, L., and Fruchart, D.**, *J. Magn. Magn. Mater.*, 86, 349, 1990a.
427. **Malaman, B., Venturini, G., Le Caer, G., Pontonnier, L., Fruchart, D., Tomala, K., and Sanchez, J. P.**, *Phys. Rev.*, B41, 4700, 1990b.
428. **Malaman, B., Venturini, G., Blaise, A., Amoretti, G., and Sanchez, J. P.**, *J. Magn. Magn. Mater.*, 104–107, 1359, 1992.
429. **Malik, S. K.**, *Int. Conf. Strongly Correlated Electron Systems*, Sendai, 1992, 9P–85, p. 170.
430. **Malik, S. K. and Adroja, D. T.**, *J. Magn. Magn. Mater.*, 102, 42, 1991.
431. **Malik, S. K., Sankar, S. G., Rao, V. U. S., and Obermyer, R.**, *AIP Conf. Proc.*, 34, 87, 1975.
432. **Malik, S. K., Sankar, S. G., Rao, V. U. S., and Obermyer, R.**, *AIP Conf. Proc.*, 34, 87, 1976.
433. **Malik, S. K., Umarjii, A. M., and Shenoy, G. K.**, *Phys. Rev.*, B32, 4426, 1985a.
434. **Malik, S. K., Umarjii, A. M., and Shenoy, G. K.**, *Phys. Rev.*, B31, 6971, 1985b.
435. **Malik, S. K., Nagarajan, R., Adroja, D. T., and Dwight, A. E.**, *Proc. Int. Conf. Valence Fluctuations*, Bangalore, India, 1987, F11.
436. **Malik, S. K., Dunlap, B., Ślaski, M., Adroja, D. T., and Dwight, A. E.**, *Conf. Abstr. Int. Conf. Magnetism*, Paris, 1988, 2P G–10, p. 156.
437. **Malik, S. K., Zhang, Z.-Y., Wallace, W. E., and Sankar, S. G.**, *J. Magn. Magn. Mater.*, 78, L6, 1989.
438. **Marazza, R., Rossi, D., and Ferro, R.**, *J. Less-Common Met.*, 75, P25, 1980a.
439. **Marazza, R., Rossi, D., and Ferro, R.**, *Gazz. Chim. Ital.*, 110, 357, 1980b.
440. **Martin, O. E. and Girgis, K.**, *J. Magn. Magn. Mater.*, 37, 228, 1983.
441. **Martin, O. E., Girgis, K., and Fischer, P.**, *J. Magn. Magn. Mater.*, 37, 231, 1983.

442. **Marusin, E. P., Bodak, O. I., Tsokol, A. O., and Fundamenskii,** *Sov. Phys. Crystallogr.*, 30, 338, 1985.
443. **Matsuura, Y., Hirosawa, S., Yamamoto, H., Fujimura, S., and Sagawa, M.,** *Appl. Phys. Lett.*, 46, 308, 1985.
444. **Matsuura, Y., Hirosawa, S., Yamamoto, H., Fujimura, S., and Osamura, K.,** *Jpn. J. Appl. Phys.*, 24, L635, 1985b.
445. **Mattis, D. C.,** *The Theory of Magnetism*, Harper & Row, New York 1965.
446. **Mc Guire, F. R., Argyle, B. E., Shafer, W. M., and Smart, J. S.,** *J. Appl. Phys.*, 34, 1345, 1963.
447. **McCall, W. M., Narasimhan, K. S. V. L., and Butera, R. A.,** *J. Appl. Phys.*, 44, 4724, 1973.
448. **Medvedeva, I. V.,Ganin, A. A., Shcherbakova, Ye. V., Yermolenko, A. S., and Bersener, Yu. S.,** *J. Alloys Comp.*, 178, 403, 1992.
449. **Melamud, M., Pinto, H., Felner, I., and Shaked, H.,** *J. Appl. Phys.*, 55, 2034, 1984.
450. **Methfessel, S. and Mattis, D. C.,** *Magnetic Superconductors*, Springer-Verlag, Berlin, 1968, 208.
451. **Meyer, I. and Felner, I.,** *J. Less-Common Met.*, 29, 25, 1972.
452. **Miedema, A. R.,** *J. Less-Common Met.*, 46, 67, 1976.
453. **Mittag, M., Rosenberg, M., and Buschow, K. H. J.,** *J. Magn. Magn. Mater.*, 82, 109, 1989.
454. **Miyako, Y., Kuwai, T., Taniguchi, T., Kawarazaki, S., Amitsuka, H., Paulsen, C. C., and Sakakibara, T.,** *J. Magn. Magn. Mater.*, 108, 190, 1992.
455. **Moodenbaugh, A. R., Cox, D. E., and Braun, H. F.,** *Phys. Rev.*, B25, 4702, 1982.
456. **Moodenbaugh, A. R., Cox, D. E., Vining, C. B., and Segre, C. K.,** *Phys. Rev.*, B29, 271, 1984.
457. **Moreau, J. M., le Roy, J., and Paccard, P.,** *Acta Cryst.*, B28, 2446, 1982.
458. **Morsen, E., Mosel, B. D., Muller-Warmuth, W., Reehuis, W., and Jeitschko, W.,** *J. Phys. Chem. Solids*, 49, 785, 1988.
459. **Moze, O., Caciuffo, R., Li, H.-S., Hu, B.-P., Coey, J. M. D., Osborn, R., and Taylor, A. D.,** *Phys. Rev.*, B42, 1940, 1990a.
460. **Moze, O., Ibberson, R. M., and Buschow, K. H. J.,** *J. Phys.*, 2, 1677, 1990b.
461. **Moze, O., Ibberson, R. M., and Buschow, K. H. J.,** *Solid State Commun.*, 78, 473, 1991.
462. **Mugnoli, A., Albinati, A., and Hawat, A. W.,** *J. Less-Common Met.*, 97, L1, 1984.
463. **Murgai, V., Raen, S., Cupta, L. C., and Parks, R. D.,** *Valence Instabilities*, Wachter P. and Boppart, H., Eds., North-Holland, Amsterdam, 1982, 537.
464. **Nakamura, H., Kitaoka, Y., Asayama, K., Onuki, Y., and Komatsubara, T.,** *J. Magn. Magn. Mater.*, 76–77, 467, 1988a.
465. **Nakamura, H., Kitaoka, Y., Yamada, H., and Asayama, K.,** *J. Magn. Magn. Mater.*, 76–77, 517, 1988b.
466. **Nambudipad, N., Stang, I., and Luders, K.,** *Solid State Commun.*, 60, 625, 1986.
467. **Narasimhan, K. S. B. L., Rao, V. U. S., Wallace, W. E., and Pop, I.,** *AIP Conf. Proc.*, 29, 594, 1975.
468. **Narasimhan, K. S. B. L., Rao, V. U. S., Bergner, R. L., and Wallace, W. E.,** *J. Appl. Phys.*, 46, 4957, 1976.
469. **Néel, L.,** *Ann. Physique* (Paris), 3, 137, 1948.
470. **Nguyen, V. N., Tcheou, R., Rossat-Mignod, J., and Ballestracci, R.,** *Solid State Commun.*, 45, 209, 1983.
471. **Niarchos, D., Zonganelis, G., Kostikas, A., and Simppoulos, A.,** *Solid State Commun.*, 59, 389, 1986.
472. **Niihara, K. and Yajima, S.,** *Chem. Lett.* (Japan), 5, 875, 1972.
473. **Niihara, K., Katajama, Y., and Yajima, S.,** *Chem. Lett.* (Japan), 6, 613, 1973.
474. **Nikitin, S. A., Ivanova, T. I., Nekrasova, O. V., Torchinova, R. S., Popov, Yu., Korjasova, O. N., and Kluchnikova, E. A.,** *Sov. Fiz. Met. Met.*, 64, 1071, 1987.
475. **Noakes, D. R. and Shenoy, G. H.,** *Phys. Lett.*, 91A, 35, 1982.

476. **Noakes, D. R., Umarji, A. M., and Shenoy, G. H.,** *J. Magn. Magn. Mater.*, 39, 309, 1983.

477. **Noakes, D. R., Shenoy, G. K., Niarchos, D., Umarji, A. M., and Aldred, A. T.,** *Phys. Rev.*, B27, 4317, 1983b.

478. **Nojiri, H., Motokawa, M., Nishiyama, K., Naganine, K., and Shigeoka, T.,** *J. Magn. Magn. Mater.*, 104–107, 1311, 1992.

479. **Nowik, I.,** *Hyperfine Interactions*, 13, 89, 1983.

480. **Nowik, I.,** *Int. Conf. Magnetism of Rare-Earths and Actinides*, Burzo, E. and Rogalski, M., Eds., Central Institute of Physics, Bucharest, 1983, 112.

481. **Nowik, I. and Felner, I.,** *Acta Phys. Polon.*, 97, 97, 1985.

482. **Nowik, I. and Felner, I.,** *Physica B*, 130, 433, 1985.

483. **Nowik, I. and Felner, I.,** *Proc. 3rd Int. Conf. Physics of Magnetic Materials*, Gorzkowski, W., Lachjowicz, H. K., and Szymczak, H., Eds., World Scientific, Singapore, 1986, 366.

484. **Nowik, I., Felner, J., and Seh, M.,** *J. Magn. Magn. Mater.*, 15–18, 1215, 1980.

485. **Nowik, I., Felner, J., and Seh, M.,** *Proc. Int. Conf. Magnetism of Rare-Earths and Actinides*, Burzo, E. and Rogalski, M., Eds., Central Institute of Physics, Bucharest, 1983, 112.

486. **Obermyer, R., Sankar, S. G., and Rao, V. U. S.,** *J. Appl. Phys.*, 50, 2132, 1979.

487. **Oesterreicher, H.,** *J. Phys. Chem. Solids*, 33, 1031, 1972.

488. **Oesterreicher, H.,** *J. Phys. Chem. Solids*, 34, 1267, 1973.

489. **Oesterreicher, H.,** *Physica Status Solidi (a)*, 34, 723, 1976.

490. **Oesterreicher, H.,** *Physica Status Solidi (a)*, 40, K139, 1977a.

491. **Oesterreicher, H.,** *J. Less-Common Met.*, 55, 199, 1977b.

492. **Oesterreicher, H.,** *Physica Status Solidi (a)*, 39, K75, 1977c.

493. **Oesterreicher, H., Corliss, L. M., and Hastings, J. M.,** *J. Appl. Phys.*, 41, 2326, 1970.

494. **Oesterreicher, H., Spada, F., and Abache, C.,** *Mater. Res. Bull.*, 19, 1060, 1984.

495. **Ohashi, K. T., Tokoyama, T., Osugi, R., and Tawara, Y.,** *IEEE Trans. Magn. MAG.*, 23, 3101, 1987.

496. **Ohashi, M., Sakurada, S., Kaneko, T., Abe, S., Yoshida, H., and Yamaguchi, Y.,** *J. Magn. Magn. Mater.*, 104–107, 1383, 1992.

497. **Okamoto, N., Nogai, H., Yoshida, H., and Tsujimura, A.,** *J. Magn. Magn. Mater.*, 70, 299, 1987.

498. **Olenitch, R. R., Akselrud, L. G., and Yarmoliuk, Ya. P.,** *Dopov. Akad. Nauk Ukr.*, RSR Ser. A(2), 84, 1981.

499. **Ondera, H., Ohashi, M., Yamauchi, H., and Yamaguchi, Y.,** *J. Magn. Magn. Mater.*, 109, 249, 1992.

500. **Palenzona, A., Cirafici, S., and Canepa, F.,** *J. Less-Common Met.*, 119, 199, 1986.

501. **Palstra, T. T. M.,** Ph.D. thesis, University of Leiden, 1986.

502. **Palstra, T. T. M., Mydosh, J. A., Nieuwenhuys, G. J., van der Kraan, A. M., and Buschow, K. H. J.,** *J. Magn. Magn. Mater.*, 36, 290, 1983.

503. **Palstra, T. T. M., Werij, M. G. C., Nieuwenhuys, G. J., Mydosh, J. A., de Boer, F. R., and Buschow, K. H. J.,** *J. Phys.*, F 14, 1961, 1984.

504. **Palstra, T. T. M., Nieuwenhuys, G. J., Mydosh, J. A., and Buschow, K. H. J.,** *Phys. Rev.*, B31, 4622, 1985.

505. **Palstra, T. T. M., Nieuwenhuys, G. J., Mydosh, J. A., Helmholdt, R. B., and Buschow, K. H. J.,** *J. Magn. Magn. Mater.*, 54–57, 995, 1986.

506. **Palstra, T. T. M., Menovsky, A. A., Nieuwenhuys, G. J., and Mydosh, J. A.,** *J. Magn. Magn. Mater.*, 54–57, 435, 1986.

507. **Palstra, T. T. M., Menovsky, A. A., Nieuwenhuys, G. J., and Mydosh, J. A.,** *Phys. Rev.*, B34, 4566, 1986a.

508. **Parks, R. D., Riehl, B., Moartensson, N. and Steglich, F.,** *Phys. Rev.*, B27, 6052, 1983.

509. Parthé, E., Chabot, B., Braun, H. F., and Engel, N., *Acta Crystallogr.*, B39, 588, 1983.
510. Pearson, W. B., *J. Less-Common Met.*, 96, 105, 1984.
511. Pearson, W. B., *J. Solid State Chem.*, 56, 278, 1985.
512. Pearson, W. S. and Villars, P., *J. Less-Common Met.*, 97, 119, 133, 1984.
513. Pecharsky, V. K., Gschneidner, K. A., Jr., and Miller, L. L., *Phys. Rev.*, B43, 10906, 1991.
514. Pellizone, M., Braun, H. F., and Muller, J., *J. Magn. Magn. Mater.*, 30, 33, 1982.
515. Pędziwiatr, A. T., *Report Institute of Nuclear Physics 1988*, No. 1417/PS, Kraków, 1988.
516. Pędziwiatr, A. T. and Wallace, W. G., *J. Less-Common Met.*, 126, 41, 1986.
517. Pędziwiatr, A. T. and Wallace, W. E., *Solid State Commun.*, 64, 1017, 1987.
518. Pędziwiatr, A. T., Jiang, S. Y., and Wallace, W. E., *J. Magn. Magn. Mater.*, 62, 20, 1986.
519. Pędziwiatr, A. T., Chen, H. Y., and Wallace, W. E., *J. Magn. Magn. Mater.*, 67, 311, 1987.
520. Pędziwiatr, A. T., Jiang, S. Y., Wallace, W. E., Burzo, E., and Pop, V., *J. Magn. Magn. Mater.*, 66, 69, 1987.
521. Pinto, H. and Shaked, H., *Phys. Rev.*, B7, 3261, 1973.
522. Pinto, H., Melamud, M., and Gurewitz, E., *Acta Crystallogr.*, A35, 533, 1979.
523. Pinto, H., Melamud, M., Gal, J., Shaked, H., and Kalvius, G. M., *Phys. Rev.*, B27, 1861, 1983.
524. Pinto, H., Melamud, M., Kuznietz, M., and Shaked, H., *Phys. Rev.*, B31, 508, 1985.
525. Pinto, R. P., Amado, M. M., Salqueiro Silva, M., Braga, M. E., Sonsa, J. B., Chevalier, B., and Etourneau, J., *J. Magn. Magn. Mater.*, 104–107, 1235, 1992.
526. Pieger, H. R., Bruck, E., Braun, E., Oster, F., Freimuth, A., Polit, B., Roden, B., and Wohlleben, D., *J. Magn. Magn. Mater.*, 63–64, 107, 1987.
527. Plumer, M. L. and Caille, A., *Phys. Rev. B*, 46, 203, 1992.
528. Pourarian, E., *Proc. 11th Int. Workshop Rare-Earth Magnets and their Application* Sankar, S. G., Ed., Carnegie Mellon University, Pittsburgh, 1990, 381.
529. Pourarian, F., Huang, M. Q., and Wallace, W. E., *J. Less-Common Met.*, 120, 63, 1986.
530. Pourarian, F., Jiang, S. Y., Sankar, S. G., and Wallace, W. E., *J. Appl. Phys.*, 63, 3972, 1988.
531. Pourian, F., Malik, S. K., Boltich, E. B., Sankar, S. G., and Wallace, W. E., *IEEE Trans. Magn. MAG–25*, 3315, 1989.
532. Quezel, S., Rossat-Mignod, J., Chevalier, B., Lejay, P., and Etourneau, J., *Solid State Commun.*, 49, 685, 1984.
533. Quezel, S., Rossat-Mignod, J., Chevalier, B., Zhong, W.-X., and Etourneau, J., *C.R. Acad. Sci. Paris*, 301, 919, 1985.
534. Quezel, S., Burlet, P., Jacoud, J. L., Regnault, L. P., Rossat-Mignod, J., Vettier, C., Lejay, P., and Flouquet, J., *J. Magn. Magn. Mater.*, 76–77, 403, 1988.
535. Rambabu, D. and Malik, S. K., *Proc. Int. Conf. Valence Fluctuations*, Bangalore, India, 1987, A24.
536. Rauchschwalbe, U., *Physica*, 147B, 1, 1987.
537. Rebelsky, L., Reilly, K., Horn, S., Borges, H., Thompson, J. D., Willis, J. O., Aikin, R., Caspari, R., and Bredl, C. D., *J. Appl. Phys.*, 63, 3405, 1988.
538. Rechenberg, H. R. and Sanchez, J. P., *Solid State Commun.*, 62, 461, 1987.
539. Rechenberg, H. R., Peduan-Filho, A., Missell, F. P., Deppe, P., and Rosenberg, M., *Solid State Commun.*, 59, 541, 1986.
540. Rechenberg, H. R., Boge, M., Jeandey, C., Oddan, J. L., and Sanchez, J. P., *Solid State Commun.*, 64, 277, 1987.
541. Rhodes, P. R. and Wohlfarth, E. P., *Proc. Roy. Soc.* (London), 273, 247, 1963.

542. **Rieger, W. and Parthé, E.,** *Monatshfte fur Chemie,* 100, 439, 1969.
543. **Rogl, P., Chevalier, B., Besnus, M. J., and Etourneau, J.,** *J. Magn. Magn. Mater.,* 80, 305, 1989.
543a. **Rogl, P., Hiebl, K., and Sienko, M. J.,** 7th Intern. Conf. Solid Compounds Transition Elements, Paper II, 4A, Grenoble, 1982.
544. **Romaka, V. A., Zarechniuk, O. S., Rychal, R. M., Yarmoluk, J. P., and Skolozdra, R. V.,** *Sov. Fiz. Met. Met.,* 54, 410, 1982a.
545. **Romaka, V. A., Grin, J. N., Yarmoluk, J. P., Zarechniuk, O. S., and Skolozdra, R. V.,** *Sov. Fiz. Met. Met.,* 54, 691, 1982b.
546. **Rosenberg, M., Mittag, M., and Buschow, K. H. J.,** *J. Appl. Phys.,* 63, 3586, 1988.
547. **Rosenberg, M., Deppe, P., Erdmann, K., and Sinnemann, Th.,** *Concerted European Action on Magnets (CEAM),* Mitchell, I. V., Coey, J. M. D., Givord, D., Harris, I. R., and Hanitsh, R., Eds., Elsevier, London, 1989, 276.
548. **Rossat-Mignod, J.,** *J. Phys.,* 40, C5–95, 1979.
549. **Rossat-Mignod, J.,** *Systematics and the Properties of the Lanthanides,* Sinha, S. P., Ed., Reidel, 1983, 255.
550. **Rossat-Mignod, J., Burlet, P., Villain, J., Bartholin, H., Wang-Tcheng-Si, Florence, D., and Vogt, O.,** *Phys. Rev.,* B16, 440, 1977.
551. **Rossat-Mignod, J., Burlet, P., Quezel, S., Effantin, J. M., Vogt, O., and Bartholin, H.,** *Ann. Chim. Fr.,* 7, 471, 1982.
552. **Rossat-Mignod, J., Effantin, J. M., Vettier, G., and Vogt, O.,** *Physica B,* 130, 555, 1985.
553. **Routsi, Ch. D., Yakinthos, J. K., and Gamari-Seale, H.,** *J. Magn. Magn. Mater.,* 98, 257, 1991.
554. **Routsi, Ch. D., Yakinthos, J. K., and Gamari-Seale, H.,** *J. Magn. Magn. Mater.,* 117, 79, 1992a.
555. **Routsi, Ch. D., Yakinthos, J. K., and Gamari-Seale, H.,** *J. Magn. Magn. Mater.,* 110, 317, 1992b.
556. **Ruderman, M. A. and Kittel, C.,** *Phys. Rev.,* 96, 99, 1954.
557. **Rundqvist, S. and Nawapong, P.,** *Acta Chem. Scand.,* 20, 2250, 1966.
558. **Rupp, B., Rogl, P., and Hullinger, F.,** *J. Less-Common Met.,* 135, 113, 1987.
559. **Sagawa, M., Fujimura, S., Yamamoto, H., Matsuua, Y., and Hiraga, K.,** *IEEE Trans. Magn.,* MAG20, 1584, 1984.
560. **Sagawa, M., Fujimura, S., Yamamoto, H., Matsuura, Y., and Hirosawa, S.,** *J. Appl. Phys.,* 57, 4094, 1985.
561. **Sakurada, S., Kaneko, T., Abe, S., Yashida, H., Ohaski, M., Kido, G., and Nakagawa, Y.,** *J. Magn. Magn. Mater.,* 90–91, 53, 1990.
562. **Sakurai, J., Yamamoto, Y., and Komura, Y.,** *J. Phys. Soc. Jpn.,* 57, 24, 1988.
563. **Sakurai, J., Yamaguchi, Y., Mibu, K., and Shinyo, T.,** *J. Magn. Magn. Mater.,* 84, 157, 1990.
564. **Sakurai, J., Nakatani, S., Adam, A., and Fujiwara, H.,** *J. Magn. Magn. Mater.,* 108, 143, 1992.
565. **Sales, B. C. and Wohlleben, D. K.,** *Phys. Rev.,* 35, 1240, 1975.
566. **Sales, B. C. and Viswanathan, R.,** *J. Low. Temp.,* 23, 449, 1976.
567. **Sampathkumaran, E. V. and Vijayaraghavan, R.,** *Phys. Rev. Lett.,* 56, 2861, 1986.
568. **Sampathkumaran, E. V., Gupta, L. C., and Vijayarghavan, R.,** *Phys. Rev. Lett.,* 43, 1189, 1979.
569. **Sampathkumaran, E. V., Gupta, L. C., and Vijayarghavan, R.,** *J. Phys. C, Solid State Phys.,* 12, 4323, 1979a.
570. **Sampathkumaran, E. V., Vijayarghavan, R., Gopalakrisharan, K. V., Pilley, R. G., and Devare, H. G.,** *Valence Fluctuations in Solides* Felicov, I. M., Hanke, W., and Maple, M. B., Eds., North-Holland, Amsterdam, 1981, 193.
571. **Sampathkumaran, E. V., Changhule, R. S., Gopalakrisharan, K. V., Malik, S. K., and Vijayarghavan, R.,** *J. Less-Common Met.,* 92, 35, 1983.

572. Sampathkumaran, E. V., Frank, E. V., Kalkowski, G., Kaindl, G., Domke, M., and Wortmann, G., *Phys. Rev.*, B29, 5702, 1984.

573. Sampathkumaran, E. V., Kalkowski, G., Laubschat, C., Kaindl, G., Domke, M., Schmiester, G., and Wortmann, G., *J. Magn. Magn. Mater.*, 47–48, 212, 1985.

574. Sampathkumaran, E. V., Dhar, S. K., and Malik, S. K., *Proc. Int. Conf. Valence Fluctuations,* Bangalore, India, 1987, B13.

575. Sampathkumaran, E. V., Dhar, S. K., Nambudripad, N., Vijayarghavan, R., Kuentzler, R., and Dossmann, Y., *J. Magn. Magn. Mater.*, 76–77, 645, 1988.

576. Sampathkumaran, E. V., Das, E. V., Vijayarghavan, R., Hirota, K., and Ishikawa, M., *Solid State Commun.*, 78, 971, 1991.

577. Sampathkumaran, E. V., Das, E. V., Vijayarghavan, R., Yamamoto, H., and Ishikawa, M., *Solid State Commun.*, 83, 609, 1992a.

578. Sampathkumaran, E. V., Das, E. V., Vijayarghavan, R., Hirota, K., and Ishikawa, M., *J. Magn. Magn. Mater.*, 108, 85, 1992b.

579. Sanchez, J. P., *Concerted European Action on Magnets (CEAM)*, Mitchell, I. V., Coey, J. M. D., Givord, D., Harris, I. R., and Hanitsch, R., Eds., Elsevier, London, 1989, 300.

580. Sanchez, J. P., Tomala, K., and Kmieć, R., J. de Phys., C8-435, 1988.

581. Sanchez, J. P., Tomala, K., Łątka, K., *J. Magn. Magn. Mater.*, 99, 95, 1991.

582. Sato, K., Isikawa, Y., and Mori, K., *J. Magn. Magn. Mater.*, 104–107, 1435, 1992.

583. Sauer, Ch., Galiski, G., Zinn, W., and Abd-Elmaguid, M. M., *Solid State Commun.*, 62, 265, 1987.

584. Schafer, W., Will, G., Gal, J., and Suski, W., *J. Less-Common Met.*, 149, 237, 1989.

585. Schlabitz, W., Baumann, J., Neumann, G., Plumacher, D., and Reggentin, K., *Crystalline Electric Field Effects in f-electron Magnetism*, Guertin, R. P., Suski, W., and Zołnierczuk, Z., Eds., Plenum Press, New York, 1982, 289.

586. Schobinger-Papamantellos, P. and Buschow, K. H. J., *J. Less-Common Met.*, 171, 321, 1991.

587. Schobinger-Papamantellos, P. and Buschow, K. H. J., *J. Alloys. Comp.*, 185, 51, 1992; 187, 73, 1992.

588. Schobinger-Papamantellos, P. and Hullinger, F., *J. Less-Common Met.*, 146, 327, 1989.

589. Schobinger-Papamantellos, P., Niggli, A., Kotsanidis, P. A., and Yakinthos, J. K., *J. Phys. Chem. Solids*, 45, 695, 1984.

590. Schobinger-Papamantellos, P., Routsi, Ch., and Yakinthos, J. K., *J. Phys. Chem. Solids*, 44, 875, 1983.

591. Sechovsky, V. and Havela, L., *J. Magn. Magn. Mater.*, 104–107, 7, 1992.

592. Sekizawa, K., Takano, Y., Takigami, H., and Takahashi, Y., *J. Less-Common Met.*, 127, 99, 1987.

593. Segre, C. K. and Braun, H. F., *Phys. Lett.*, A85, 372, 1981.

594. Segre, C. K., Croft, M., Hodges, J. A., Murgai, V., Gupta, L. C., and Parks, R. D., *Phys. Rev. Lett.*, 42, 1947, 1982.

595. Segre, C. K., Croft, M., Hodges, J. A., Murgai, V., Gupta, L. C., and Parks, R. D., *Phys. Rev. Lett.*, 49, 1947, 1982.

596. Severing, A., Holland-Moritz, E., Rainford, B. D., Culverhouse, S. R., and Frick, B., *Phys. Rev.*, B39, 2557, 1989.

597. Shaked, H., Pinto, M., and Felner, I., *AIP Conf. Proc.* (Modulated Structure), 53, 295, 1979.

598. Shcherbakova, Ye. V., Yermolenko, A. S., and Makarova, G. M., *Fiz. Met. Metalloved.*, 68, 1130, 1989.

599. Shelton, R. N., Braun, H. F., and Misick, E., *Solid State Commun.*, 52, 797, 1984.

600. Shelton, R. N., Hausermann-Berg, L. S., Johnson, M. J., Klavius, P., and Yang, D., *Phys. Rev.*, B34, 199, 1986.

601. Shigeoka, T., *J. Sci. Hiroshima Univ.*, A48(2), 103, 1984.

602. **Shigeoka, T., Fujii, H., Fujiwara, H., Yagasaki, K., and Okamoto, T.,** *J. Magn. Magn. Mater.*, 31–34, 209, 1983.

603. **Shigeoka, T., Iwata, N., Fujii, H., and Okamoto, T.,** *J. Magn. Magn. Mater.*, 53, 58, 1985.

604. **Shigeoka, T., Iwata, N., Fujii, H., and Okamoto, T.,** *J. Magn. Magn. Mater.*, 63–64, 1343, 1986.

605. **Shigeoka, T., Iwata, N., Fujii, H., Okamoto, T., and Hashimoto, Y.,** *J. Magn. Magn. Mater.*, 70, 239, 1987.

606. **Shigeoka, T., Iwata, N., and Fujii, H.,** *J. Magn. Magn. Mater.*, 76–77, 189, 1988.

607. **Shigeoka, T., Iwata, N., Hashimoto, Y., Andoh, Y., and Fujii, H.,** *Physica,* B156–157, 741, 1989.

608. **Shigeoka, T., Fujii, H., Yanenabu, K., Sugiyama, K., and Date, M.,** *J. Phys. Soc. Jpn.*, 58, 394, 1989a.

609. **Shigeoka, T., Saeki, M., Iwata, N., Takabatake, T., and Fujii, H.,** *J. Magn. Magn. Mater.*, 90–91, 557, 1990.

610. **Shigeoka, T., Iwata, N., and Fujii, H.,** *J. Magn. Magn. Mater.*, 104–107, 1229, 1992.

611. **Shigeoka, T., Kawano, S., Iwata, N., and Fujii, H.,** *Physica,* B180–181, 82, 1992.

612. **Shoemaker, C. B. and Shoemaker, D. P.,** *Acta Crystallogr.*, 18, 900, 1965.

613. **Shoemaker, C. B., Shoemaker, D. P., and Fruchart, R.,** *Acta Crystallogr.*, C40, 1665, 1984.

614. **Siek, S. and Szytuła, A.,** *J. Phys.*, 40, C5–162, 1979.

615. **Siek, S., Szytuła, A., and Leciejewicz, J.,** *Physica Status Solidi* (a), 46, K101, 1978.

616. **Siek, S., Szytuła, A., and Leciejewicz, J.,** *Solid State Commun.*, 39, 863, 1981.

617. **Sinha, V. K., Cheng, S. F., Wallace, W. E., and Sankar, S. G.,** *J. Magn. Magn. Mater.*, 80, 281, 1989; 81, 227, 1989.

618. **Sill, R. L. and Esau, E. D.,** *J. Appl. Phys.*, 55, 1844, 1984.

619. **Sill, R. L. and Hitzman, C. J.,** *J. Appl. Phys.*, 53, 2061, 1981.

620. **Sima, V., Smetana, Z., Sechovsky, V., Grossinger, R., and Franse, J. J. M.,** *J. Magn. Magn. Mater.*, 31–34, 201, 1983.

621. **Sinnema, S., Radwański, R. J., Franse, J. J. M., de Mooij, D. B., and Buschow, K. H. J.,** *J. Magn. Magn. Mater.*, 44, 333, 1984.

622. **Sinnemann, Th., Rosenberg, M., and Buschow, K. H. J.,** *J. Less-Common Met.*, 146, 223, 1989.

623. **Sinnemann, Th., Rosenberg, M., and Buschow, K. H. J.,** *Hyperfine Interactions,* 50, 675, 1989a.

624. **Skolozdra, R. V. and Koretskaya, O. E.,** *Ukr. Fiz. Zh.*, 29, 606, 1981.

625. **Skolozdra, R. V. and Komarovskaja, L. P.,** *Ukr. Fiz. Zh.* (USSR), 27, 1834, 1982.

626. **Skolozdra, R. V. and Komarovskaja, L. P.,** *Ukr. Fiz. Zh.* (USSR), 28, 1093, 1983.

627. **Skolozdra, R. V., Koretskaya, O. E., and Gorelenko, Yu. K.,** *Ukr. Fiz. Zh.* (USSR), 27, 263, 1982.

628. **Skolozdra, R. V., Koretskaya, O. E., and Gorelenko, Yu. K.,** *Izv. Akad. Nauk SSSR Neorg. Mater.*, 20, 604, 1984.

629. **Skolozdra, R. V., Mendzyk, V. M., Gorelenko, Yu. K., and Tkachuk, V. D.,** *Sov. Fiz. Met. Met.*, 52, 966, 1981.

630. **Skrabek, E. A. and Wallace, W. E.,** *J. Appl. Phys.*, 34, 1356, 1963.

631. **Ślaski, M. and Szytuła, A.,** *J. Less-Common Met.*, 87, L1, 1982.

632. **Ślaski, M., Szytuła, A., and Leciejewicz, J.,** *J. Magn. Magn. Mater.*, 39, 268, 1983.

633. **Ślaski, M., Szytuła, A., Leciejewicz, J., and Zygmunt, A.,** *J. Magn. Magn. Mater.*, 46, 114, 1984.

634. **Ślaski, M., Kurzyk, J., Szytuła, A., Dunlap, B. D., Sungaila, Z., and Umezawa, A.,** *J. Phys.*, 49, C8–427, 1988.

635. **Slebarski, A.,** *J. Less-Common Met.*, 72, 231, 1980.

636. **Slebarski, A.,** *J. Magn. Magn. Mater.*, 66, 107, 1987.

637. **Slebarski, A. and Zachorowski, W.,** *J. Phys.*, F14, 1553, 1984.

638. Smit, H. H. A., Thiel, R. C., and Buschow, K. H. J., *J. Phys.*, F18, 295, 1988.
639. Smith, J. L. and Kmetko, E. A., *J. Less-Common Met.*, 90, 83, 1983.
640. Solzi, M., Pareti, L., Moze, O., and David, W. I. F., *J. Appl. Phys.*, 64, 5084, 1989.
641. Spada, F., Abache, C., and Oesterreicher, H., *J. Less-Common Met.*, 99, L21, 1984.
642. Stearns, M. B., *Phys. Rev.*, B4, 4081, 1971.
643. Stadelmaier, H. H., El-Masry, N. A., and Stallard, S. R., *J. Appl. Phys.*, 57, 4149, 1985.
644. Stanley, H. B., Lynn, J. W., Shelton, R. N., and Klavius, P., *Phys. Rev. J. Appl. Phys.*, 61, 3371, 1987.
645. Stefański, P., *thesis*, Institute of Molecular Physics Polish Academy of Sciences, Poznań, 1991.
646. Stefański, P. and Wrzeciono, A., *Physica Status Solidi*, 1989.
647. Stefański, P., Kowalczyk, A., and Wrzeciono, A., *J. Magn. Magn. Mater.*, 81, 155, 1989.
648. Stefański, P., Szlaferek, A., and Wrzeciono, A., *Phys. Status Solidi* (b), 156, 657, 1989.
649. Steglich, F., *J. Phys. Chem. Solids*, 50, 225, 1989.
650. Steglich, F., Aarts, J., Bredl, C. D., Lieke, W., Meschede, D., Franse, W., and Schaffer, J., *Phys. Rev. Lett.*, 43, 1892, 1979.
651. Steglich, F., Geibel, C., Horn, S., Ahlheim, U., Lang, M., Sparn, G., Loidl, A., Krimmel, A., and Assmus, W., *J. Magn. Magn. Mater.*, 90–91, 383, 1990.
652. Stevens, K. W. H., *Proc. Phys. Soc.*, A65, 209, 1952.
653. Stewart, G. R., *Rev. Mod. Phys.*, 56, 755, 1984.
654. Stewart, A. M. and Coles, B. R., *J. Phys.*, E4, 458, 1974.
655. Stewart, G. A. and Zukrowski, J., *Crystalline Electric Fields Effects in f-electron Magnetism*, Guertin, R. P., Suski, W., and Zołnierek, Z., Eds., Plenum Press, New York, 1982, 319.
656. Strzeszewski, J., Nazareth, A., and Hadjipanayis, G. C., *J. Appl. Phys.*, 63, 3978, 1988.
657. Szpunar, B. and Szpunar, J. A., *J. Appl. Phys.*, 57, 4130, 1985.
658. Szytuła, A., *J. Less-Common Met.*, 157, 167, 1990.
659. Szytuła, A., *Handbook Magnetic Materials*, Vol. 6, Buschow, K. H. J., Ed., North-Holland, Amsterdam, 1991, chap. 2, p. 85.
660. Szytuła, A., *J. Alloys Comp.*, 178, 1, 1992.
661. Szytuła, A. and Szott, I., *Solid State Commun.*, 40, 199, 1981.
662. Szytuła, A. and Siek, S., *Proc. VII Int. Conf. Solid Compounds Transition Elements*, Grenoble, 1982a, IIA11.
663. Szytuła, A. and Siek, S., *J. Magn. Magn. Mater.*, 27, 49, 1982b.
664. Szytuła, A. and Leciejewicz, J., *Handbook on the Physics and Chemistry of Rare Earths*, Gschneidner, K. A., Jr. and Eyring, L., Eds., North-Holland, Amsterdam, 1989, chap. 83, p. 133.
665. Szytuła, A. and Zygmunt, A., *Acta Phys. Polon.*, in press.
666. Szytuła, A., Leciejewicz, J., and Bińczycka, H., *Physica Status Solidi (a)*, 52, 76, 1980.
667. Szytuła, A., Bażela, W., and Leciejewicz, J., *Solid State Commun.*, 48, 1053, 1983.
668. Szytuła, A., Ślaski, M., Ptasiewicz-Bąk, H., Leciejewicz, J., and Zygmunt, A., *Solid State Commun.*, 52, 396, 1984.
669. Szytuła, A., Leciejewicz, J., and Ślaski, M., *Solid State Commun.*, 58, 683, 1986.
670. Szytuła, A., Budkowski, A., Ślaski, M., and Zach, R., *Solid State Commun.*, 57, 813, 1986a.
671. Szytuła, A., Leciejewicz, J., Bażela, W., Ptasiewicz-Bąk, H., and Zygmunt, A., *J. Magn. Magn. Mater.*, 69, 299, 1987.

672. Szytuła, A., Oleś, A., Perrin, M., Ślaski, M., Kwok, W., Sungaila, Z., and Dunlap, B. D., *J. Magn. Magn. Mater.*, 69, 305, 1987a.

673. Szytuła, A., Ptasiewicz-Bąk, H., Leciejewicz, J., and Bażela, W., *Solid State Commun.*, 66, 309, 1988.

674. Szytuła, A., Bażela, W., and Leciejewicz, J., *J. Phys.*, 49, C8–383, 1988a.

675. Szytuła, A., Ślaski, M., Kurzyk, J., Dunlap, B. D., Sungailia, Z., and Umezawa, A., *J. Phys.*, 49, C8–437, 1988b.

676. Szytuła, A., Oleś, A., Allain, Y., and Andre, G., *J. Magn. Magn. Mater.*, 75, 298, 1988c.

677. Szytuła, A., Ptasiewicz-Bąk, H., Leciejewicz, J., and Bażela, W., *J. Magn. Magn. Mater.*, 80, 189, 1989.

678. Szytuła, A., Oleś, A., and Perrin, M., *J. Magn. Magn. Mater.*, 86, 377, 1990.

679. Szytuła, A., Radwański, R. J., and de Boer, F. R., *J. Magn. Magn. Mater.*, 104–107, 1237, 1992.

680. Szytuła, A., Leciejewicz, J., and Małetka, A., *J. Magn. Magn. Mater.*, 118, 302, 1992.

681. Tebble, R. S. and Craik, D. J., *Magnetic Materials*, Wiley-Interscience, New York, 1969, 61.

682. Tagawa, H., Inabla, K., Sakurai, J., and Kamura, K., *Solid State Commun.*, 66, 993, 1988.

683. Takabatake, T., Nakazawa, Y., Isikawa, M., Sakakibara, T., Koga, K., and Oguno, I., *J. Magn. Magn. Mater.*, 76–77, 87, 1988.

684. Takanayagi, S., Woods, S. B., Wada, N., Watanabe, T., Onuki, Y., Kobori, A., Komatsubara, T., Imai, M., and Asano, H., *J. Magn. Magn. Mater.*, 76–77, 281, 1988.

685. Takano, Y., Ohhata, K., and Sekizawa, K., *J. Magn. Magn. Mater.*, 66, 187, 1987.

686. Takano, Y., Ohhata, K., and Sekizawa, K., *J. Magn. Magn. Mater.*, 70, 242, 1987a.

687. Takano, Y., Takigami, H., Kanno, K., and Sekizawa, K., *J. Magn. Magn. Mater.*, 104–107, 1367, 1992.

688. Takeda, K., Konishi, K., Deguchi, H., Iwata, N., and Shigeoka, T., *J. Magn. Magn. Mater.*, 104–107, 901, 1992.

689. Takegahara, K., Variase, A., and Kasuya, T., *J. Phys.* (Paris), 41, C5–327, 1980.

690. Thompson, J. D., Willis, J. O., Godart, C., Mac Laughlin, D. E., and Gupta, L. C., *Solid State Commun.*, 56, 169, 1985.

691. Thuy, N. P. and Franse, J. J. M., *J. Magn. Magn. Mater.*, 54–57, 915, 1986.

692. Thuy, N. P., Hong, N. M., Hien, T. D., and Franse, J. J. M., *Proc. 11th Int. Workshop Rare Earth Magnets and their Applications*, Dankar, S. G., Ed., Carnegie Mellon University, Pittsburgh, 1990, 60.

693. Tokuhara, K., Ohtsu, Y., Ono, F., Yamada, O., Sagawa, M., and Matsuura, Y., *Solid State Commun.*, 56, 333, 1985.

694. Tomala, K., Sanchez, J. P., and Kmieć, R., *J. Phys.: Condens. Matter.*, 1, 9231, 1989.

695. Tomala, K., Blaise, A., Kmieć, R., and Sanchez, J. P., *J. Magn. Magn. Mater.*, 117, 275, 1992.

696. Tran, V. H. and Troć, R., *J. Magn. Magn. Mater.*, 102, 74, 1991.

697. Uemura, Y. J., Kossler, W. J., Yu, X. H., Schone, H. E., Kempton, J. R., Stronach, C. E., Barth, S., Gygax, F. N., Hitti, B., Schenck, A., Baines, C., Lankford, W. E., Onuki, Y., and Komatsubara, T., *Phys. Rev.*, B39, 4726, 1989.

698. Umarji, A. M., Noakes, D. R., Voccaro, P. J., Shenoy, G. K., Aldred, A. T., and Niarchos, D., *J. Magn. Magn. Mater.*, 36, 61, 1983.

699. Umarji, A. M., Shenoy, G. K., and Noakes, D. R., *J. Appl. Phys.*, 55, 2297, 1984.

700. Umarji, A. M., Malik, S. K., and Shenoy, G. K., *Solid State Commun.*, 53, 1029, 1985.

701. Umarji, A. M., Godart, C., Gupta, L. C., and Vijayaraghavan, R., *Pramana — J. Phys.*, 27, 321, 1986.
702. Urbain, G., Weiss, P., and Tromba, F., *Compt. Rend.*, Acad. Sci., 200, 1232, 1935.
703. Vaishanova, D. P., Kimball, C. W., Umarij, A. M., Malik, S. K., and Shenoy, G. K., *J. Magn. Magn. Mater.*, 49, 286, 1985.
704. Van Noort, H. M., de Mooij, D. B., and Buschow, K. H. J., *J. Less-Common Met.*, 113, L9, 1985.
705. Venturini, G. and Roques, B., *J. Less-Common Met.*, 146, 24, 1989.
706. Venturini, G., Malaman, B., Meot-Meyer, M., Fruchart, D., le Caer, G., Malterre, D., and Roques, B., *Revue de Chimie Minerale*, 23, 162, 1986.
707. Venturini, G., Malaman, B., Pontonnier, L., and Fruchart, D., *Solid State Commun.*, 67, 193, 1988.
708. Venturini, G., Malaman, B., Pontonnier, L., Bacmann, M., and Fruchart, D., *Solid State Commun.*, 66, 597, 1988a.
709. Venturini, G., Malaman, B., and Roques, B., *J. Less-Common Met.*, 146, 271, 1989.
710. Venturini, G., François, M., Malaman, B., and Roques, B., *J. Less-Common Met.*, 160, 215, 1990.
711. Venturini, G., Chafik El Idrissi, B., and Malaman, B., *J. Magn. Magn. Mater.*, 94, 35, 1991.
712. Venturini, G., Malaman, B., Tomala, K., Szytuła, A., and Sanchez, J. P., *Phys. Rev.*, B46, 207, 1992.
713. Venturini, G., Welter, R., and Malaman, B., *J. Appl. Comp.*, 185, 99, 1992a.
714. Verhoef, R., De Boer, F. R., Zhang, Z.-D., and Buschow, K. H. J., *J. Magn. Magn. Mater.*, 75, 319, 1988.
715. Verhoef, R., de Boer, F. R., Franse, J. J. M., Denissen, C. J. M., Jacobs, T. H., and Buschow, K. H. J., *J. Magn. Magn. Mater.*, 80, 41, 1989.
716. Vining, C. B. and Shelton, R. N., *Phys. Rev.*, B27, 2800, 1983.
717. Vinokurova, L., Ivanov, V., and Szytuła, A., *J. Magn. Magn. Mater.*, 99, 193, 1991.
718. Wachter, P., *C.R.C. Crit. Rev.*, 9, 2219, 1971.
719. Wallace, W. E., *Rare Earth Intermetallics*, Academic Press, New York, 1973.
720. Wallace, W. E., *Prog. Solid State Chem.*, 16, 127, 1986.
721. Wang, F. E. and Gilfrich, J., *Acta Crystallogr.*, 21, 746, 1986.
722. Wang, X.-Z., Chevalier, B., Berlureau, T., Etourneau, J., Coey, J. M. D., and Cadogan, J. M., *J. Less-Common Met.*, 138, 235, 1988.
723. Weidner, P., Sandra, R., Appl, L., and Shelton, R. N., *Solid State Commun.*, 53, 115, 1985.
724. Weitzer, F., Hiebl, K., and Rogl, P., *Solid State Commun.*, 82, 353, 1992.
725. Welter, R., Venturini, G., and Malaman, B., *10th Int. Conf. Solid Comp. Trans. Elements*, Munster, 1991, P–229–FR, SA.
726. Welter, R., Venturini, G., and Malaman, B., *J. Less-Common Met.*, 189, 49, 1992.
727. Westerholt, K. and Bach, H., *Phys. Rev. Lett.*, 47, 1925, 1981.
728. Williams, A. R., Morizzi, V. L., Malozemoff, A. P., and Tekura, K., *IEEE Trans. Magn. MAG.*, 19, 1983.
729. Xing, F. and Ho, W.-W., *Proc. 10th Int. Workshop Rare Earth Magnets*, Vol. I, Kyoto, 1989, 209.
730. Yakinthos, J. K., *J. Magn. Magn. Mater.*, 46, 300, 1985.
731. Yakinthos, J. K., *J. Phys.*, 47, 673, 1986.
732. Yakinthos, J. K., *J. Phys.*, 47, 1239, 1986a.
733. Yakinthos, J. K. and Ikonomou, P. F., *Solid State Commun.*, 34, 777, 1980.
734. Yakinthos, J. K. and Gamari Seale, H., *Z. Phys.*, B48, 251, 1982.
735. Yakinthos, J. K. and Roudaut, E., *J. Magn. Magn. Mater.*, 68, 90, 1987.
736. Yakinthos, J. K., Routsi, Ch., and Ikonomou, P. F., *J. Less-Common Met.*, 72, 205, 1980.

737. **Yakinthos, J. K., Routsi, Ch., and Schobinger-Papamantellos, P.,** *J. Magn. Magn. Mater.,* 30, 355, 1983.
738. **Yakinthos, J. K., Routsi, and Schobinger-Papamantellos, P.,** *J. Phys. Chem. Solids,* 45, 689, 1984.
739. **Yakinthos, J. K., Kotsanidis, P. A., Schafer, W., and Will, G.,** *J. Magn. Magn. Mater.,* 81, 163, 1989.
740. **Yakinthos, J. K., Kotsanidis, P. A., Schafer, W., and Will, G.,** *J. Magn. Magn. Mater.,* 89, 299, 1990.
741. **Yakinthos, J. K., Kotsanidis, P. A., Schafer, W., and Will, G.,** *J. Magn. Magn. Mater.,* 102, 71, 1991.
742. **Yamauchi, H., Yamada, M., Yamaguchi, Y., Yamamoto, H., Hirosawa, S., and Sagawa, M.,** *J. Magn. Magn. Mater.,* 54–57, 575, 1986.
743. **Yang, Y.-C., Kebe, B., James, W. J., Deportes, J., and Yelon, W. B.,** *J. Appl. Phys.,* 52, 2077, 1981.
744. **Yang, Y.-C., Kong, L.-S., Sun, S.-H., Gu, D.-M., and Cheng, B.-P.,** *J. Appl. Phys.,* 63, 3702, 1988.
745. **Yang, Y.-C., Sun, S.-H., Zhang, Z.-Y., Luo, T., and Gao, J.-J.,** *Solid State Commun.,* 68, 175, 1988a.
746. **Yang, Y.-C., Sun, S.-H., and Cheng, B.-P.,** *Ann. Phys. Sci.,* 38, 1429, 1989a.
747. **Yang, Y.-C., Sun, S.-H., Kong, L.-S., and Zha, Y.-B.,** *Sci. China,* A32, 1398, 1989b.
748. **Yang, Y.-C., Ge, S.-L., Zhang, X.-D., Kong, L.-S., Pan, Q., Hou, Y.-T., and Yang, J.-L.,** *Proc. 11th Int. Workshop Rare-Earth Magnets and Their Application,* Vol. II Sankar, S. F., Ed., Pittsburgh, October, 1990, 190.
748a. **Yang, Y.-C., Zhang, X.-D., Kong, L.-S., Pan, Q., Ge, S.-L., Yang, J.-L., Ding, Y.-F., Zhang, B.-S., Ye, C.-T., and Jin, L.,** *Solid State Commun.,* 78, 313, 1991.
749. **Yarovetz, V. I.,** *Autoreferat Dis. Kand. Khim.,* thesis, (Abstr.; in Russian), (Nauk, Lvov), 1978, 24.
750. **Yarovetz, V. I. and Gorelenko, Y. K.,** *Vestn. Lvov Univ. Ser. Khim.,* 23, 20, 1981.
751. **Yermolenko, A. S., Shcherbakova, Ye. V., Andreev, A. V., and Baranov, N. V.,** *Fiz. Z. Met. Metalloved.,* 65, 749, 1988.
752. **Yosida, K.,** *Phys. Rev.,* 106, 893, 1957.
753. **Yosida, K. and Watabe, A.,** *Progr. Theor. Phys.,* 28, 361, 1962.
754. **Zerguine, M. I.,** *Thesis,* Grenoble, 1988.
755. **Zhang, L. Y. and Wallace, W. E.,** *J. Less-Common Met.,* 149, 371, 1989.
756. **Zhang, Z.-D., Sun, X. K., Chuang, Y. C., de Boer, F. R., and Radwański, R. J.,** *J. Phys.* (Paris), 49, C8–569, 1988.
757. **Zhang, L. Y., Pouraian, F., and Wallace, W. E.,** *J. Magn. Magn. Mater.,* 71, 203, 1988a.
758. **Zhang, L. Y., Pouraian, F., and Wallace, W. E.,** *J. Magn. Magn. Mater.,* 74, 101, 1988b.
759. **Zhang, L. Y., Zheng, Y., and Wallace, W. E.,** *Proc. 11th Int. Workshop Rare Earth Magnets and Their Applications,* Vol. II Sankar, S. G., Ed., Pittsburgh, October, 1990, 219.
760. **Zheng, Ch. and Hoffmann, R.,** *J. Solid State Chem.,* 72, 58, 1988.
761. **Zhong, X.-F. and Ching, W. Y.,** *J. Appl. Phys.,* 67, 4768, 1990.
762. **Zouganelis, G., Kostikas, A., Simopoulos, and Niarchos, D.,** *J. Magn. Magn. Mater.,* 75, 91, 1991a.
763. **Zygmunt, A. and Szytuła, A.,** *Acta Magnetica,* Suppl. 84, 193, 1984.
764. **Zygmunt, A., Bażela, W., Leciejewicz, J., Małetka, K., and Szytuła, A.,** *J. Less-Common Met.,* in press.
765. **Moodenbaugh, A. R., Cox, D. E., and Vining, C. B.,** *Phys. Rev.,* 32, 3103, 1985.
766. **André, G., Bourée-Vigneron, F., Oleś, A., and Szytuła, A.,** *J. Magn. Magn. Mater.,* 86, 387, 1990.
767. **Bourée-Vigneron, F.,** *Physica Scripta,* 44, 27, 1991.

INDEX

A

AFI magnetic ordering, 136, 138, see also
    Magnetic ordering
AFIII magnetic ordering, 117, 132, 143, see also
    Magnetic ordering
AIB_2-type structure, 14–15, 90–93, see also
    Crystal structure types
Anisotropic next-nearest neighbor-Ising (ANNNI)
    model, 3–4, 177–178, see also Exchange
    interactions
Anisotropy
    cerium pnictides, 3–4
    crystal electric field relation, 64, 75–76
    europium compounds, 153
    exchange interactions, 174
    neodymium compounds, 184
    rare earth compounds
        RMn_2X_2 constants, 163, 164, 167
        RPd_2Si and RPd_2Ge bonding, 26, 27
        RT_2X_2, 119
        RT_4B, 240
        $RT_{4+x}Al_{8-x}$, 230
        RT_6X_6, 224
        RT_{12-x}, 193, 199
        $R_2Co_{14}B$, 207, 209
        $R_2Fe_{14}B$, 207
        $R_2Fe_{14}C$, 207
        $R_2T_{14}X$, 213, 218, 219, 221
    $ThCr_2Si_2$-type structures, 177
ANNNI model, see Anisotropic next-nearest
    neighbor-Ising model
Antiferromagnetic ordering, see also
        Antiferromagnets; Ferromagnetic ordering;
        Ferromagnets; Magnetic ordering
    $CaIn_2$-type structures, 95
    cerium compounds, 2, 98, 142, 143, 145
    crystal electric field relation, 76
    europium compounds, 4–6, 151–153
    $Fe_2P(ZrNiAl)$-type structures, 84, 89
    gadolinium compounds, 123, 124
    lanthanide elements, 62
    lanthanum compounds, 95, 231–233
    neodymium, 1
    PbFCl-type structures, 95–96
    rare earth compounds
        RFe_2X_2, 125, 126
        RMn_2X_2, 161, 162, 164–171
        RTX_2, 99, 103, 104
        RTX_3, 109
        RT_2X, 105, 108, 109
        RT_2X_2, 114–115, 117, 119, 121, 156, 159
        RT_2X_2 with heavy lanthanides,
            126–129, 132, 133, 135, 136, 139

        $RT_{4+x}Al_{8-x}$, 224, 229, 230
        RT_6X_6, 223–224
        $R_2Fe_{14}B$, 202
        R_2RhSi_3, 93
        $R_2Rh_3Si_5$, 114
        $R_2T_{14}X$, 216
    samarium compounds, 122
Antiferromagnets, see also Antiferromagnetic
        ordering; Ferromagnetic ordering;
        Ferromagnets; Magnetic ordering
    AIB_2- and Ni_2In-type structures, 91, 93
    $CaIn_2$-type structures, 94
    $MgCu_2$-type structures, 83
    PbFCl-type structures, 95–96
    rare earth compounds
        RMn_2X_2, 161
        RTX_2, 101, 103
        RT_2X_2, 121, 129
        RT_6X_6, 224
        R_2RhSi_3, 93
Atomic moment, 193, 196
Atomic positions
    AIB_2 and related structures, 14, 15
    $CaCu_5$-type structures, 57
    $CaIn_2$- and $GaGeLi$-type structures, 15
    $CeCu_2$-type structures, 21
    Fe_2-P-type structures, 12
    $LaPtSi$-type structures, 16
    $MgAgAs$-type structures, 10
    $MgCu_2$-type structures, 9
    $MgZn_2$-type structures, 12
    $Nd_2Fe_{14}B$, 48, 49
    $NdFe_{10}Mo_2$, 45
    PbFCl-type structures, 17
    rare earth compounds
        RTX_2, 24, 26
        RT_6X_6-type, 52, 53
        $R_2T_3Si_5$-type, 28

B

$BaAl_4$, 29
$BaNiSn_3$-type structure, 29, 31, 109

C

$CaAl_2Si_2$-type structure, 158
$CaBe_2Ge_2$-type structure, 29–31, 43, 44, 180
$CaCu_5$-type structures, 55–57
$CaIn_2$-type structure, 94–95
CEF, see Crystal electric field (CEF) model
Cerium
    $CaCu_5$-type structures, 55

CeAg$_2$In, 107
CeAgGe, 99
CeCuSi, 90–91
CeCu$_2$-type structure, 21, 98–99
CeCu$_2$Si$_2$, neutron scattering, 81, 82
CeGeAl, 95
CeMn$_2$(Si$_{1-x}$Ge$_x$)$_2$, 168
CeNiSi$_2$-type structure, 22, 23, 25, 101, 103
CePbSb, 95
CePtSi$_2$, 103
Ce(Rh$_{1-x}$Ru$_2$)$_2$Si$_2$, 143, 144
CeSiAl, 95
CeT$_2$Be$_2$, 146
CeT$_2$Ge$_2$, 146, 148, 151
CeT$_2$Si$_2$, 43, 146, 148, 151, 152
CeT$_2$Sn$_2$, 146, 147, 149, 151, 152
CeT$_2$X$_2$
   crystal field schemes, 179–181
   magnetic properties, 140–151
CeTSi$_2$, 104
Ce$_2$Ni$_{17}$Si$_5$-type structures, 54
Fe$_2$P(ZrNiAl)-type structure, 84, 89
Kondo effect, 7
magnetic ordering of pnictides, 3–4
magnetic properties of alloys, 2
Pauling metallic radii, 12
Chemical pressure, 150
Chemical shift, 175
Coercivity, 241
Critical distance, 169
Crystal electric field (CEF)
   cerium compounds, 147, 149, 179–180
   dysprosium compounds, 185
   erbium compounds, 186–187
   free ion interactions, 70–73
   holmium compounds, 185–186
   hyperfine interactions, 78–80
   level, 80–82
   magnetic moment relation, 73–76, 173
   magnetic ordering role, 245
   magnetic part of molar capacity, 77–79
   neodymium compounds, 184, 185
   neutron scattering by transitions, 80–82
   overview, 67–69
   praseodymium compounds, 180–184
   rare earth compounds
      RMn$_2$X$_2$, 159, 161, 166, 167
      RTX, 191–192
      RTX$_2$, 101, 104, 190, 191
      RTX$_3$, 109
      RT$_2$X, 105, 107, 192
      RT$_2$X$_2$, 115, 158, 189–191
      RT$_2$X$_2$ with heavy lanthanides, 132, 135, 137
      RT$_4$B, 240
      R$_{1+t}$T$_4$B$_4$, 241
      R$_2$Fe$_3$Si$_5$, 110–113
      R$_2$T$_{14}$X, 220–223
      splitting in R$_2$T$_{14}$X, 216

      substituted R$_2$T$_{14}$X, 212
   terbium compounds, 184–185
   thulium compounds, 187
   transport properties, 76–77
Crystal structure types
   AlB$_2$, 14–15, 90–93
   BaNiSn$_3$, 29, 31, 109
   CaAl$_2$Si$_2$, 158
   CaBe$_2$Ge$_2$, 29–31, 43, 44, 180
   CaCu$_5$, 55–57
   CaIn$_2$, 15, 94–95
   CeCu$_2$, 21, 22, 28, 98, 99
   CeNiSi$_2$-type structures, 22, 23, 25, 101, 103,
      179
   Ce$_2$Ni$_{17}$Si$_5$-type structure phases, 54
   CoSn-derivate structure types, 52–53
   Fe$_2$-P, 12, 13
   Fe$_2$P(ZrNiAl), 84–89
   Fe$_3$C, 26, 27, 108
   GaGeLi, 15
   HfFe$_6$Ge$_6$, 52–53
   HfFe$_6$Sn$_6$, 223
   LaIrSi, 11
   LaIrSi (ZrOS), 84
   LaPtSi, 16, 95
   MgAgAs, 10, 11, 83
   MgCu$_2$, 9, 10, 83
   MgZn$_2$, 12, 13, 89–90
   NaZn$_{13}$, 54, 55
   NdCo$_4$, 57–59
   Ni$_2$In, 14–15, 90–93
   PbFCl, 17, 95–98
   R$_2$T$_3$Si$_5$, 28
   R$_2$T$_{14}$X-type structure, 47–51
   SrNi$_{12}$B$_6$(EuN$_{12}$B$_6$), 55
   TbFeSi$_2$, 23, 99
   TbNiC$_2$, 24, 26
   ThCr$_2$Si$_2$, 177
   ThMn$_{12}$, 44–48
   TiNiSi type, 18–21, 96–98
   YPd$_2$Si, 27
   ZrBeSi-type structure, 18
Crystal structures
   RMn$_6$Sn$_6$ compounds, 52
   RT$_2$X$_2$ phase compounds, 29–30
   R$_2$T$_3$Si$_5$, 28
Curie temperature, see also Curie-Weiss law
   AlB$_2$- and Ni$_2$In-type structures, 90, 91
   CeCu$_2$-type structures, 99
   CeT$_2$X$_2$ compounds, 146
   de Gennes function comparison in R$_2$Fe$_{14}$B
      compounds, 206
   LaIrSi(ZrOS)-type structures, 84
   LaPtSi-type structures, 95
   La(T$_{1-x}$)$_{13}$ compounds, 232, 234
   MgAgAs-type structures, 83
   PbFCl-type structures, 95–98
   rare earth compounds

RTX$_2$, 103–105
RTX$_3$, 109
RT$_2$X$_2$, 156, 158
RT$_2$X$_2$ with heavy lanthanides, 132, 136, 138
RT$_4$B, 239
RT$_{4+x}$Al$_{8-x}$, 229
RT$_9$Si$_2$, 235
RT$_{12}$B$_6$, 237
RT$_{12-x}$, 196
R$_{1+\epsilon}$T$_4$B$_4$, 241
substituted R$_2$T$_{14}$X, 210–212
SmT$_2$X$_2$ compounds, 122
YbT$_2$X$_2$ compounds, 155
Curie-Weiss law, see also Curie temperature
    AlB$_2$- and Ni$_2$In-type structures, 90, 91
    CaIn$_2$-type structures, 94
    CeCu$_2$-type structures, 99
    CeT$_2$X$_2$ compounds, 140, 146, 147
    Fe$_2$P(ZrNiAl)-type structures, 84, 89
    LaIrSi(ZrOS)-type structures, 84
    LaPtSi-type structures, 95
    MgCu$_2$-type structures, 83
    PbFCl-type structures, 96–98
    rare earth compounds
        RTX$_2$, 103–105
        RTX$_3$, 109
        RT$_2$X, 108
        RT$_2$X$_2$, 156, 158
        RT$_2$X$_2$ with heavy lanthanides, 127, 132, 134, 136, 138
        R$_2$T$_3$Si$_5$, 114
    SmT$_2$X$_2$ compounds, 122
    YbT$_2$X$_2$ compounds, 155

D

de Gennes factor
    lanthanide elements, 63
    PbFCl-type structures, 98
    rare earth compounds
        RCu$_2$Si$_2$, 188, 191
        RTX$_2$, 101, 103, 190, 191
        RT$_2$X$_2$ with heavy lanthanides, 132
        R$_2$Fe$_{14}$B, 206
        R$_2$Fe$_3$Si$_5$, 110, 111
    RKKY model relation, 173, 245
Density of states (DOS), 169, 172, 173
Diamagnetic properties, 229
Dipole-dipole interaction, 110, 112
Domain-wall nucleation process, 241
Doniach model, 146, 150
DOS, see Density of states
Dysprosium
    DyAlGa, 91, 92
    DyCo$_2$Si$_2$, 75–76
    DyFeAl, 90
    DyFe$_{10}$Cr$_2$, 221
    DyMn$_2$Ge$_2$, 166

DyT$_2$Ge$_2$, 126
DyT$_2$Si$_2$, 35, 42, 126
DyT$_2$X$_2$ compounds, 185
DyTiFe$_{11-x}$Co$_x$, 198, 201
    magnetic ordering, 3

E

Effective magnetic moment, 91, 97, 172–173, see also Magnetic moment
Effective moments, 94
Effective paramagnetic moment, 142, see also Paramagnetic moment
Electric resistivity, see Resistivity, electrical
Electron charge transfer, 200
Electronic energy bands, 169
Electrostatic field, 219, 220
Electrostatic potential, 68
Energy levels, 180–183, 186
Enthalpy, 47
Entropy, 135
Erbium
    ErFe$_{10}$Cr$_2$, 221
    ErT$_2$Ge$_2$, 128
    ErT$_2$Si$_2$, 128
    ErT$_2$X$_2$, 186–187
    ErTiFe$_{11-x}$Co$_x$, 198, 201
    magnetic ordering, 3
Europium
    chalcogenides, 4–6
    EuPdSi, 84
    EuSe, 6
    EuTGa, 99
    EuT$_2$Ge$_2$, 42, 153
    EuT$_2$Si$_2$, 35, 153
    EuT$_2$X$_2$, 151–154
Exchange integrals, 4, 62, 92, 136, 149
Exchange interactions
    binary lanthanide -3d- transition metal compounds, 66–67
    cerium pnictides, 4
    crystal electric field theory, 68, 74, see also Crystal electric field
    europium chalcogenides, 5–6
    gadolinium compounds, 123
    lanthanide elements, 62, 64
    lanthanum compounds, 232
    magnetic moment relation, 172, 173
    praseodymium compounds, 182
    rare earth compounds
        metallics, 173–178
        RT$_4$B, 239
        RT$_{4+x}$Al$_{8-x}$, 230
        RT$_{12}$B$_6$, 237
        RT$_{12-x}$, 198, 215–218
        R$_{1+\epsilon}$T$_4$B$_4$, 241
        R$_2$T$_{14}$X, 207, 211, 212, 217, 218

F

Fe, see Iron
Fermi energy, 7, 140, 155, 215
Fermi level, 147, 169
Fermi surface
  cerium compounds, 4, 43
  effective magnetic moment relation, 172, 173
  lanthanide elements, 62
  RKKY theory and, 174, 176
Ferrimagnets, 91, 93, 99, 240
Ferromagnetic ordering, see also Antiferromagnetic
      ordering; Antiferromagnets; Ferromagnets;
      Magnetic ordering
  AlB$_2$- and Ni$_2$In-type structures, 90
  CaIn$_2$-type structures, 95
  CeCu$_2$-type structures, 99
  EuT$_2$X$_2$ compounds, 153
  GdT$_2$X$_2$ compounds, 123
  LaPtSi-type structures, 95
  La(T$_{1-x}$)$_{13}$ compounds, 231
  MgZn$_2$-type structures, 89–90
  rare earth compounds
    RMn$_2$X$_2$, 159, 161–166, 168–171
    RTX$_2$, 99, 104
    RT$_2$X$_2$, 117, 120, 156
    RT$_2$X$_2$ with heavy lanthanides, 128–129
    RT$_4$B, 239, 240
    RT$_{4+x}$Al$_{8-x}$, 224, 229
    RT$_6$X$_6$, 223–224
    RT$_9$Si$_2$, 235
    RT$_{12}$B$_6$, 236
    R$_{1+\epsilon}$T$_4$B$_4$, 241, 242
Ferromagnetism, 1, 64, 142, 198, see also
      Antiferromagnetic ordering;
      Antiferromagnets; Ferromagnets; Magnetic
      ordering
Ferromagnets, see also Antiferromagnetic ordering;
      Antiferromagnets; Ferromagnets; Magnetic
      ordering
  Fe$_2$P(ZrNiAl)-type structures, 84
  PbFCl-type structures, 95
  rare earth compounds
    RTX$_2$, 103
    RT$_2$X$_2$, 121, 137
    R$_2$RhSi$_3$, 93
    RT$_4$B, 240
Free-ion moment, 121
Free ion state, 69–73, 74

G

Gadolinium
  crystal electric field splitting, 68, 69
  GaGeLi, 15
  GdAiSn, 94
  GdCiSi, 176
  Gd(Co$_{1-x}$B$_x$) series, 56

GdCuGe, 91
GdCuSn, 94
GdFe$_{12-x}$M$_x$, 222, 223
GdMn$_2$Ge$_2$, 164
GdRu$_2$Si$_2$, 124
GdTAl, 84
GdTSn, 84
GdT$_2$Ge$_2$, 42, 123
GdT$_2$Si$_2$, 123, 172–174
GdT$_2$X$_2$, 122–124
GdTiFe$_{11-x}$Co$_x$, 198, 201
historical background, 1
magnetic ordering, 3
Germanides, 23

H

Hall effect, 76
Heat capacity, 108, 120, 135, 140, 144
Heat of formation, 243
Heavy fermion effect
  cerium compounds, 140–142, 144,
      148–150
  Fe$_2$P(ZrNiAl)-type structures, 89
  lanthanides, 6–8
  LaPtSi-type structures, 95
  RTX$_2$ compounds, 103, 104
Heisenberg ferromagnet, 5, 6, see also
      Ferromagnets
Heisenberg model, 66
Helicoidal magnetic ordering, 133, 136–137, 245,
      see also Antiferromagnetic ordering;
      Antiferromagnets; Ferromagnets; Magnetic
      ordering
Heusler alloys, 25, 27, 105, 192
HfFe$_6$Ge$_6$-type structure, 52–53
HfFe$_6$Sn$_6$-type structure, 223
Holmium
  HoAlGa, 91
  HoCoSi$_2$, 14, 23, 25
  HoFe$_5$Al$_7$, 229–230
  HoPd$_2$Sn, 107
  HoT$_2$X$_2$ compounds, 127, 185–186
  HoTiFe$_{11-x}$Co$_x$, 198, 201
  magnetic ordering, 3
  Pauling metallic radius, 14
Hund state, 76
Hund's rule, 61
Hybrides, 212–214, see also Individual entries
Hybridization model, 7, 8
Hydrogenation, 47, 51, 199–200, 212–214
Hyperfine field, 79, 101, 135, see also Hyperfine
      interactions
Hyperfine interactions
  CaIn$_2$-type structures, 94
  crystal electric field theory, 78–80
  europium compounds, 153
  gadolinium compounds, 124

lanthanide elements, 62
lanthanum compounds, 233
rare earth compounds, 138, 156, 165, 200

I

Interaction integrals, 136
Interatomic distances
    CaBe$_2$Ge$_2$-type structures, 43
    CeCu$_2$-type structures, 22
    lanthanum compounds, 234
    neodymium compounds, 48, 50
    rare earth compounds
        RMn$_2$X$_2$, 171
        RT$_2$X$_2$, 30, 33–35, 244
        RT$_{12-x}$, 198
    terbium compounds, 24, 26, 52–53
    TiNiSi-type structures, 20–21
Interatomic radii, 14
Ionization state, 68, see also Crystal electric field
Iron compounds
    Fe$_2$-P-type structure, 12
    Fe$_2$P (ZrNiAl)-type structure, 85–89
    Fe$_3$C-type structure, 26, 27, 108
    substitution, 209–212
Ising linear-chain model, 75
Isomer shift, 151, 152

K

KKR method, 169
Kondo effect
    cerium compounds, 141, 144, 145, 147–151
    rare earth compounds, 108
Kondo lattice, 6, 7, 8

L

Lanthanum
    La(Al$_2$Fe$_{1-x}$), 232
    La(Co$_{1-x_i}$x)$_{13}$, 236
    LaCu$_2$Si$_2$, 81, 82
    La(Fe$_{0.88-x}$Co$_x$Al$_{0.12}$)$_{13}$ system, 233
    La(Fe$_{1-x}$Al$_x$)$_{13}$, 236
    La(Fe$_{1-x}$Si$_x$)$_{13}$, 236
    LaFe$_{13}$, 55
    LaIrSI type crystals, 11
    LaIrSi (ZrOS)-type structure, 84
    LaMnSi$_2$, 23, 25
    LaPtSi-type structure, 95
    La$_2$Pt$_2$Ge$_2$, 43
    La(T$_{1-x}$)$_{13}$ phases, 231–234
    NaZn$_{13}$-type structure, 54
Lande factors, 74
Lanthanide elements, 1, 2
Lanthanides, see also Individual entries
    binary
        heat of formation, 243

magnetic interactions of -3d- transition metal
        compounds, 65–67
    physical properties, 6–8
    heavy, magnetic properties, 125–140
    magnetic interactions of elements, 61–65
    Miedema theorem, 243–244
    ternary, crystal structures
        CaCu$_5$-type structures, 55–57
        Ce$_2$Ni$_{17}$Si$_5$-type structure phases, 54
        NaZn$_{13}$-stype structure phases, 54, 55
        NdCo$_4$-type structures and polytypes, 57–59
        RTX phases, 9–22, see also RTX phases
        RTX$_2$ phases, 22–26
        RT$_2$X phases, 24–27
        RT$_2$X$_2$ phases, 28–44
        RT$_6$X$_6$ compounds with CoSn-derivate
            structure types, 52–53
        R$_2$T$_3$Si$_5$ phases, 27–28
        R$_2$T$_{14}$X-type structure, 47–51
        SrNi$_{12}$B$_6$(EuN$_{12}$B$_6$)-type structure phases, 55
        ThMn$_{12}$-type structure, 44–48
    physical parameters of ternary, 8
Laplace equation, 70–71
Lattice parameters
    CaBe$_2$Ge$_2$-type structures, 43, 44
    CaCu$_5$-type structures, 57
    europium selenium, 6
    Fe$_2$-P-type structures, 13
    LaPtSi-type structures, 16
    rare earth compounds
        RMn$_2$X$_2$, 161, 162
        RNiSi$_2$ and RNiGe$_2$, 23, 24
        RPd$_2$Si and RPd$_2$Ge, 26, 27
        RTX, 9, 10
        RT$_2$X$_2$, 31–33
        RT$_{12-x}$Al$_x$, 46
        R$_2$T$_{14}$X, 49–51
    TiNiSi-type structures, 18–20
LMTO method, 43
Local density of states (LDOS), 215, see also
        Density of states

M

Magnetic effective moment, see Effective magnetic
        moment
Magnetic fields, 73, 74, 134
Magnetic moment(s)
    AIB$_2$- and Ni$_2$In-type structures, 90, 91
    binary lanthanide -3d- transition metal
        compounds, 65, 67
    CaIn$_2$-type structures, 95
    CeCu$_2$-type structures, 98
    cerium
        pnictides, 2, 4
        CeT$_2$X$_2$, 142–144, 148–151, 180, 182
    crystal electric field theory, 73–76
    erbium compounds, 186–187

Fe$_2$P(ZrNiAl)-type structures, 84
lanthanide compounds, 244
lanthanide elements, 1, 61, 64
LaPtSi-type structures, 95
lanthanum compounds, 231
MgAgAs-type structures, 83
MgZn$_2$-type structures, 90
PbFCl-type structures, 96–98
rare earth compounds
   RMn$_2$X$_2$, 159, 161, 164–165, 168, 170
   RTX$_2$, 99–101, 104
   RT$_2$Si$_2$, 188, 189
   RT$_2$X, 107, 108
   RT$_2$X$_2$, 114, 115, 119, 121, 158
   RT$_2$X$_2$ with heavy lanthanides,
     126–133, 135, 139
   RT$_4$B, 240
   RT$_{4+x}$Al$_{8-x}$, 224, 229–231
   RT$_9$Si$_2$, 235
   RT$_{12}$B$_6$, 236
   RT$_{12-x}$, 195–197, 200, 214–215
   R$_2$Co$_3$Si$_5$, 114
   R$_2$Co$_{14}$B, 209
   R$_2$Fe$_3$Si$_5$, 112
   R$_2$Fe$_{14}$B, 203–206
   R$_2$Fe$_{14}$C, 207
   R$_2$RhSi$_3$, 93
terbium compounds, 184
ternary compounds, 171–173
thulium compounds, 187
Magnetic ordering, see also Antiferromagnetic
   ordering; Antiferromagnets; Ferromagnetic
   ordering; Ferromagnets
CaIn$_2$-type structures, 94
cerium compounds, 140, 145, 146, 152
europium compounds, 4–5
LaIrSi(ZrOS)-type structures, 84
lanthanide elements, 1, 3
lanthanum compounds, 231–233
LaPtSi-type structures, 95
MgAgAs-type structures, 83
PbFCl-type structures, 97, 98
rare earth compounds
   heavy, 2, 3
   RFe$_{12-x}$Al$_x$, 144
   RMn$_2$X$_2$, 161, 171, 244
   RTX$_2$, 103, 190, 191
   RTX$_3$, 109, 110
   RT$_2$X, 108
   RT$_2$X$_2$, 114
   RT$_2$X$_2$ with heavy lanthanides, 132, 134
   RT$_{4+x}$Al$_{8-x}$, 224
   RT$_6$X$_6$, 223–224
   R$_{1+\epsilon}$T$_4$B$_4$, 241, 242
   R$_2$Co$_{14}$B, 209
   R$_2$Fe$_3$Si$_5$, 112
   R$_2$RhSi$_3$, 93
ytterbium compounds, 155

Magnetic phase diagrams
cerium compounds, 143, 144, 151
gadolinium compounds, 123
lanthanum compounds, 232, 233
rare earth compounds
   RMn$_2$X$_2$, 161–162, 164, 166, 168
   RT$_2$X$_2$, 118, 120, 121
   RT$_2$X$_2$ with heavy lanthanides, 131, 134, 137,
     138
   RT$_{12}$B$_6$, 237, 238
   RT$_{12-x}$, 198, 200, 201
   R$_2$T$_{14}$X, substituted, 211–213
Magnetic phases, 6, see also Magnetic phase
   diagrams
Magnetic properties, see also Individual entries
crystal electric field, see also Crystal electric
   field (CEF)
   RTX compounds, 191–192
   RTX$_2$ compounds, 190, 191
   RT$_2$X$_2$ compounds, 179–191
   RT$_2$X compounds, 192
exchange interactions in RTX compounds,
   173–178
lanthanide elements, 61–65
lanthanide -3d- transition metal binaries, 65–67
magnetic moment in RTX compounds, 171–173
RMn$_2$X$_2$, 159–171
RTX phases
   AlB$_2$ and Ni$_2$In-type structures, 90–93
   CaIn$_2$-type structure, 94–95
   CeCu$_2$-type structure, 98–99
   Fe$_2$P (ZrNiAl)-type structure, 85–89
   LaIrSi (ZrOS)-type structure, 84
   LaPtSi-type structure, 95
   MgAgAs-type structure, 83
   MgCu$_2$-type structure, 83
   MgZn$_2$-type structure, 89–90
   PbFCl-type structure, 95–96
   R$_2$RhSi$_3$ compounds, 93–94
   TiNiSi-type structure, 96–98
RTX$_2$ phases, 99–105
RTX$_3$ phases, 109–110
RT$_2$X phases, 105–109
RT$_2$X$_2$ phases
   CeT$_2$X$_2$, 140–151
   EuT$_2$X$_2$, 151–154
   GdT$_2$X$_2$, 122–124
   heavy lanthanides, 125–140
   other compounds, 156–159
   RT$_2$Si$_2$ and RT$_2$Ge$_2$, 114–122
   SmT$_2$X$_2$, 122
   YbT$_2$X$_2$, 155–156
   R$_2$T$_3$Si$_5$ phases, 110–114
Magnetic structure
AlB$_2$- and Ni$_2$In-type structures, 91, 92
crystal electric field anisotropy relation, 64
PbFCl-type structures, 96
R$_2$RhSi$_3$ compounds, 94

RT$_{12-x}$ compounds, 195
Magnetic susceptibility, see Susceptibility, magnetic
Magnetization
  cerium compounds, 141, 142, 145
  dysprosium compounds, 75–76, 185
  high-field of RT$_2$X$_2$ compounds, 116
  neodymium compounds, 184
  PbFCl-type structures, 96, 97
  praseodymium compounds, 182–183
  rare earth compounds
    RMn$_2$X$_2$, 159, 161, 163, 164, 169
    RT$_2$X$_2$, 119–121
    RT$_2$X$_2$ with heavy lanthanides, 128, 129, 136–139
    RT$_4$B, 239, 240
    RT$_{4+x}$Al$_{8-x}$, 229–231
    RT$_9$Si$_2$, 235
    RT$_{12}$B$_6$, 238
    RT$_{12-x}$, 193, 196, 200
    R$_{1+\epsilon}$T$_4$B$_4$, 241, 242
    R$_2$Co$_{14}$B, 207, 209
    R$_2$Fe$_{14}$B, 202, 203, 207
    R$_2$Fe$_{14-x}$T$_x$B, 209–210
    R$_2$T$_{14}$X, 213, 214, 219–222
    substituted R$_2$T$_{14}$X, 211
  YbT$_2$X$_2$ compounds, 155
Magnetocrystalline anisotropy, see also Anisotropy
  PbFCl-type structures, 97
  RMn$_2$X$_2$ compounds, 159, 161
  RT$_2$X$_2$ compounds, 119
  RT$_{12-x}$ phases, 218–223
  R$_2$T$_{14}$X compounds, 216
Magnetoresistance, 95
Magnetostriction, 61, 131
MALCAO method, 43
Mean-field approximation, 180
Mean-field theory, 215–216, 218
Metamagnetic behavior, 99
Metamagnetic process, 129, 133
Metamagnetic transition
  cerium compounds, 141
  lanthanum compounds, 232
  RT$_2$X compounds, 109
  RT$_2$X$_2$ compounds, 114, 120, 121
    with heavy lanthanides, 138, 139
MgAgAs-type structure, 10–11, 83
MgCu$_2$ type crystals, 9–10, 83
MgZn$_2$-type structure, 12, 13, 89–90
Micromagnetic behavior, 136
Micromagnetic state, 232, 234
Microscopic exchange, 216
Miedema theorem, 243–244
Mixed valency
  CeNiSi$_2$-type structures, 23
  cerium compounds, 2, 141
  europium compounds, 151, 154
  Fe$_2$P(ZrNiAl)-type structures, 89

Heusler alloys, 105
  lanthanides, 6–8
  rare earth compounds, 42, 46
  ytterbium compounds, 155
Modulated antiferromagnetic structure, 119, see also Antiferromagnetic ordering
Modulated magnetic structure, 119, 120, 136, 142, see also Magnetic structure
Mössbauer spectroscopy
  Fe$_2$P(ZrNiAl)-type structures, 84
  CeCu$_2$-type structures, 99
  dysprosium compounds, 185
  europium compounds, 151–154
  gadolinium compounds, 124
  lanthanide elements, 1
  lanthanum compounds, 233
  magnetic moment for R$_2$T$_{14}$X compounds, 214
  MgZn$_2$-type structures, 90
  praseodymium compounds, 181
  rare earth compounds
    RMn$_2$X$_2$, 165
    RTX$_2$, 101
    RTX$_3$, 109
    RT$_2$X, 108
    RT$_2$X$_2$, 41–42, 114, 156
    RT$_2$X$_2$ with heavy lanthanides, 132, 134, 138, 139
    RT$_4$B, 240
    RT$_{4+x}$Al$_{8-x}$, 229, 230
    RT$_6$X$_6$, 224
    RT$_{12-x}$, 195, 197
    R$_{1+\epsilon}$T$_4$B$_4$, 241
    R$_2$Fe$_{14}$B, 203, 204
    R$_2$Fe$_{14}$C, 207
    R$_2$Fe$_3$Si$_5$, 113
  ThMn$_{12}$ phase compounds, 45, 47
  ytterbium compounds, 155
Molar capacity, 77–78, see also Crystal electric field
Molecular field, 166, 216, 239, 240, see also Molecular field
Molecular field theory
  crystal electric field theory comparison, 68, 75
  europium chalcogenides, 5–6
  lanthanide elements, 62
  -3d- transition binary metallics, 65–66
Muon spin relaxation, 140

N

NaZn$_{13}$-type structure phases, 54, 55
Néel temperature
  AlB$_2$- and Ni$_2$In-type structures, 91
  CaIn$_2$-type structures, 94, 95
  CeCu$_2$-type structures, 98, 99
  cerium compounds, 4, 141, 144–147
  europium selenium, 6
  europium compounds, 152, 153

Fe$_2$P(ZrNiAl)-type structures, 89
gadolinium compounds, 123
lanthanum compounds, 233
MgCu$_2$-type structures, 83
PbFCl-type structures, 96–98
rare earth compounds
  RMn$_2$X$_2$, 169, 170
  RTX$_2$, 101, 103
  RT$_2$X, 108, 109
  RT$_2$X$_2$, 115, 121, 156, 158
  RT$_2$X$_2$ with heavy lanthanides, 127, 129, 132–134, 136, 137
  RT$_4$B, 239
  RT$_6$X$_6$, 224
  R$_2$Fe$_3$Si$_5$, 110, 112
Neodymium
antiferromagnetic ordering, 1
NdAlGa, 91
Nd(Co$_{1-x}$Fe$_2$Si$_2$, 237
Nd(Co$_{1-x}$Fe$_x$)$_9$Si$_2$, 235
NdCo$_2$Ge$_2$, 118, 130
NdCo$_2$Si$_2$, 118
NdCo$_4$B$_4$-type structure, 57–59
NdCo$_9$Si$_2$, 235
NdCu$_2$Si$_2$, 120
NdFeSi$_2$, 100
NdFe$_{10}$Mo$_2$, 45
NdFe$_{12}$B$_6$, 55
NdMnSi$_2$, 99–101
NdNiSi$_2$, 103
NdRuSi$_2$, 23–24
NdRu$_2$Ge$_2$, 120, 177–178
NdRu$_2$Si$_2$, 177–178
NdT$_2$Ge$_2$, 115, 116
NdT$_2$Si$_2$, 115, 116
NdT$_2$X$_2$, 184, 185
Nd$_2$Fe$_{14}$B
  crystal field electric, 220, 221
  molecular field analysis, 65
  R$_2$T$_{14}$X structure prototype, 47–49, 202
Nd$_2$RhSi$_3$, magnetic ordering, 93, 94
Nd$_5$Fe$_{18}$B$_{18}$, 58, 59
Neutron diffraction
AlB$_2$- and Ni$_2$In-type structures, 90, 91
CaIn$_2$-type structures, 95
CeCu$_2$-type structures, 98, 99
cerium compounds, 140, 142, 143
gadolinium compounds, 124
heavy rare earths, 2, 3
magnetic moment determination for R$_2$T$_{14}$X compounds, 214
MgZn$_2$-type structures, 90
PbFCl-type structures, 96–98
rare earth compounds
  RMn$_2$X$_2$, 161, 165, 167
  RTX$_2$, 99, 103, 104
  RT$_2$X, 105, 108, 109
  RT$_2$X$_2$, 31, 33, 114, 119–121, 156, 158–159

RT$_2$X$_2$ with heavy lanthanides, 125–130, 133, 135–136
RT$_{4+x}$Al$_{8-x}$, 224, 228
RT$_6$X$_6$, 224
RT$_{12-x}$, 195–198
R$_2$Fe$_3$Si$_5$, 112, 113
R$_2$Fe$_{14}$B, 203, 204
R$_2$RhSi$_3$, 93
ThMn$_{12}$ phase compounds, 45, 47
Neutron inelastic scattering method, 192
Neutron scattering, 1, 80–82
Ni$_2$In-type structures, 14–15, 90–93
Nitrogenation, 200
NMR, see Nuclear magnetic resonance
Nuclear magnetic resonance (NMR), 42, 158

O

Orbital momentum, 79
Orthorhombicity parameter, 18, 19

P

Paramagnetic Curie temperature, 96, see also Curie temperature
Paramagnetic moments, 84, 91, 98, see also Magnetic moments
Paramagnetic ordering, 83, 139, 146, see also Antiferromagnetic ordering; Ferromagnetic ordering; Magnetic ordering; Van Vleck paramagnetism
Paramagnets, see also Pauli paramagnets
  CaIn$_2$-type structures, 95
  PbFCl-type structures, 97, 98
  rare earth compounds, 93, 109, 120
Pauli paramagnets
  AlB$_2$- and Ni$_2$In-type structures, 91
  cerium compounds, 141, 149
  lanthanum compounds, 234
  PbFCl-type structures, 96, 97
  rare earth compounds
    RMn$_2$X$_2$, 159
    RTX$_2$, 101
    RT$_2$X, 108
    RT$_2$X$_2$, 114, 158
    R$_2$Co$_3$Si$_5$, 113–114
  TiNiSi-type structures, 96
Pauling metallic radii, 12, 14, 15, 23
PbFCl-type structure, 17, 95–98
Photoemission spectra, 43
Point-charge model, 68, 72, 73, 183–184, 220–221
Praseodymium
  PrCo$_2$Si$_2$, 116, 118
  PrFeSi$_2$, 23, 25
  PrMnSi$_2$, 99, 100, 103
  PrT$_2$Ge$_2$ compounds, 114, 115
  PrT$_2$Si$_2$ compounds, 114, 115

PrT_2X_2 compounds, 180–184
Pressure, see also Curie temperature; Néel
　　temperature
　curie temperature in RMn_2X_2 compounds, 161,
　　162
　lanthanum compound magnetic properties, 233,
　　234
　Néel temperature in CeT_2X_2 phases, 146, 147
Prometheum, historical background, 1
Promotion model, 7

Q

Quadrupole interactions, 135, 154, 188, 190
Quadrupole scattering, 77
Quadrupole splitting, 41–42

R

Rare earth compounds
　RAlGa, 91
　RAl_2Si_2, 158
　RAu_2Si_2, 139
　RCoAl, 90
　$RCoSi_2$, 101
　$R(Co_{1-x}Fe_x)_9Si_2$, 237
　RCo_2B_2, 156
　RCo_2Si_2, 35
　RCo_6Ge_6, 53
　RCo_9Si_2, 54, 237
　$RCo_{10}SiC_{0.5}$, 54
　$RCo_{12}B_6$, 55, 238
　RCr_4Al_8, 227
　RCr_6Al_6, 228
　RCuAl, 12, 13
　RCuSi, 14, 90
　RCu_2Ge_2, 42
　RCu_2Si_2, 42
　RCu_4Al_8, 226
　RCu_6Al_6, 228
　$RFeSi_2$, 23
　RFe_2Si_2, 174, 188, 190
　RFe_4Al_8, 226
　RFe_5Al_7, 227
　RFe_6Sn_6, 53, 224, 225, 227–228, 231
　$RFe_{10}Cr_2$, 195, 197
　$RFe_{10}M_2$, 217
　$RFe_{10}SiC_{0.5}$, 54
　$RFe_{10}V_2$, 221, 222
　$RFe_{10-x}Co_xV$, 198, 199
　$RFe_{11}Ti$, 195, 197, 223
　$RFe_{12-x}Al_x$, 244
　$RFe_{12-x}M_x$, 45, 46, 217–219
　RIr_2Si_2, 138
　$RMnSi_2$, 23, 99
　RMn_2Ge_2, 160
　RMn_2Si_2, 160
　RMn_2X_2, 159–171, 244

RMn_4Al_8, 227
RMn_6Al_6, 228
RMn_6Ge_6, 225
RMn_6Sn_6, 52, 223, 225
RMn_{12}, 226
RNiAl, 12, 13
$RNiGa_2$, 105
RNiGe, 96
RNiSn, 97
$RNiX_2$, 100, 101
RNi_2P_2, 158
RNi_9Si_2, 54
$RNi_{12}B_6$, 55
ROs_2Si_2, 137
RPbSb compounds, 95
RPdSn compounds, 98
RPd_2Ge, 26, 27
RPd_2Si, 26, 27
RPd_2Si_2, 137, 138
RPtSn, 97
RPt_2Si_2, 43, 44
RTX
　crystal electric field, 191–192
　magnetic data, 85–89
　AlB_2 and Ni_2In-type structures, 14–15, 90–93
　$CaIn_2$-type structure, 15–16, 94–95
　$CeCu_2$-type structure, 21–22, 98–99
　Fe_2-P-type structures, 12, 13, 85–89
　LaIrSi (ZrOS)-type structure, 11, 84
　LaPtSi-type structures, 16, 95
　MgAgAs-type structure, 10–11, 83
　$MgCu_2$-type structure, 9–10, 83
　$MgZn_2$-type structure, 12–13, 89–90
　PbFCl-type structures, 16–17, 95–96
　R_2RhSi_3 compounds, 93–94
　TiNiSi-type structures, 17–21, 96–98
RTX_2
　crystal electric field, 190, 191
　magnetic properties, 99–105
　ternary lanthanides, 22–25, see also
　　Lanthanides, ternary
RTX_3, 109–110
RT_2Ge_2, 114–122
RT_2Si_2, 80, 81, 114–122
RT_2X, 105–109, 174–175, 192
RT_2X_2
　antiferromagnetic structure, 176
　crystal electric field, 179–190, 191
　interatomic distances, 244
　CeT_2X_2, 140–151
　EuT_2X_2, 151–154
　GdT_2X_2, 122–124
　heavy lanthanides, 125–140
　other compounds, 156–159
　RT_2Si_2 and RT_2Ge_2, 114–122
　SmT_2X_2, 122
　ternary lanthanides, 29–44, see also
　　Lanthanides, ternary

YbT$_2$X$_2$, 155–156
RT$_4$B, 239–241
RT$_{4+x}$Al$_{8-x}$, 224–231
RT$_6$X$_6$, 52–53, 223–225
RT$_9$Si$_2$, 234–236
RT$_{12}$B$_6$, 236–238
RT$_{12-x}$, 193–202
RT$_{12-x}$M$_x$, 47
R$_{1+\epsilon}$T$_4$B$_4$, 241–242
R$_2$Co$_3$Si$_5$, 113–114
R$_2$Co$_{14}$B, 209, 210
R$_2$Co$_{14}$BH$_x$, 214
R$_2$Fe$_3$Si$_5$ compounds, 110–113
R$_2$Fe$_{14}$B
  hydrogenation and magnetic properties, 213
  magnetic moments, 214
  magnetic phases, 209, 210
  molecular field modeling, 216–217
  light rare earths interaction, 48
R$_2$Fe$_{14}$C, 48, 209, 210
R$_2$Fe$_{14}$X, 219
R$_2$Fe$_{14-x}$Co$_x$B, 211, 212
R$_2$Fe$_{14-x}$T$_x$B, 209–210
R$_2$RhSi$_3$, 14–15, 93–94
R$_2$Rh$_3$Si$_5$ compounds, 114
R$_2$T$_3$Si$_5$, 28–29, 110–114
R$_2$T$_{14}$X
  exchange interaction models, 215–218
  magnetic moments, 214–215
  magnetic properties, 202–209, 212–214
  magnetocrystalline anisotropy, 218–223
  substitutions, 209–212
  ternary lanthanides, 47–51, see also
    Lanthanides, ternary
R$_{2-x}$Y$_x$Co$_{14}$B, 213
RTiFe$_{11}$, 200, 202
RTiFe$_{11-x}$Co$_x$, 47, 198, 201
Resistivity
  CeT$_2$X$_2$ compounds, 145, 146
  crystal electric field theory, 76–77
  electric
    CaIn$_2$-type structures, 95
    Fe$_2$P(ZrNiAl)-type structures, 89
    Kondo effect, 7
    lanthanum compounds, 233, 234
    MgCu$_2$-type structures, 83
    PbFCl-type structures, 98
    RMn$_2$X$_2$ compounds, 163, 168
    RT$_2$X compounds, 108, 109
    RT$_2$X$_2$ compounds, 115, 120
    RT$_2$X$_2$ with heavy lanthanides, 129, 131–133
  magnetic, 98
Rhodes-Wohlfarth plot, 169, 171
RKKY interactions, see also Individual entries
  cerium compounds, 4, 146, 149, 150
  RT$_2$X$_2$ compounds with heavy lanthanides, 138
  R$_2$Fe$_3$Si$_5$ compounds, 110
RKKY model, see also Individual entries

de Gennes factor relation, 173
  exchange interactions relation, 174–176
  Fe$_2$P(ZrNiAl)-type structures, 84
  gadolinium compounds, 123
  lanthanide elements, 61, 62
  magnetic ordering schemes relation, 245
Rock-salt crystal structure, 2, 4
Russel-Saunders coupling, 70

S

Saturation effect, 142
Saturation magnetization
  lanthanum compounds, 232
  rare earth compounds, 94, 163, 198–199, 211,
    212
Saturation moment, 120
Scandium compounds, 28
Schottky heat, 78, 79, 105
Shielding effect, 183
Silicides, 14–15, 93, see also Individual entries
Slater-Pauling curve, 198
Samarium compounds, 122, 161, 162
Specific heat
  CeCu$_2$-type structures, 98
  cerium compounds, 140, 143, 145, 147
  crystal electric field theory, 78, 79
  dysprosium compounds, 185
  Fe$_2$P(ZrNiAl)-type structures, 89
  holmium compounds, 185
  LaPtSi-type structures, 95
  PbFCl-type structures, 98
  rare earth compounds
    RMn$_2$X$_2$, 168
    RT$_2$X, 105
    RT$_2$X$_2$, 115, 156
    RT$_2$X$_2$ with heavy lanthanides, 129, 132, 135,
      136
    R$_2$Fe$_3$Si$_5$, 112
    R$_2$Fe$_{14}$B, 207
    R$_2$T$_{14}$X, 215
Spherical harmonics, 71–72
Spin compensation effect, 236, 238
Spin flop, 114
Spin fluctuation temperature, 8
Spin glass behavior, 83, 136, 229
Spin reorientation
  RT$_4$B compounds, 240
  RT$_6$X$_6$ compounds, 224
  RT$_9$Si$_2$ compounds, 235
  RT$_{12-x}$M$_x$ phases, 195, 200
  R$_2$Co$_{14}$B compounds, 209
  R$_2$Fe$_{14}$B compounds, 206–207
  R$_2$T$_{14}$X compounds, 213, 214, 221–223
  substituted R$_2$T$_{14}$X compounds, 211, 212
Strontium compounds, 55, 238, see also
    Lanthanides, ternary
Stannides, 23

Stevens coefficients, 64, 72–73, 75, 242
Stevens factor, 199, 220, 221, 240
Stoner condition, 8, 169
Superconductivity
  cerium compounds, 140
  Heusler alloys, 105
  LaIrSi(ZrOS)-type structures, 84
  LaPtSi-type structures, 95
  rare earth compounds
    RTX_2, 104
    RTX_3, 109
    RT_2X, 108
    RT_2X_2, 114
    $R_2Fe_3Si_5$, 113
  YbT_2X_2 compounds, 155
Superexchange mechanisms, 4–6, see also
        Exchange interactions
Susceptibility
  crystal electric field theory, 74, 76
  magnetic, AlB_2- and Ni_2In-type structures, 90
    $CaIn_2$-type structures, 94, 95
    $CeCu_2$-type structures, 99
    cerium compounds, 141–147
    effective magnetic moments relation, 171
    europium compounds, 152–154
    $Fe_2P(ZrNiAl)$-type structures, 89
    LaIrSi(ZrOS)-type structures, 84
    LaPtSi-type structures, 95
    MgAgAs-type structures, 83
    $MgCu_2$-type structures, 83
    neodymium compounds, 184
    PbFCl-type structures, 97, 98
    praseodymium compounds, 183
    RMn_2X_2 compounds, 161
    RTX_2 compounds, 101, 104–105
    RT_2X compounds, 108
    RT_2X_2 compounds, 115, 122, 156
    RT_2X_2 compounds with heavy lanthanides,
        126–127, 129, 131–132, 134, 136, 138
    $R_2Co_3Si_5$ compounds, 113–114
    samarium compounds, 122
    ytterbium compounds, 155

T

Terbium compounds
  magnetic ordering, 3
  $TbAg_2In$, 105, 107
  TbAlGa, 91
  TbCiSi, 91
  $TbCoSi_2$, 100, 101
  $TbCo_2B_2$, 178
  $TbFeSi_2$-type structure, 23, 99
  $TbFe_{10}Cr$, 221
  $TbFe_{10.5-x}Co_xW_{12}$, 198, 200
  $TbFe_{12-x}Co_x$, 198, 200
  $TbMn_2Ge_2$, 23, 25
  $TbNiC_2$, 24, 26, 100

$TbNiSi_2$, 103
$TbNi_2Si_2$, 131, 134
$Tb(Rh_{1-x}Ru_2)Si_2$, 137
TbT_2Ge_2 compounds, 125
TbT_2Si_2 compounds, 125
TbT_2X_2 compounds, 184–185
$TbTiFe_{11-x}Co_x$, 198, 201
Ternary compounds, 3d-rich, see also Lanthanides,
        ternary
  exchange interaction models, 215–218
  hybrides, 212–214
  $La(T_{1-x})_{13}$ phases, 231–234
  magnetic moments, 214–215
  magnetocrystalline anisotropy, 218–223
  RT_4B, 239–241
  $RT_{4+x}Al_{8-x}$, 224–231
  RT_6X_6 phases, 223–224
  RT_9Si_2, 234–236
  $RT_{12}B_6$ series, 236–238
  RT_{12-x} phases, 193–202
  $R_{1+x}T_4B_4$, 241–242
  $R_2Co_{14}B$, 207–209
  $R_2Fe_{14}B$, 202–207
  $R_2Fe_{14}C$, 207, 208
  substitution compounds, 209–212
Thermal conductivity, 76
Thermal hysteresis, 229
Thermoelectric power, 76, 95, 133
Thorium
  anisotropy, 177
  $CeNiSi_2$-type structure relation, 22, 23, 25, 101,
        103, 179
  magnetic behavior of compounds, 244
  magnetic moment, 195–197
  magnetic ordering, 3
  physical effects in, 8
  RIr_2Si_2 silicides, 138
  RT_2X_2 phases, 29–31, 35, 156, 158, 244
  RT_xAl_{12-x} magnetic properties, 226–228
  ternary lanthanides, 44–47, see also
        Lanthanides, ternary
Thulium
  TmT_2Ge_2, 129
  TmT_2Si_2, 129
  TmT_2X_2, 187
  $Tm_{2-x}Dy_2Fe_{14}B$, 212, 213
TiNiSi-type structure, 18, 19, 20–21, 96–98
Transition metals, 1
Transport, 77–76, 103

V

Van Vleck paramagnetism, 105, 122, 142, 156,
        158, see also Paramagnetic ordering;
        Paramagnets

W

Wigner-Seitz cell boundary, 243

X

XAS, see X-ray absorption spectroscopy
X-ray absorption spectroscopy (XAS), 42, 154, 155
X-ray diffraction
  $NdCo_4B_4$-type structures, 58
  RT_2X_2 phase compounds, 31, 33, 41
  RT_4B compounds, 241
  $ThMn_{12}$ phase compounds, 45

Y

Yttrium compounds
  $Y(Co_{1-x}Fe_x)_4B$, 237
  YCo_6Ge_6, 53
  $YCo_{12-x}V_x$, 196, 198
  $YFe_{12-x}Al_x$, 45

$YIrGe_2$, 24, 26
YPd_2Si, 27
$YTiFe_{11}N_x$, 47
$YTiFe_{11-x}Co_x$, 198, 201
$YZnSi$, 91
$Y_2Fe_{14}B$, 203, 214
$Y_2Fe_{14}XH_x$, 213
Y_2PdSi_3, 94
Ytterbium compounds
  $YbTGa$, 99
  YbT_2Si_2, 156
  YbT_2X_2, 155–156
  historical background, 1
  Kondo effect, 7

Z

ZrBeSi-type structure, 18